TI-89 Graphing Calculator For Dummies®

Cheat Sheet

P9-BTM-696

Hard-to-Find Keys

- ∫ **(Integration):** Press [2nd][7].
- ***d* (differentiation):** Press [2nd][8].
- ′ **(prime):** Press [2nd][=].
- θ **(theta):** Press [◆][^].
- < **(less than):** Press [2nd][0].
- > **(greater than):** Press [2nd][.].
- ° **(degrees):** Press [2nd][|].
- ∠ **(angle):** Press [2nd][EE].
- ∞ **(infinity):** Press [◆][CATALOG].
- ***i* (complex number *i*):** Press [2nd][CATALOG].

TI-92 and Voyage 200 Keystrokes

- **Home:** Press [◆]Q.
- **Catalog:** Press [2nd][2].
- **Custom:** Press [2nd][3].
- **Units menu:** Press [◆]P.
- ′ **(prime):** Press [2nd]B.
- θ **(theta):** The key to the right of M.
- ° **(degrees):** Press [2nd]D.
- ∠ **(angle):** Press [2nd]F.
- ∞ **(infinity):** Press [2nd]J.
- ***i* (complex number *i*):** Press [2nd]I.

Shortcut Keystrokes

- @: Press [◆][STO▸].
- !: Press [◆][÷].
- ≠: Press [◆][=].
- ≤: Press [2nd][0][=].
- ≥: Press [2nd][.][=].
- Σ: Press [◆][(][↑][3].
- σ: Press [◆][(][alpha][3].

How Do You Do It?

- **Approximate an answer:** Press [◆][ENTER].
- **Abort a calculation:** Press [ON].
- **Use the previous graphing window:** Press [F2]⏶⏶⏵[1].

TI-89 Graphing Calculator For Dummies®

Cheat Sheet

Where Are They?

- **Absolute value:** MATH Number menu ([2nd][5][1]).
- **log(*x*):** Catalog ([CATALOG][4][2nd][▼][2nd][▼][▲]).
- **and:** MATH Test menu ([2nd][5][8]).
- **≤, ≥:** MATH Test menu ([2nd][5][8]).
- **Trigonometric functions:** MATH Trig menu ([2nd][5][alpha][=]).
- **Hyperbolic functions:** MATH Hyperbolic menu ([2nd][5][alpha][)]).
- **Cross and Dot products:** MATH Matrix Vector ops menu ([2nd][5][4][▲]).
- **Polar, cylindrical, and spherical coordinates:** MATH Matrix Vector ops menu ([2nd][5][4][▲]).
- **σ:** CHARACTER Greek menu ([2nd][+][1]).
- **$\bar{x}$, $\bar{y}$:** CHARACTER Math menu ([2nd][+][2]).

What's the Proper Format?

- **Indefinite integral:** ∫(function, variable).
- **Definite integral:** ∫(function, variable, lower limit, upper limit).
- **Derivative:** *d*(function, variable).
- **nth derivative:** *d*(function, variable, *n*).
- **Real solutions to an equation:** Solve(equation, variable).
- **System of two equations:** Solve(equat1 and equat2, {var1, var2}).
- **Complex solutions to an equation:** cSolve(equation, variable).
- **Solutions to a differential equation:** deSolve(1st or 2nd order differential equation, independent variable, dependent variable).

For Dummies: Bestselling Book Series for Beginners

TI-89 Graphing Calculator FOR DUMMIES®

TI-89 Graphing Calculator FOR DUMMIES®

by C. C. Edwards

WILEY
Wiley Publishing, Inc.

TI-89 Graphing Calculator For Dummies®
Published by
Wiley Publishing, Inc.
111 River Street
Hoboken, NJ 07030-5774

www.wiley.com

Published by Wiley Publishing, Inc., Indianapolis, Indiana

Published simultaneously in Canada

For general information on our other products and services, please contact our Customer Care Department within the U.S. at 800-762-2974, outside the U.S. at 317-572-3993, or fax 317-572-4002.

For technical support, please visit www.wiley.com/techsupport.

Wiley also publishes its books in a variety of electronic formats. Some content that appears in print may not be available in electronic books.

Library of Congress Control Number: 2005923785

ISBN-13: 978-0-7645-8912-6

ISBN-10: 0-7645-8912-1

Manufactured in the United States of America

10 9 8 7 6 5 4 3 2

1B/RV/QU/QW/IN

WILEY

About the Author

C. C. Edwards has a Ph.D. in mathematics from the University of Wisconsin, Milwaukee, and is currently teaching mathematics on the undergraduate and graduate levels. She has been using technology in the classroom since before Texas Instruments came out with its first graphing calculator, and she frequently gives workshops at national and international conferences on using technology in the classroom. She is the author of *TI-83 Plus Graphing Calculator for Dummies* and *TI-84 Plus Graphing Calculator for Dummies* and she has written 40 activities for the Texas Instruments Explorations Web site. She was an editor of *Eightysomething!,* a newsletter that used to be published by Texas Instruments. She still hasn't forgiven TI for canceling that newsletter.

Just barely five feet tall, CC, as her friends call her, has three goals in life: to be six inches taller, to have naturally curly hair, and to be independently wealthy. As yet, she is nowhere close to meeting any of these goals. When she retires, she plans to become an old-lady carpenter.

Dedication

This book is dedicated to Julia Hamp Edwards, my grandmother. She was a very wise and loving person who saw to it that I got to go to college.

Author's Acknowledgments

I'd like to thank the folks at John Wiley & Sons who worked behind the scenes on the creation of this book. I don't come in contact with many of these people until the book is in the editing stage, which takes place after I write the acknowledgments. But from past experience I know that they do an extremely good job. Unfortunately, I can't tell you about it at this time. Their names appear in the Publisher's Acknowledgments.

The Wiley people I've been dealing with so far are the same people I worked with on the *TI-83 Plus Graphing Calculator For Dummies* and *TI-84 Plus Graphing Calculator For Dummies* books: Melody Layne, acquisitions editor, and Christopher Morris, project editor. As usual, Melody has given me many superb ideas for the book and has been a great liaison with Texas Instruments. And Chris, as before, has given me extremely good criticism. It's hard to disagree with Chris because he's usually right on target.

On the home front, I thank all my academic colleagues and students who inspire me. To name names, they are Ioana Mihaila, Bogdan Mihaila, Olcay Akman, Fusun Akman, Menassie Ephrem, Tom Hoffman, Karen Thompson, Caroline Lowery, Cassandra Lim, and many, many others. I also thank Ginny Levson, who provided me with the musical background that kept me sane while writing this book — it was a CD, in case you're wondering. And, as corny as it sounds, I thank my border collies, Nip and Tess, for putting up with fewer walks and less play time.

Publisher's Acknowledgments

We're proud of this book; please send us your comments through our online registration form located at www.dummies.com/register/.

Some of the people who helped bring this book to market include the following:

Acquisitions, Editorial, and Media Development

Project Editor: Christopher Morris

Acquisitions Editor: Melody Layne

Copy Editor: Virginia Sanders

Technical Editor: Dr. Douglas Shaw, University of Northern Iowa

Editorial Manager: Kevin Kirschner

Media Development Supervisor: Richard Graves

Editorial Assistant: Amanda Foxworth

Cartoons: Rich Tennant (www.the5thwave.com)

Composition Services

Project Coordinator: Adrienne Martinez

Layout and Graphics: Joyce Haughey, Stephanie D. Jumper, Barry Offringa, Julie Trippetti, Erin Zeltner

Proofreaders: Leeann Harney, Jessica Kramer, Joseph Niesen, Carl Pierce

Indexer: TECHBOOKS Production Services

Special Help *Emily Bain*

Publishing and Editorial for Technology Dummies

Richard Swadley, Vice President and Executive Group Publisher

Andy Cummings, Vice President and Publisher

Mary Bednarek, Executive Acquisitions Director

Mary C. Corder, Editorial Director

Publishing for Consumer Dummies

Diane Graves Steele, Vice President and Publisher

Joyce Pepple, Acquisitions Director

Composition Services

Gerry Fahey, Vice President of Production Services

Debbie Stailey, Director of Composition Services

Contents at a Glance

Table of Contents

Introduction

Do you know how to use the TI-89, TI-89 Titanium, TI-92 Plus, or Voyage 200 graphing calculator to do each of the following?

- Solve equations and systems of equations
- Factor polynomials
- Evaluate derivatives and integrals
- Graph functions, parametric equations, polar equations, and sequences
- Create Stat Plots and analyze statistical data
- Multiply matrices
- Solve differential equations and systems of differential equations
- Transfer files between two or more calculators
- Save calculator files on your computer
- Add applications to your calculator so it can do even more than it could when you bought it

If not, this is the book for you. Contained within these pages are straightforward, easy-to-follow directions that tell you how to do everything listed here — and much, much more.

About This Book

Although this book doesn't tell you how to do *everything* the calculator is capable of doing, it gets pretty close. It covers more than just the basics of using the calculator, paying special attention to warning you of the problems you could encounter if you knew only the basics of using the calculator.

This is a reference book. It's process-driven, not application-driven. I don't give you a problem to solve and then tell you how to use the calculator to solve that particular problem. Instead, I give you the steps you need to get the calculator to perform a particular task, such as constructing a histogram.

Using This Book with a TI-92 Plus or a Voyage 200

Although this book isn't expressly written for the TI-92 Plus and Voyage 200, it does tell you how to use these calculators, but you must keep in mind three minor differences between these calculators and the TI-89:

- The TI-92 Plus and Voyage 200 have QWERTY keyboards, the TI-89 doesn't. So when I give directions telling you to press [ALPHA] to enter a letter on the TI-89, on a TI-92 Plus or Voyage 200 you'd simply press the letter on the QWERTY keyboard.
- There are some symbols on the TI-89 keypad that are located in different places on the TI-92 Plus and Voyage 200 keypads. But this is no big deal the Cheat Sheet that comes with this book lists these symbols and their locations on the TI-92 Plus and Voyage 200. Just refer to the Cheat Sheet when, for example, I tell you to press [◆][^] to enter θ and you notice that θ isn't above the [^] key on the TI-92 Plus and Voyage 200.
- The TI-89 has only five Function keys on the keypad just below the calculator's screen, [F1] through [F5]; the TI-92 and Voyage 200 have eight Function keys on the keypad below this screen. On the TI-89 you, for example, access function F6 by pressing [2nd][F1], on the TI-92 and Voyage 200 you access this function by pressing the F6 key. So in this book when you see directions telling you to press [2nd][F1], [2nd][F2], or [2nd][F3] you should respectively press the F6, F7, or F8 key on a TI-92 or Voyage 200 calculator.

Conventions Used in This Book

When I refer to "the calculator," I am referring to the TI-89 and the TI-89 Titanium. With three minor differences, the instructions I give for "the calculator" can also be used on a TI-92 or Voyage 200. I explain these differences in the preceding section.

When I want you to press a key on the calculator, I use an icon for that key. For example, if I want you to press the ENTER key, I say "Press [ENTER]." If I want you to press a series of keys, such as the MODE key and then the Right Arrow key, I say, "Press [MODE][▶]." You press all keys on the calculator one at a time. On the calculator, there is no such thing as holding down one key while you press another key.

Becoming handy with the location of the keys on the calculator is tricky enough, but remembering the locations of the secondary functions (the color-coded functions above most keys) is even more of a challenge. So when I want you to access one of those functions, I give you the actual keystrokes. For example, if I want you to access the MATH menu, I tell you to press [2nd][5]. This is a simpler method than that of the manual that came with your calculator it would say, "Press [2nd][MATH]," and then make you hunt for the location of the secondary function MATH. The same principle holds for using key combinations to enter specific characters; for example, I tell you to press [ALPHA][0] to enter the less-than symbol.

When I want you to use the Arrow keys, but not in any specific order, I say, "Press ◀▶▲▼." If I want you to use only the Up and Down Arrow keys, I say, "Press ▲▼."

What You Don't Have to Read

Of course, you don't have to read anything you don't want to. The only items in this book that you really don't need to read are the items next to Technical Stuff icons. These items are designed for the curious reader who wants to know, but doesn't really need to know, why something happens.

Other items that you might not need to read are the paragraphs that follow the steps in a procedure. These paragraphs give you extra help in case you need it. The steps themselves are in **bold**; the explanatory paragraphs are in a normal font.

Foolish Assumptions

My nonfoolish assumption is that you know (in effect) nothing about using the calculator, or you wouldn't be reading this book. My foolish assumptions are as follows:

- You own, or have access to, a TI-89, TI-89 Titanium, TI-92 Plus, or Voyage 200 calculator.
- If you want to transfer files from your calculator to your computer, I assume that you have a computer and know the basics of how to operate it.

How This Book Is Organized

The parts of this book are organized by tasks that you would like to have the calculator perform.

Part I: Making Friends with the Calculator

This part describes the basics of using the calculator. It addresses such tasks as adjusting the contrast and getting the calculator to perform basic arithmetic operations.

Part II: Doing Algebra and Trigonometry

This part tells you how to solve equations, factor polynomials, find partial fraction decompositions, and evaluate trigonometric functions. This part also tells you how to deal with complex numbers and how to find complex solutions to equations.

Part III: Graphing and Analyzing Functions

In this part, think visual. Part III tells you how to graph and analyze functions and how to create a table for the graph. This part also tells you how to find critical points and inflection points.

Part IV: Working with Sequences, Parametric Equations, and Polar Equations

This part describes how you can graph and analyze parametric equations, polar equations, and sequences.

Part V: Doing Calculus

This part is loaded with information on how to do almost anything related to calculus, vector calculus, and differential equations. It tells you how to evaluate integrals, derivatives, limits, dot products, and cross products, and how to graph in 3D. It also tells you how to solve differential equations and how to graph slope fields. It even tells you how to find Taylor polynomials and how to convert between rectangular, cylindrical, and spherical coordinates.

Part VI: Dealing with Matrices

Part VI gives you the basics on how to add, subtract, multiply, invert, and transpose matrices. And then it delves deeper into the world of matrices by telling you how to use matrices to solve systems of equations and how to find eigenvalues and eigenvectors.

Part VII: Dealing with Probability and Statistics

It's highly probable that Part VII tells you how to deal with probability and statistics. In particular, my statistics show that it tells you how to generate random numbers, evaluate permutations and combinations, analyze one- and two-variable data, and find a regression equation that models your data.

Part VIII: Communicating with PCs and Other Calculators

Your calculator joins the information superhighway. Part VIII describes how you can save calculator files on a computer and how you can transfer files from one calculator to another.

Part IX: The Part of Tens

Part IX contains a plethora of wonderful information. This part tells you about the many wonderful applications you can put on your calculator, and it describes the most common errors and error messages that you might encounter.

Icons Used in This Book

This book uses four icons to help you along the way. Here's what they are and what they mean:

The text next to this icon tells you about shortcuts and other ways of enhancing your use of the calculator.

The text next to this icon tells you something you should remember so you don't run into problems later. Usually it's a reminder to enter the appropriate type of number so you can avoid an error message.

There is no such thing as crashing the calculator. But this icon warns you of those *few* times when you can do something wrong on the calculator and be totally baffled because the calculator gives you confusing feedback either no error message or a cryptic error message that doesn't tell you the true location of the problem.

This is the stuff you don't need to read unless you're really curious.

Where to Go from Here

You don't have to read this book from cover to cover. You don't even have to start reading at the beginning of a chapter. When you want to know how to get the calculator to do something, just start reading at the beginning of the appropriate section. The index and table of contents should help you find whatever you're looking for.

Custom menus for the TI-89 calculator are available at the Wiley Web site at `www.dummies.com/go/ti-89fd`.

Part I

Making Friends with the Calculator

The 5th Wave By Rich Tennant

In this part . . .

This part takes you once around the block with the basics of using the calculator. In addition to showing you how to use the calculator to evaluate arithmetic expressions, I discuss the elementary calculator functions — including multi-use keys, menus, modes, and the CATALOG. I also cover the fundamentals of using and combining expressions, including the order of operations and storing and recalling variables. In addition, I explain how to use the numerous functions housed in the MATH Number menu to perform tricks such as finding quotients and remainders when doing long division.

Chapter 1

Coping with the Basics

In This Chapter

- Turning the calculator on and off
- Using the keyboard
- Using the menus
- Setting the mode of the calculator
- Using the CATALOG

TI-89 graphing calculators are loaded with many useful features. With them, you can graph and investigate functions, parametric equations, polar equations, and sequences. You can also produce 3D graphs, contour maps, slope fields, and direction fields. These calculators can even factor expressions and solve systems of equations. And if that's not enough to keep you busy, a TI-89 can integrate, differentiate, evaluate limits, solve differential equations, analyze statistical data, and manipulate matrices. You can even turn this calculator into an e-book reader!

But if you've never used a graphing calculator before, you might at first find it a bit intimidating. After all, it contains several dozen menus, some of which contain three or four submenus. But getting used to working with the calculator really isn't hard. After you get familiar with what the calculator is capable of doing, finding the menu that houses the command you need is quite easy. And you have this book to help you along the way.

When to Change the Batteries

The convenience of battery power has a traditional downside: What if the batteries run out of juice at a crucial moment, such as during a final exam? Fortunately, the calculator gives you some leeway. When your batteries are low, you see the BATT warning message displayed to the right of the last line of the screen. After you see this message for the first time, the calculator should, according to the manufacturer, continue to function just fine for at least one week. When the batteries are so low that you might not make it through that final exam, the calculator highlights the BATT warning message, as illustrated in the second picture in Figure 1-1. One exception exists: If you attempt to transfer data or an application from a PC or another calculator to a calculator that has low batteries, the calculator with the low batteries displays a warning message telling you to change the batteries and refuses to make the transfer. (Part VIII explains how to transfer data and applications.)

Because you've likely put batteries into countless toys, you should have no trouble opening the cover on the back of the calculator and popping in four AAA batteries. Above the AAA battery chamber is a panel that opens to the compartment containing the backup battery. The lid of the panel indicates the type of battery housed in the compartment. The manufacturer recommends that you replace this battery every three or four years. So mark your calendar!

Turning the Calculator On and Off

Press [ON] to turn the calculator on. (The [ON] key is the last key in the left column of keys on the keyboard.) The first time you turn on the TI-89 Titanium, you see the application screen, as shown in the first picture in Figure 1-1. On the TI-89, you see the Home screen, as in the second picture in this figure.

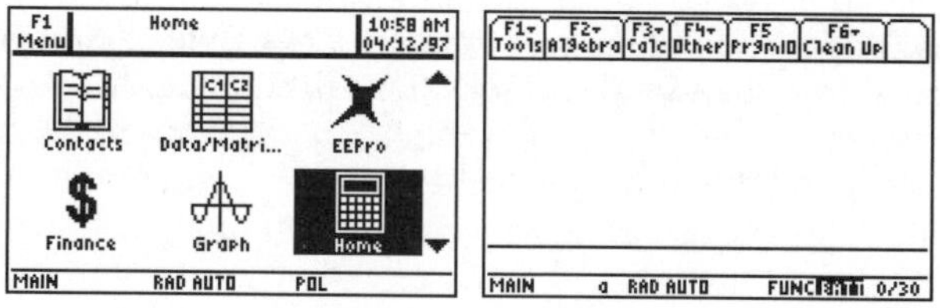

Figure 1-1: The TI-89 Titanium Application screen (left) and Home screen (right).

To turn the calculator off, press [2nd] and then press [ON]. (The [2nd] key is the second key from the top in the left column of keys.) Pressing [2nd][ON] to turn off the calculator also exits any application you were using (such as the Graph application) so that when you turn the calculator back on, you're confronted with the Application screen on the TI-89 Titanium or the Home screen on the TI-89.

TIP

If you want to turn the calculator off without exiting your current application, press [◆] and then press [ON]. (The [◆] key is the third key from the top in the left column of keys.) The next time you turn the calculator on, it will be exactly as you left it.

To prolong the life of the batteries, the calculator automatically turns itself off after several minutes of inactivity. But don't worry — when you press [ON], all your work appears on the calculator just as you left it before the calculator turned itself off.

In some types of light, the screen can be hard to see. To increase the contrast, press and hold [◆] and repeatedly press [+] until you have the desired contrast. To decrease the contrast, press and hold [◆] and repeatedly press [−].

What Is the Home Screen?

The Home screen is the screen where most of the action takes place as you use the calculator — it's where you evaluate expressions and execute commands. This is also the screen you usually return to after you've completed a task such as entering a matrix in the Data/Matrix editor or graphing a function.

REMEMBER

Press [HOME] to go to the Home screen from any other screen. If you want to clear the contents of the Home screen, press [F1][8].

The second picture in Figure 1-1 (shown previously) displays the Home screen layout. The first line is the Toolbar, which houses all the neat commands you can use. Under that is the large space, called the history area, where the calculator displays the results of whatever it is you told the calculator to do. The second-to-last line is the command line — this is where you enter the commands telling the calculator to do something. And the last line is the status line, where the calculator tells you what it's up to. For example, the third entry (RAD) on this line in Figure 1-1 tells you that the calculator is expecting all angles to be measured in radians.

Using the Keyboard

The row of keys under the calculator screen contains the function keys you use to select the Toolbar menus at the top of the screen. The next three rows, for the most part, contain editing keys, menu keys, and Arrow keys. The Arrow keys (◀▶▲▼) control the movement of the cursor. The remaining rows contain, among other things, the keys you typically find on a scientific calculator.

Using the [2nd] key

Above and to the left of most keys is a secondary key function written in the same color as the [2nd] key. To access that function, first press [2nd] and then press the key. For example, π is above the [^] key, so to use π in an expression, press [2nd] and then press [^].

Because hunting for the entities written above the keys can be tedious, in this book I use only the actual keystrokes. For example, I make statements like, "Enter π into the calculator by pressing [2nd][^]." Most other books would state, "π is entered into the calculator by pressing [2nd][π]."

When the [2nd] key is active and the calculator is waiting for you to press the next key, you see 2nd displayed to the left on the last line of the screen.

Using the [♦] key

Above and to the right of most keys on the top half of the keyboard and the last three keys in the right column is a secondary key function written in the same color as the [♦] key. To access that function, first press [♦] and then press the key. For example, θ is above the [^] key, so to use θ in an expression, press [♦] and then press [^].

Hunting for the entities written above the keys can be a pain, so for this book I tell you only the actual keystrokes. For example, I make statements like, "Enter θ into the calculator by pressing [♦][^]." Most other books would state, "θ is entered into the calculator by pressing [♦][θ]."

When the [♦] key is active and the calculator is waiting for you to press the next key, the calculator displays a diamond to the left on the last line of the screen.

Using the [ENTER] key

You use the [ENTER] key to evaluate expressions and to execute commands. After you have, for example, entered an arithmetic expression (such as 5 + 4), press [ENTER] to evaluate that expression. In this context, the [ENTER] key functions as the equal sign. I explain entering arithmetic expressions in Chapter 2.

Using the Arrow keys

The Arrow keys (◀, ▶, ▲, and ▼) control the movement of the cursor. These keys are in a circular pattern in the upper-right corner of the keyboard. As expected, ▶ moves the cursor to the right, ◀ moves it to the left, and so on. When I want you to use the Arrow keys — but not in any specific order — I refer to them all together, as in "Use ◀▶▲▼ to place the cursor on the entry." Or I simply write, "use the Arrow keys."

Using the [ALPHA] and [↑] keys to write text

You wouldn't want to write a novel on the calculator, but you often need to write text in order to give a name to things such as the variables you define and the files you store in the calculator. You enter these names by using the [ALPHA] and [↑] keys and the letters of the alphabet appearing above and to the right of most keys on the bottom half of the keyboard. The letters *x*, *y*, *z*, and *t* have their own keys: [X][Y][Z][T].

TIP

On the TI-89, the ALPHA key, [alpha], is in lowercase letters.

To enter a single lowercase letter, first press [ALPHA] and then press the key corresponding to that letter; for an uppercase letter, first press [↑] and then press the key corresponding to the letter. For example, the letter G is above the [7] key; to enter a lowercase *g*, press [ALPHA] and then press [7]; to enter an uppercase *G*, press [↑] and then press [7].

WARNING!

The calculator *is not* case sensitive. If, for example, you save a number with the name "A" and then save another number with the name "a," the calculator overwrites the number stored in "A" with the number you are saving in "a."

Because hunting for letters on the calculator can become annoying, I tell you the exact keystrokes needed to create them. For example, if I want you to enter the lowercase letter *g*, I say, "Press [ALPHA][7] to enter the letter *g*." Most other books would say "Press [ALPHA][G]" and leave it up to you to figure out where that letter is on the calculator.

You must press [ALPHA] before entering each letter. However, if you want to enter many letters, first press [2nd][ALPHA] to lock the calculator in Alpha mode. Then all you have to do is press the keys for the various letters. When you're finished, press [ALPHA] to take the calculator out of Alpha mode. For example, to enter Exam into the calculator, press [2nd][ALPHA][↑][÷][X][−][5] and then press [ALPHA] to tell the calculator that you're no longer entering letters. To enter more than one uppercase letter, press [↑][ALPHA], enter the letters, and then press [ALPHA] to take the calculator out of Alpha mode.

TIP

When the calculator is in Alpha mode and is waiting for you to enter a letter, you see the letter *a* displayed to the left on the last line of the screen, as seen in the second picture in Figure 1-1. When you press [2nd][ALPHA] to lock the calculator in Alpha mode, the *a* at the bottom of the screen is highlighted. If the calculator is waiting for you to enter an uppercase letter, you see ↑ displayed. And when you press [↑][ALPHA] to enter several uppercase letters, 🄰 appears at the bottom of the screen.

REMEMBER

Keys to remember

The following keystrokes are invaluable:

- [ESC]: This is the equivalent of the Escape key on a computer. It gets you out of whatever you're doing. For example, when you're in a menu and decide you really don't want to be in that menu, press [ESC] to exit the menu. Or if you have moved the cursor into the history area and want to return to the command line, press [ESC].
- [ENTER]: This key is used to execute commands and to evaluate expressions. When evaluating expressions, it's the equivalent of the equal sign.
- [CLEAR]: This is the erase key. If you're entering something into the calculator and change your mind, press this key two times — once to erase the characters to the right of the cursor and a second time to erase the rest of the entry. If you want to erase an entry in the history area of the Home screen, use the Arrow keys to highlight that entry and then press [CLEAR].
- [←]: This is equivalent to the Backspace key on a computer — it erases the character to the left of the cursor.

Recalling and Editing Entries

Being able to edit a current or previous entry is a real timesaver. The first picture in Figure 1-2 provides an example. The first entry in the history area found the lcm (least common multiple) of 4, 6, and 14. If you now wanted to find the lcm of 4, 6, and 35, wouldn't it be a lot easier to edit this entry by changing the 14 to a 35 than it would be to key in the whole entry from scratch? This section tells you how to do this.

Recalling an entry to the command line

You can edit an entry only when it's on the command line (the second line from the bottom of the screen). For example, in the first picture in Figure 1-2, the last entry evaluated is gcd(84, 35). Because this entry remains on the command line, you can, for example, edit it to find gcd(84, 24) by simply changing 35 to 24. (I explain editing later in this section.) The first line in this picture found the lcm of 4, 6, and 14. If you want to edit this entry to find, for example, the lcm of 4, 6, and 35, you would first recall the entry to the command line and then edit it. To recall an entry to the command line, repeatedly press ⊖ to highlight the entry you want to place on the command line, as illustrated in the second picture in Figure 1-2. Then press [ENTER] to place the entry on the command line, as in the third picture in Figure 1-2.

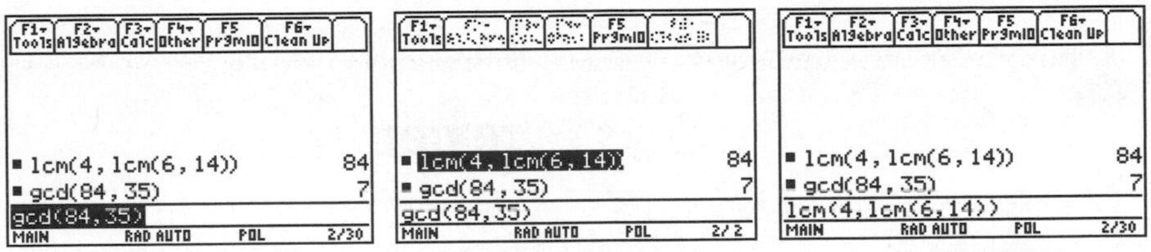

Original screen Highlight entry Press [ENTER]

Figure 1-2: Recalling entries to the command line.

TIP

You can use the ▲ key to scroll past the top of the screen. If you want to return to the command line from the history area without placing a new entry on it, press [ESC].

Editing entries on the command line

The calculator offers four ways to edit an entry on the command line:

- **Deleting the entire entry:** Press [CLEAR] one or two times to erase an entry on the command line. If the entry is highlighted or if the cursor is at the left of the entry, pressing [CLEAR] one time erases the entry; in all other situations, you need to press [CLEAR] twice to erase the entry.
- **Erasing part of an entry:** To erase a single character, use the ◀▶ keys to place the cursor to the right of the character you want to erase and then press [←] to delete that character.
- **Inserting characters:** Because the insert mode is the default, to insert a character, just use the ◀▶ keys to place the cursor where you want to insert the character and then key in the character.
- **Keying over existing characters:** You can press [2nd][←] to toggle the calculator between the Insert and Type Over modes. When the calculator is in Insert mode, the cursor appears as a vertical line placed between two characters; in Type Over mode, the cursor is a square that highlights a character. Keying in a character while in Type Over mode does exactly what you'd expect — it replaces the highlighted character with the character you key in. When you're finished keying over characters, press [2nd][←] to put the calculator back in Insert mode.

The BUSY Indicator

When the calculator is busy performing calculations, you see the BUSY indicator highlighted at the right of the last line of the screen.

If the calculator takes too long to graph a function, evaluate an expression, or execute a command, and you want to abort the process, press [ON]. When confronted with the Break error message, press [ESC] or [ENTER].

Accessing Applications

The most often used applications, such as Home, Graph, and Table, have their own key or key combination. For example, press [HOME] to go to the Home screen; to go to the Graph screen, press [◆][F3].

To access the other applications, press [APPS]. On the TI-89 Titanium, you see a screen similar to the one in the first picture in Figure 1-1; on the TI-89, you see a numbered list of applications. To select an application on the TI-89 Titanium, use the Arrow keys to highlight the desired application and then press [ENTER]. On the TI-89, you select an application by keying in its number or by using the ▲▼ keys to highlight the application and then pressing [ENTER].

There isn't enough space in this book to tell you how to use all of these applications, but I do tell you how to use Home, Graph, and Table, as well as the Data/Matrix editor. For a brief explanation of what some of the other applications can do, see Chapter 23.

Using Menus

You can find most functions and commands in the menus housed in the calculator — and just about every chapter in this book refers to them. This section is designed to give you an overview of how to find and select menu items.

Accessing a menu

Each application has its own set of menus in the Toolbar at the top of the screen, as illustrated in the second picture in Figure 1-1 appearing earlier in this chapter. These menus are specific to the application and aren't always available in other applications. In addition to the Toolbar menus, the calculator has several menus that are always available no matter what application you're using. MATH and UNITS are examples of such menus.

Each menu, whether it be on the Toolbar or not, has its own key or key combination. For example, to access the F1 Tools menu in the Toolbar, press [F1]. To access the MATH menu (which you see in the first picture in Figure 1-3), press [2nd][5].

An arrow appearing to the right of a menu item means that that item houses a submenu. This is illustrated in the first picture in Figure 1-3, in which each item in the menu houses a submenu. Only the last five items in the menu in the third picture in Figure 1-3 house submenus.

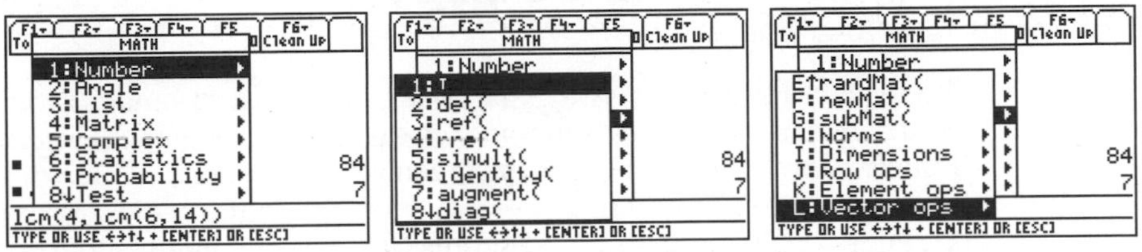

Figure 1-3: The MATH menu and the MATH Matrix submenu.

Scrolling a menu

If a menu contains more items than can be displayed on the screen, a down or up arrow appears after the number or letter of the last or the first item on the screen. This is illustrated in the first two pictures in Figure 1-3, in which the down arrow to the right of the number 8 indicates that there are more items at the bottom of the menu than can fit on the screen. To see these menu items, repeatedly press ▼ until they come into view. To get quickly to the bottom of a menu from the top of the menu, press ▲. The result of doing this is illustrated in the third picture of Figure 1-3, in which you now see an up arrow to the right of E, indicating that there are more items at the top of this menu than can be displayed on the screen. To quickly get from the bottom to the top of the menu screen, press ▼.

Selecting menu items

To select a menu item from a menu, key in the number (or letter) of the item or use ▼▲ to highlight the item and then press [ENTER].

TIP

Press [ESC] to exit a menu or submenu without selecting a menu item.

Setting the Mode

The Mode menu, which you access by pressing [MODE], is the most important menu on the calculator — among other things, it tells the calculator how you want numbers and graphs to be displayed. The three screens (pages) that constitute Mode menu are shown in Figure 1-4. Menu items that are illegible, such as the third item in the second picture in Figure 1-4, are modes that aren't available at the current time.

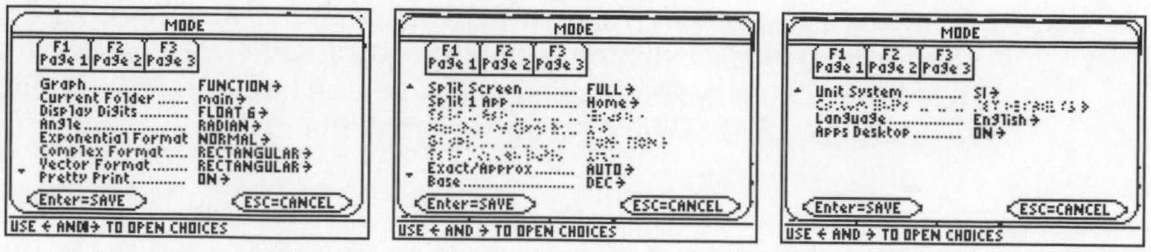

Figure 1-4: The three pages of the Mode menu.

You set a mode the same way you select an item from a menu (which I explain in the preceding section). For example, to change the Angle mode from radians to degrees, press [F1] if you're not already on page one of the Mode menu, repeatedly press ⊖ until you highlight RADIAN in the Angle option, press ⊙ to display the options in the Angle menu, and then press [2] to select the DEGREE option. I illustrate this procedure in Figure 1-5.

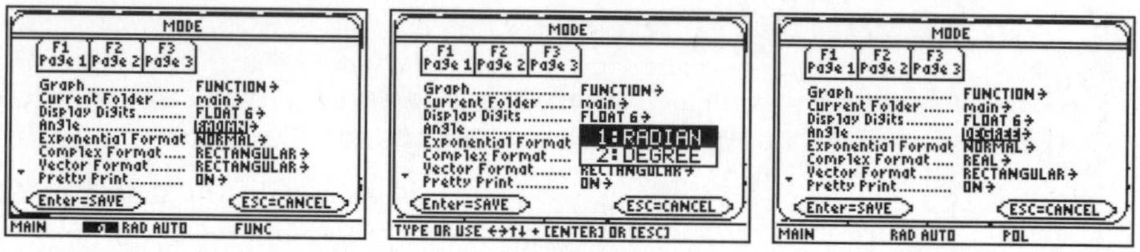

Highlight Angle option — Press ⊙ — Select option

Figure 1-5: Changing the Angle mode.

Here are your choices for the items in the Mode menu. When you finish making your selections, press [ENTER] to save your new Mode settings.

- **Graph:** This setting tells the calculator what type of objects you plan to graph. Your choices are FUNCTION, PARAMETRIC, POLAR, SEQUENCE, 3D, and DIFF EQUATIONS. These types of graphs are explained later in this book.

  The current Graph mode of the calculator is displayed on the status line. For example, FUNC on the last line in the first picture of Figure 1-5 indicates that the calculator is set to graph functions of one variable; POL on the last line in the third picture of Figure 1-5 tells you that the calculator is set to graph polar equations.

- **Current Folder:** This setting tells the calculator where it can find the variables and files you have stored in the calculator. The folders in which you save variables and files work the same way computer folders do. It is usually sufficient to save all your variables and files in the default MAIN folder. I don't cover creating folders in this book. The active folder appears on the left end of the status line, as in the first picture in Figure 1-5 (shown earlier).

- **Display Digits:** The full value of a number sometimes contains too many digits to display on the screen. This option tells the calculator how many digits of a number to display of the screen. FIX *n* tells the calculator to round numbers so that *n* digits appear after the decimal point, and FLOAT *n* tells the calculator to round numbers so that a total of *n* digits appear. The FLOAT option varies the number of decimal places depending on the value of the number. If a number can't be displayed by using the Display Digits setting, it is displayed in scientific notation.

The default Display Digits setting is FLOAT 6, which is just fine for normal use. With this setting, 12.3456789 is rounded to 12.3457, a total of six digits. But if you're working with money, set this option to FIX 2 so that all numbers are rounded to two decimal places.

- **Angle:** This setting tells the calculator the units it should use when you enter an angle. Your options are RADIAN and DEGREE. The current Angle mode of the calculator is displayed on the status line. For example, RAD on the last line in the first picture of Figure 1-5 (shown earlier) indicates that the calculator is set to measure all angles in radians.

  If you plan to enter some angles in radians and some in degrees, set the Angle mode to RADIAN. When you want to enter an angle in degrees, just enter the angle and then press [2nd][I] to tell the calculator to measure that particular angle in degrees instead of radians.

  If you're planning on graphing trigonometric functions, put the calculator in Radian mode. Reason: Most trig functions are graphed for $-2\pi \le x \le 2\pi$. That is approximately $-6.28 \le x \le 6.28$. That's not a bad value for the limits on the *x*-axis. But if you graph in Degree mode, you would need $-360 \le x \le 360$ for the limits on the *x*-axis. This is doable . . . but trust me, it's easier to graph in RADIAN mode.

- **Exponential Format:** This setting controls how the calculator displays numbers. Your options are NORMAL, SCIENTIFIC, and ENGINEERING. In NORMAL mode, the calculator displays numbers in the usual numeric fashion that you used in elementary school — provided it can display the number by using the current Display Digits mode; if it can't, the number is displayed in scientific notation.

  In SCIENTIFIC mode, numbers are displayed by using scientific notation, and in ENGINEERING mode, numbers appear in engineering notation. The three modes are illustrated in Figure 1-6.

  In scientific and engineering notation, the calculator displays E*n* to denote multiplication by 10^n.

Normal Scientific Engineering

Figure 1-6: Normal, scientific, and engineering notations.

- **Complex Format:** This mode tells the calculator how to display complex numbers. In REAL mode, the calculator displays an error message when confronted with a complex result, as illustrated in the first picture in Figure 1-7. In the RECTANGULAR and POLAR modes, real numbers are displayed in the normal fashion and complex numbers are displayed in the $a + bi$ rectangular format or in the $re^{i\theta}$ polar format, as in the last two pictures in Figure 1-7.

  Because you don't want to get an error message when the calculator encounters a complex result, set the Complex mode to RECTANGULAR or POLAR.

- **Vector Format:** You have three choices, namely RECTANGULAR, CYLINDRICAL, and SPHERICAL. In RECTANGULAR mode, two- and three-dimensional vectors are displayed as $[x, y]$ and $[x, y, z]$. In the CYLINDRICAL and SPHERICAL modes, two-dimensional vectors are displayed in $[r, \theta]$ polar format. Three-dimensional vectors are displayed by using cylindrical or spherical coordinates, depending on the selected mode.
- **Pretty Print:** You have two options, OFF or ON. When Pretty Print is set to ON, the entries displayed on the calculator screen look . . . well, they look pretty, the way they do in a textbook. When it's set to OFF, the entries displayed on the screen look the same as those you enter on the command line. For example, to square 4, you enter 4^2 on the command line. With Pretty Print ON, the calculator displays 4^2; when it's OFF, the calculator displays 4^2. My recommendation is to leave Pretty Print ON.
- **Split Screen:** When this option is set to FULL, only one application is displayed on the screen. If you want to see two applications at the same time, such as a graph and a table, you can select TOP-BOTTOM to display one application at the top of the screen and the other at the bottom. You can also select LEFT-RIGHT to display one application on the left side of the screen and the other on the right. I explain split screens in Chapters 5 and 6.

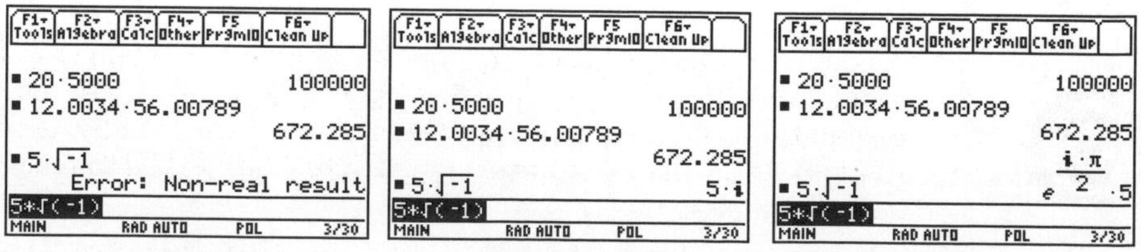

REAL RECTANGULAR POLAR

Figure 1-7: Displaying numbers by using the three Complex mode settings.

- **Split 1 App, Split 2 App, Number of Graphs, and Graph 2:** When the Split Screen mode is set to FULL, only the Split 1 App option is available. This option tells the calculator what application is to appear on the screen. In FULL screen mode, this option is equivalent to selecting an application from the Application menu, as I describe earlier in this chapter.

  If the Split Screen mode is set to either TOP-BOTTOM or LEFT-RIGHT, Split 1 App tells the calculator which application is to appear at the top or at the left of the screen; Split 2 App tells the calculator which other application is to appear on the screen. Set the Number of Graphs option to 2 only if both parts of the screen are to display graphs; otherwise, set it to 1.

  When the Number of Graphs is set to 2, the mode for the first graph is set by the first (Graph) option in the Mode menu; you use the Graph 2 option to set the graph mode for this second graph. This, for example, allows you to view a parametric graph and a polar graph on the same screen.

- **Exact/Approximate:** You have three choices, AUTO, EXACT, and APPROXIMATE. As expected, in EXACT mode, the calculator gives only the exact value of all results, and in APPROXIMATE mode, all results are displayed as decimal approximations. In EXACT mode, if the calculator can't simplify the expression you entered, it simply redisplays the exact same expression you entered. AUTO mode, however, is the best of both worlds because it displays the exact value when possible; otherwise, it displays the approximate value.

  TIP: Set the Exact/Approximate mode to AUTO. Then if the calculator gives you an exact result and you want to approximate it, simply press [◆][ENTER].

- **Base:** If you aren't a computer scientist or an engineer, you want to set this option to DEC so that your numbers are displayed in base 10. HEX displays numbers in base 16 and BIN in base 2.

- **Unit System and Custom Units:** Select SI or ENG/US if you want results involving units to use the scientific (metric) or English units. With these modes, the Custom Units option is not available. Select CUSTOM if you want to specify your own basic units of measure. Then define these units of measure in the Custom Units option.

- **Language:** Well, if you're reading this book, you might as well set this to English.

  TIP: A common prank is to set someone's calculator to a language he can't read. If this happens to you, press [MODE][F3][▼][▶][1][ENTER] to reset the calculator to English. If the prankster also set the Unit System mode to CUSTOM, you have to press [MODE][F3][▼][▼][▶][1][ENTER] to reset the calculator to English.

- **Apps Desktop:** This item is available only on the TI-89 Titanium. When Apps Desktop is set to ON, applications are displayed in picture form, as in the first picture in Figure 1-1. When it's set to OFF, applications appear in list format.

Using the CATALOG

The calculator's CATALOG houses every command and function used by the calculator. However, it's easier to use the keyboard and the menus to access these commands and functions than it is to use the CATALOG. Several exceptions exist, however; for example, the log function isn't housed in any of the calculator's menus. To use this command, you have to key it into the calculator or select it from the CATALOG. If you have to use the CATALOG, here's how to do it:

1. **If necessary, use the Arrow keys to place the cursor at the location where you want to insert a command or function found in the CATALOG.**
2. **Press [CATALOG] and enter the first letter in the name of the command or function.**

   The calculator is automatically placed in Alpha mode, so to enter the letter, just press the key corresponding to that letter. (I explain Alpha mode and entering text earlier in this chapter in the section titled "Using the [ALPHA] and [↑] keys to write text.") After entering the letter, the calculator displays the CATALOG options that begin with that letter.
3. **Repeatedly press ▼▲ to move the indicator to the desired command or function.**
4. **Press [ENTER] to select the command or function.**

Setting the Clock

The clock on the TI-89 is visible only in the Calendar and Planner applications; on the TI-89 Titanium, it is also visible on the Application screen. (The upper-right corner of the first picture in Figure 1-1, appearing earlier in this chapter, shows a clock in desperate need of setting.) To set the clock, follow these steps:

1. **Press [HOME] if you aren't already on the Home screen and then press [F1]▲X to choose Clock from the Tools menu.**

   TIP: If Clock doesn't appear as an item in the Tools menu on your TI-89, you need to upgrade the operating system of your calculator if you want it to be endowed with a clock. You find out how to do this in Chapter 21.
2. **Press ▶ and then either press [1] for a 12-hour clock or press [2] for a 24-hour clock.**

3. **Press ▼ and use the number keys to enter the current hour. Then press ▼ and key in the minutes.**
4. **Press ▼▶[1] to select AM or press ▼▶[2] to select PM.**
5. **Press ▼▶ and then press the key corresponding to the number of the date format of your choice.**
6. **Press ▼ and use the number keys to enter the current year.**
7. **Press ▼▶ and then press the key corresponding to the number of the current month.**
8. **Press ▼ and use the number keys to enter the current day of the month.**
9. **Press ▼▶[2] to turn on the clock and then press [ENTER] to save your settings.**

Chapter 2

Doing Basic Arithmetic

In This Chapter

- Entering and evaluating arithmetic expressions
- Obeying the order of operations
- Using the MATH Number menu
- Storing and recalling variables
- Combining expressions

When you use the calculator to evaluate an arithmetic expression such as $5^{10} + 4^6$, the format in which the calculator displays the answer depends on how you've set the mode of the calculator. Do you want answers displayed in scientific notation? Do you want all numbers rounded to two decimal places? Setting the mode of the calculator affords you the opportunity to tell the calculator how you want these, and other questions, answered. (I explain setting the mode in Chapter 1.) When you're doing basic, real-number arithmetic, the mode is set as described in Chapter 1.

Entering and Evaluating Expressions

Arithmetic expressions are evaluated on the Home screen. If the Home screen isn't already displayed on the calculator, press [HOME] to display it. If you want to clear the contents of the Home screen, press [F1][8].

You enter arithmetic expressions in the calculator the same way you would write them on paper if you were restricted to using the division sign (/) for fractional notation. This restriction sometimes requires parentheses around the numerator or the denominator, as illustrated in the first picture in Figure 2-1, in which I used parentheses for the second entry but not the first.

[−] versus [(-)]

The subtraction ([−]) key and the negation ([(-)]) key are entirely different. They are not interchangeable. Use the [−] key to indicate subtraction; use the [(-)] key before a number to identify that number as negative. If you improperly use [(-)] to indicate a subtraction problem, or if you improperly use [−] to indicate that a number is negative, you get an error message. You can see the use of these two symbols in the second picture in Figure 2-1.

REMEMBER

When entering numbers, don't use commas. For example, enter the number 1,000,000 in the calculator as 1000000.

After entering the expression, press [ENTER] to evaluate it. The calculator displays the answer to the right of the expression, as shown in Figure 2-1.

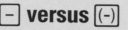

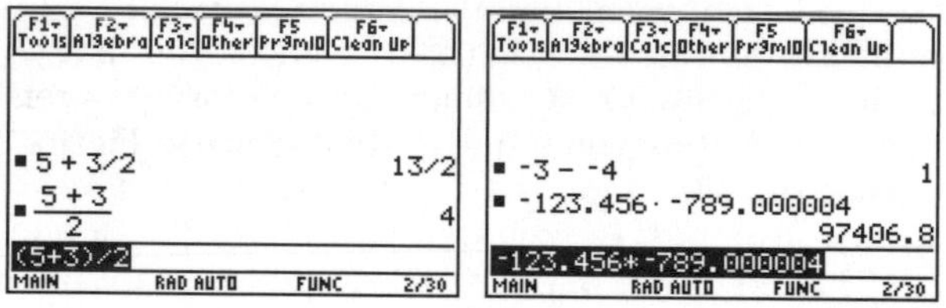

Figure 2-1: Evaluating arithmetic expressions.

Finding Important Keys

Starting with the fifth row of the calculator, you find the functions commonly used on a scientific calculator. Here's what they are and how you use them:

- **π:** To enter the transcendental number π in an expression, press [2nd][^], as shown in the first line of Figure 2-2.
- **The logarithmic and exponential functions:** The natural log (ln *x*) and exponential (e^x) functions are located above the [X] (letter X) key in the left column of keys. To enter the natural log function in an expression, press [2nd][X]; for the exponential function, press [◆][X]. After entering the function, use the number keys to enter the argument and then press [)] to close the parentheses. This is illustrated in the second line of Figure 2-2, where *e*(1) is used to enter the transcendental number *e* in an expression.

There is no special key for the common log (log *x*) function. It must be written as text in Alpha mode or selected from the CATALOG. (I cover writing text in Alpha mode and using the CATALOG in Chapter 1.) These are the keystrokes for entering **log** in Alpha mode: [2nd][ALPHA][4][-][7][ALPHA]. Evaluating log(10) appears in the third line of Figure 2-2.

- **The trigonometric and inverse trigonometric functions:** You find the most often used trigonometric functions and their inverses in the fifth row of the keyboard. These functions require that the argument of the function be enclosed in parentheses. To remind you of this, the calculator provides the first parenthesis for you. You must supply the argument and the closing parenthesis. For example, the keystrokes for the first entry in Figure 2-2 are: [◆][Y][2nd][Z][2nd][^][)][)][ENTER].

  TIP

  The other trigonometric functions and their inverses are housed in the MATH Trig menu. The section in Chapter 3 on evaluating sec, csc, cot, $\sec^{-1}$, $\csc^{-1}$, and $\cot^{-1}$ tells you how to use this menu.

- **The square-root function:** The square-root function is located above the [×] (times) key in the right column of keys. As with all functions, the square-root function requires that its argument be enclosed in parentheses. To remind you of this, the calculator provides the first parenthesis for you. You must supply the argument and the closing parenthesis (as in the entry on the command line of the picture in Figure 2-2).

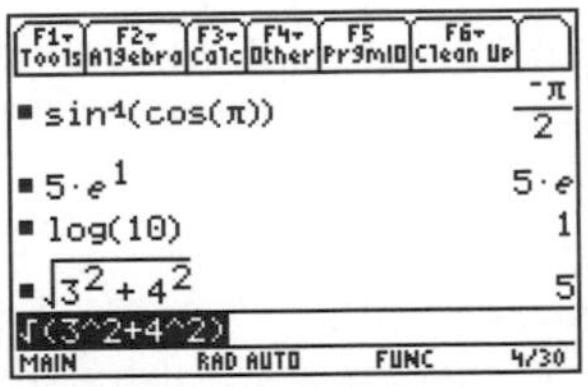

Figure 2-2: Examples of arithmetic expressions.

Using the MATH Number Menu

If you want to evaluate an arithmetic expression and you need a function other than those listed in the preceding section, you most likely can find that function in the MATH Number menu.

TIP

The trigonometric functions and their inverses are housed in the MATH Trig menu. The section in Chapter 3 on evaluating trig functions tells you how to use this menu.

TIP

The hyperbolic functions are housed in the MATH Hyperbolic menu. The section in Chapter 12 about dealing with hyperbolic functions tells you how to use this menu.

To display the MATH Number menu, press [2nd][5][▶]. The functions housed in the MATH Number menu are shown in the two pictures in Figure 2-3. After selecting a function from this menu, you must supply the argument and then press [)] to close the parenthetical expression. Here's what these functions do:

- **exact:** This function converts a decimal to a fraction.
- **abs:** This is the absolute value function.
- **round:** This function rounds a number to a specified number of decimal places. Its format is round(*number, digits*). For example, round(40/29, 2) returns 1.38 (40/29 rounded to two decimal places).
- **iPart and fPart:** The iPart function returns the integer part of a number and fPart returns the fractional part of the number. For example, iPart(40/29) = 1 and fPart(40/29) = 11/29.
- **floor and ceiling:** The floor function of a number is the largest integer that is less than or equal to that number; the ceiling function is the smallest integer that is greater than or equal to the number. For example, floor(12.35) = 12 and ceiling(12.35) = 13.
- **sign:** This function tells you whether the argument is negative or positive by returning –1 if it's negative and 1 if it's positive. If the argument is 0, the calculator returns either ±1 or sign(0), depending on the Mode setting of the calculator. For example, sign($e - \pi$) = –1.
- **mod and remain:** Both functions, mod(n, d) and remain(n, d), find the remainder when integer n (the dividend) is divided by nonzero integer d (the divisor), but they use different formulas to find this remainder. As a consequence, there are some circumstances in which the remainder found by these functions is a negative integer. If you're doing modular arithmetic, in which it doesn't matter whether the result is negative, use the **mod** function. On the other hand, if you're doing a long division problem and want to find a remainder that is zero or a positive integer, do the following:
    - **Use mod(n, d) when the divisor d is positive.** For example, mod(11, 3) = 2 and mod(–11, 3) = 1.
    - **Use remain(n, d) when the dividend n is nonnegative and the divisor d is negative.** For example, remain(11, –3) = 2.
    - **Use pencil and paper when both the dividend n and divisor d are negative.** In this situation, both the **mod** and **remain** functions give negative results. For example, mod(–11, –3) = remain(–11, –3) = –2. As a workaround, you can add the absolute value of the divisor to the result given by either function to find the zero or positive integer remainder. For example, when –11 is divided by –3, the remainder is mod(–11, –3) + 3 = –2 + 3 = 1.

- **lcm and gcd:** These functions find the least common multiple and greatest common divisor, respectively, of two integers. For example, lcm(8, –6) = 24 and gcd(8, –6) = 2. To find the least common multiple of three integers, enter the function as lcm(*a*, lcm(*b*, *c*)). Similarly, gcd(*a*, gcd(*b*, *c*)) finds the greatest common divisor of the integers *a*, *b*, and *c*.

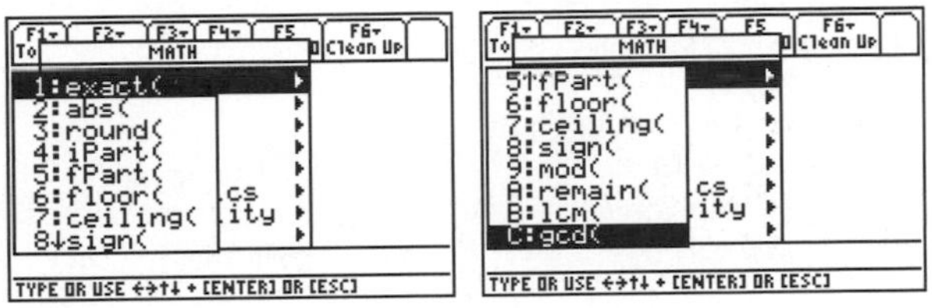

Figure 2-3: The functions in the MATH Number menu.

Understanding the Order of Operations

The order in which the calculator performs operations is the standard order that everyone uses. Spelled out in detail, here is the order in which the calculator performs operations:

1. The calculator simplifies all expressions surrounded by parentheses.
2. The calculator evaluates all functions that require an argument.

   Examples of such functions are ln, sin, and all the functions listed in the preceding section.
3. The calculator evaluates all expressions raised to an exponent.
4. The calculator negates all expressions preceded by the negation symbol — the symbol entered by pressing [(-)].

   REMEMBER: Evaluating –3^2 might not give you the expected answer. We think of –3 as being a single, negative number. So when we square it, we expect to get +9. But the calculator gets –9. This happens because, according to the order of operations, the calculator first squares the 3 and then negates the result. To avoid this potentially hazardous problem, always surround negative numbers with parentheses *before* raising them to a power.
5. The calculator evaluates all multiplication and division problems as it encounters them, proceeding from left to right.
6. The calculator evaluates all addition and subtraction problems as it encounters them, proceeding from left to right.

Well, okay, what does the phrase "x plus 1 divided by x minus 2" actually *mean* when you say it aloud? Well, that depends on how you say it. Said without pausing, it means $(x + 1)/(x - 2)$. Said with a subtle pause after the "plus" and another before the "minus," it means $x + (1/x) - 2$. The calculator can't hear speech inflection, so make good use of those parentheses when you're "talking" to the calculator.

Using a Previous Answer

You can use a previous answer in an arithmetic expression you want to evaluate. To enter this answer in your expression, use the ⊖ key to highlight the answer and then press [ENTER] to place it in your expression. This is illustrated in Figure 2-4.

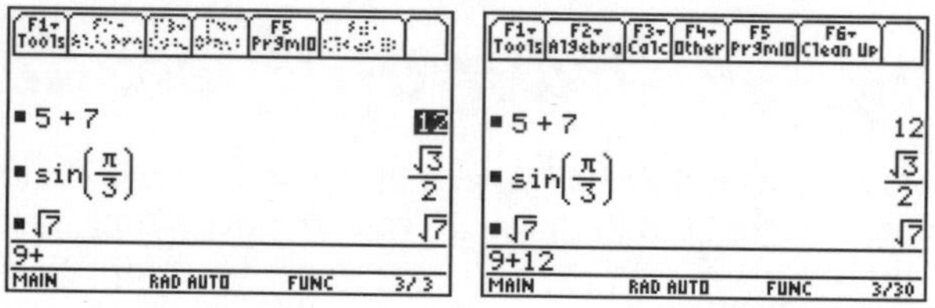

Highlight answer

Press [ENTER]

Figure 2-4: Inserting a previous answer in an expression.

If you want to start an expression with the previous answer, all you have to do is key in the first operation that is to appear after the answer — the calculator automatically knows to use the previous answer. For example if you add 1+2 and then want to divide the answer by 3, simply key in [÷][3] and then press [ENTER] to evaluate the result.

If you want to reuse a command, function, or expression that you previously entered in the calculator — but with different instructions, arguments, or variables — simply recall that command, function, or expression and then edit it. I tell you how to do this in Chapter 1.

Storing Variables

If you plan to use the same number many times when evaluating arithmetic expressions, consider storing that number in a variable. In this context, a variable is simply a name you give to the number. To store a number in a variable, follow these steps:

1. **If necessary, press HOME to enter the Home screen.**
2. **Enter the number you want to store in a variable.**

   You can store the number as an arithmetic expression. This is illustrated in the first entry in Figure 2-5. After you complete the steps for storing the number, the calculator evaluates that expression.
3. **Press STO▶.**
4. **Press ALPHA and then press the key corresponding to the letter of the variable in which you want to store the number.**

   The first picture in Figure 2-5 shows this process. The letters used for storing variables are the letters of the alphabet and the Greek letter θ.

   TIP: If you want to give your number a name consisting of more than one letter, press 2nd ALPHA and then press the keys corresponding to the letters in the name. The name can have no more than eight characters.
5. **Press ENTER to store the value.**

   This is illustrated in the first entry in Figure 2-5.

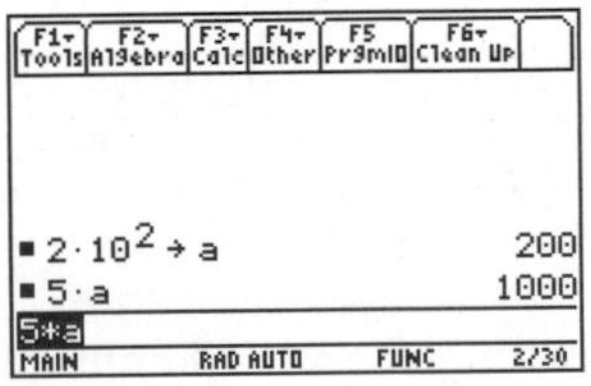

Figure 2-5: Storing a number in a variable.

TIP: The number you store in a variable remains stored in that variable until you delete the variable or until you *or the calculator* stores a new number in that variable. Because the calculator uses the letters *x*, *y*, *z*, *t*, *n*, and θ when graphing functions, parametric equations, sequences, and polar equations, the calculator might change the value stored in these variables when the calculator is in graphing mode. For example, if you store a number in the variable *x* and then ask the calculator to find the zero of the graphed function x^2, the calculator replaces the number stored in *x* with 0, the zero of x^2. So avoid storing a value in one of these variables (*x*, *y*, *z*, *t*, *n*, and θ) if you want that value to remain stored in that variable after you have graphed functions, parametric equations, sequences, or polar equations.

After you have stored a number in a variable, you can insert that number into an arithmetic expression. To do so, place the cursor where you want the number to appear, and then key in the name of the variable in which the number is stored. This is illustrated in the second entry in Figure 2-5.

WARNING!

Implied multiplication (juxtaposition)

Do not use juxtaposition (*ab*) to denote the product of two stored variables *a* and *b*. If you do, you get garbage or an error message because the calculator interprets *ab* as being a single entity having a two-letter name. My recommendation is to *always* use the [×] key when you're multiplying.

Deleting Stored Variables

If you stored a number in variable *a* and then at a later date use that variable name for something else (like getting the calculator to display the quadratic formula by having it solve the equation $ax^2 + bx + c = 0$), the calculator does something you don't want it to do — it replaces *a* with the number stored in *a*. So, after you finish using a stored variable, make sure you delete it from the memory of the calculator.

To delete all variables that have one-letter names, press [2nd][F1][1][ENTER]. To delete *all* one-letter named variables and also clear the Home screen, press [2nd][F1][2][ENTER][ENTER].

To delete a single variable, whether it has a single- or multi-character name, follow these steps:

1. **Press [2nd][−] to display the VAR-LINK screen.**
2. **Repeatedly press ⊙ to highlight the name of the variable you want to delete.**

   Variables containing numbers have the extension EXPR displayed to the right of their names.
3. **Press [F1][1][ENTER] to delete that variable from the memory of the calculator.**
4. **Press [ESC] to exit the VAR-LINK screen.**

To delete several variables on the VAR-LINK screen, use the ⊝ key to highlight one variable and press [F4] to place a check next to it. Continue this process until you have placed a check mark next to all the variables you want to delete. Then press [F1][1][ENTER] to delete the checked variables.

Combining Expressions

You can *combine* (link) several expressions or commands into one expression by using a colon to separate the expressions or commands. (You enter a colon by pressing [2nd][4].) Combining expressions is a really handy way to write mini-programs, as I detail in the "Writing a mini-program" sidebar.

Writing a mini-program

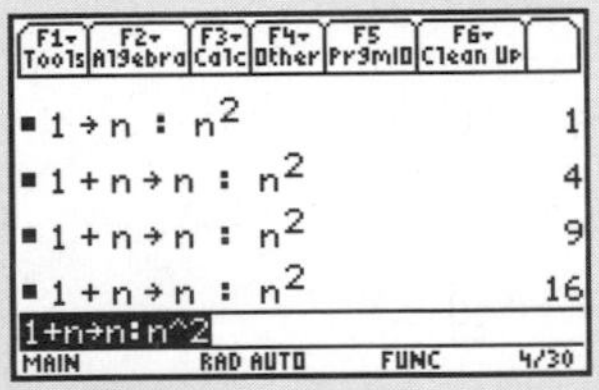

This figure depicts a program that calculates n^2 when n is a positive integer. The first line of the program initiates the program by storing the value 1 in n and calculating its square. (I explain storing a number in a variable and combining expressions earlier in this chapter.)

The next line on the left shows the real guts of the program. It increments n by 1 and then calculates the square of the new value of n. Of course, the calculator doesn't do these calculations until after you press [ENTER]. But the really neat thing about this is that the calculator continues to execute this same command each time you press [ENTER]. Because n is incremented by 1 each time you press [ENTER], you get the values of n^2 when n is a positive integer.

Part II

Doing Algebra and Trigonometry

The 5th Wave By Rich Tennant

"IT SAYS HERE IF I SUBSCRIBE TO THIS MAGAZINE, THEY'LL SEND ME A FREE DESK-TOP CALCULATOR. DESKTOP CALCULATOR?!! WHOOAA – WHERE HAVE I BEEN?!!

In this part . . .

This part introduces you to the wonders of a CAS (Computer Algebra System) calculator. It tells you how to solve equations and systems of equations, how to factor polynomials, and . . . well, it tells you how to do many other things that will just amaze you.

And if you're doing trigonometry or working with complex numbers, be sure to check out these chapters.

Chapter 3

Solving Equations, Factoring, and Other Great Stuff

In This Chapter

- Defining your own functions (user-defined functions)
- Evaluating trigonometric, logarithmic, and exponential functions
- Solving equations and systems of equations
- Solving inequalities
- Simplifying polynomial and trigonometric expressions
- Factoring polynomials

This chapter really shows off the Computer Algebra System (CAS) features of the calculator. It tells you how to get the calculator to factor polynomials, simplify algebraic and trigonometric expressions, solve systems of equations, and so on. It even tells you how to get the calculator to tell you those trigonometric formulas you might have forgotten, such as the half-angle formulas.

Evaluating Functions

Do you need to evaluate tan(π/12) or ln(2)? Or do you have your own special function that you need to evaluate at several different values of the independent variable? If so, this section tells you how to do this and much, much more.

Dealing with user-defined functions

The calculator comes equipped with many built-in functions like sin(x) and e^x, but there are often times when you want to define your own function. For example, if you want the calculator to graph $f(x) = x^3 + 2x^2 + 3x + 1$, factor $f(x)$, and evaluate $f(2)$, you'd store the definition of this function in the calculator so that each time you used the function to do a different task you could reference it by name, $f(x)$, instead of having to key in its rather long definition.

Creating a user-defined function

On the Home screen, there are two ways to save the definition of a function in the calculator so that you can later refer to it by its name. One way is to use the STO▶ key to store the definition in the function name the same way you store a number in a variable name, as I explain in Chapter 2. This is illustrated in the first line of the first picture in Figure 3-1. (You enter the arrow in this line by pressing STO▶, and you enter the letter *f* by pressing ALPHA I.) The other way to define a function on the Home screen is by using the **Define** command, as illustrated in the third line of the first picture in Figure 3-1. You enter the **Define** command by pressing F4 1, and this command requires that you enter the function in the format *function name = function definition.* (The = key is in the middle of the left column of keys.)

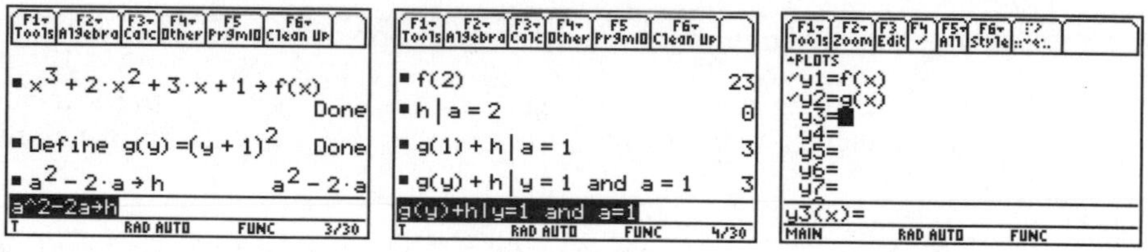

Figure 3-1: Defining and using user-defined functions.

The name you give a user-defined function can be, for example, in the form $f(x)$, $g(y)$, *harry*(a), or just plain h, as in the third function in the first picture in Figure 3-1. The advantage of naming the function h instead of *harry*(a) or $h(a)$ is that entering the function name takes fewer keystrokes. And an abbreviated function name, such as h, is acceptable to the calculator when performing basic algebraic operations such as evaluating or factoring the function. But if you also want to graph the function, you must use the $h(x)$, $h(a)$ or *harry*(y) format for the function name because the calculator's graphing program requires that you specify the argument of the function. If you ask the calculator to graph the user-defined function h, you get an error message.

Evaluating user-defined functions

Evaluating a user-defined function depends on how the function is named. To evaluate a function whose name also specifies the independent variable, such as the first two functions in the first picture in Figure 3-1, simply enter the name of the function, press (, enter the value of the independent variable, press), and then press ENTER, as illustrated in the first line of the second picture of Figure 3-1.

To evaluate a function whose name doesn't specify the independent variable, such as the third function in the first picture in Figure 3-1, enter the name of the function, press [|] (the **with** command), enter the name of the independent variable, press [=], enter the value of the independent variable, and then press [ENTER], as illustrated in the second line of the second picture of Figure 3-1. (The **with** command, [|], is under the [=] key in the left column of keys.)

The last two lines of the second picture in Figure 3-1 show examples of evaluating arithmetic combinations of user-defined functions in which one function specifies the independent variable and the other doesn't. The user-defined functions in this figure are those appearing in the first picture of this figure. In the third line in the second picture in Figure 3-1, the value of the independent variable for the first function is defined by $g(1)$. This is allowable because the name of this function — $g(y)$ — uses the independent variable. But the second function h isn't defined by using its independent variable, so it is necessary to specify the value of its variable by using the **with** command.

The last line of the second picture of Figure 3-1 illustrates a situation in which you need to use the **with** command, [|], to tell the calculator the values of two different independent variables; namely the value of the independent variable y for function g and the independent variable a for function h. This is achieved by connecting these values with the **and** command. You enter the **and** command into the calculator by pressing [2nd][5][8][8] to select this command from the MATH Test menu.

Graphing user-defined functions

The calculator is capable of graphing only the user-defined functions that specify the independent variable in the name of the function, such as the first two functions in the first picture in Figure 3-1. Moreover, when graphing functions, the calculator requires that the independent variable be denoted by x. But if your user-defined function uses some other letter to denote the independent variable, that's okay. The next paragraph tells you how to handle such a situation.

The third picture in Figure 3-1 illustrates how user-defined functions are entered in the Y= editor so that the calculator can graph them. (I explain using the Y= editor and graphing functions in Chapter 5.) If the function is defined as a function of x, simply enter the function name in the Y= editor, as illustrated in $y1$ in this picture, and then graph the function. On the other hand, if the function isn't a function of x, as with the second function in the first picture in Figure 3-1, just use the variable x instead of the variable used in defining the function, as illustrated in $y2$ in the third picture in Figure 3-1, in which function g was originally defined as a function of the variable y.

Evaluating trig functions

Some of the trigonometric functions are found on the calculator's keypad, and some are not. The sine, cosine, and tangent functions, and their inverses, appear in the fifth row of calculator keys above the [Y], [Z], and [T] keys, but the cosecant, cosine, and cotangent functions and their inverses are housed in the MATH Trig menu. This section tells you how to access and use all these trigonometric functions.

Evaluating sine, cosine, and tangent

To evaluate the sine, cosine, or tangent function on the Home screen, follow these steps:

1. **If you aren't already on the Home screen, press [HOME].**

   [HOME] is the fourth key from the top in the left column of keys.

2. **Press [2nd].**

3. **Press [Y] to evaluate sin, [Z] to evaluate cos, or [T] to evaluate tan.**

   The function appears on the command line followed by a left parenthesis, as illustrated in the first picture in Figure 3-2.

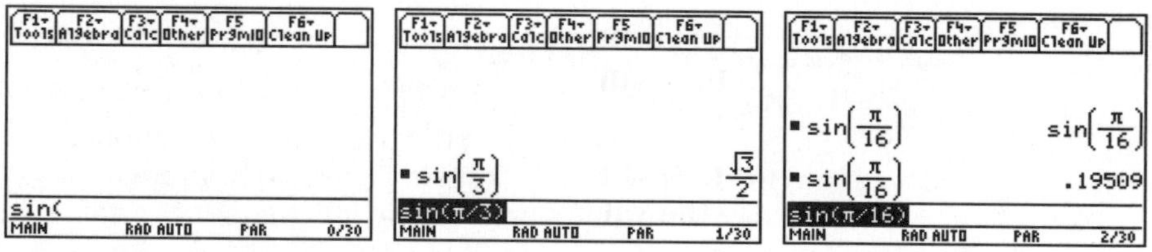

Figure 3-2: Evaluating trigonometric functions.

4. **Enter the value of the argument (angle) of the function, as illustrated in the second picture of Figure 3-2.**

   The angle you enter for the argument must agree with the Angle mode (radians or degrees) of the calculator. This mode appears at the bottom of the screen of the calculator, as illustrated in the first picture of Figure 3-2, where RAD indicates that the calculator is in Radian mode. A calculator in Degree mode displays DEG at the bottom of the screen. (I explain setting the Angle mode in Chapter 1.)

   REMEMBER

   It is possible to override the Angle mode of the calculator. If your calculator is in Radian mode and you want to enter an angle in degrees, simply enter the degree measure of the angle and then press [2nd][I] to place the degree symbol after the entry. If the calculator is in Degree mode and you enter an angle in radians, press [2nd][5][2][2] to place the radian symbol after the entry. A radian angle such as π/4 must first be enclosed in parentheses before placing the radian symbol after it.

5. **Press [)] to close the parentheses.**
6. **Press [ENTER] to evaluate the function, as illustrated in the second picture in Figure 3-2.**

   If your calculator is in Exact or Auto mode, as indicated by the word EXACT or AUTO appearing at the bottom of the screen, the calculator finds the exact value of the function when possible. If your calculator is in Approximate mode, as indicated by the word APPROX at the bottom of the screen, the calculator approximates the value of the function even when an exact value exists. You find out about setting the Mode in Chapter 1.
7. **If the calculator's answer is exactly what you entered, as in the first line of the third picture in Figure 3-2, press [◆][ENTER] to approximate the value of the function, as illustrated in the last line of this figure.**

Evaluating sin^{-1}, cos^{-1}, and tan^{-1}

To evaluate sin^{-1}, cos^{-1}, or tan^{-1} on the Home screen, follow these steps:

1. **If you aren't already on the Home screen, press [HOME].**
2. **Press [◆].**
3. **Press [Y] to evaluate sin^{-1}, press [Z] to evaluate cos^{-1}, or press [T] to evaluate tan^{-1}.**

   The function appears on the command line, followed by a left parenthesis.
4. **Enter the value of the argument of the function. Then press [)] to close the parentheses.**
5. **Press [ENTER] to find the exact value of the function or press [◆][ENTER] to approximate the value of the function.**

   If the calculator is in Radian mode, as indicated by RAD at the bottom of the screen, the exact or approximate value the calculator returns for the inverse trigonometric function is in radians; otherwise, it's in degrees.

   If you press [ENTER] and the calculator returns exactly what you entered, this means that the calculator is unable to find the exact value of the function. In this case, press [◆][ENTER] to have the calculator approximate the value. If your calculator is in Approximate mode, the calculator always returns an approximation of the value of the function even if an exact value exists. (See Chapter 1 for information on setting the Mode.)

Evaluating sec, csc, cot, sec^{-1}, csc^{-1}, and cot^{-1}

The secant, cosecant, and cotangent functions and their inverses are housed in the MATH Trig menu on the calculators that have an up-to-date operating system. All TI-89 Titanium calculators contain the MATH Trig menu, but some older, non-Titanium TI-89's do not. To see whether your TI-89 (non-Titanium) calculator has an up-to-date operating system, press [HOME] if you aren't on the Home screen and then press [F1][⊙]. If **Clock** is the last entry you see on the screen, your calculator's operating system is up-to-date enough to contain the MATH Trig menu; if the last entry is **About**, you need to upgrade the operating system in order to access the MATH Trig menu. The sidebar in this chapter, titled "Dealing with an out-of-date OS," tells how to update the operating system so that your calculator does contain the MATH Trig menu. It also tells you how to handle the secant, cosecant, and cotangent functions and their inverses if you're unable to upgrade an out-of-date operating system.

To evaluate the secant, cosecant, or cotangent functions or their inverses on the Home screen, follow these steps:

1. **If you aren't already on the Home screen, press [HOME].**
2. **Press [2nd][5]⊖ to display the second screen of the MATH menu, as illustrated in the first picture in Figure 3-3.**

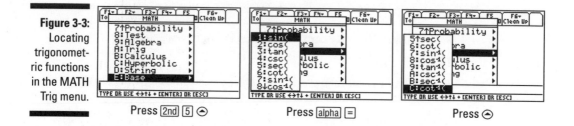

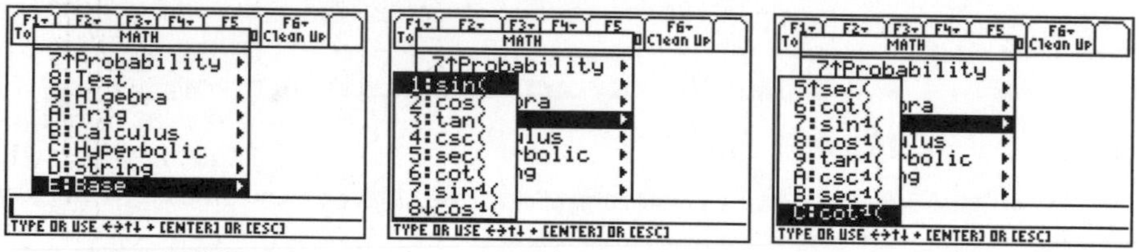

Press [2nd] [5] ⊖ Press [alpha] [=] Press ⊖

Figure 3-3: Locating trigonometric functions in the MATH Trig menu.

3. **Press ⊖⊖⊖⊖[ENTER] (or equivalently, press [ALPHA][=]) to select A: Trig from this menu.**

   The menu housing the trigonometric functions appears on the screen, as illustrated in the second picture of Figure 3-3.

   TIP: If you don't see the **A: Trig** option on this menu, your calculator has an out-of-date operating system. See the sidebar in this chapter titled "Dealing with an out-of-date OS" for information on what to do in this situation.

4. **If the function you want to evaluate doesn't appear in this menu, press ⊖ to display the rest of the menu, as illustrated in the third picture in Figure 3-3.**
5. **Use the ⊖⊖ keys to highlight the function you want to evaluate and then press [ENTER].**

   The function appears on the command line followed by a left parenthesis, as illustrated in the first picture in Figure 3-2.

6. If you're evaluating sec, csc, or cot, follow Steps 4 through 7 in the section titled "Evaluating sine, cosine, and tangent." If you're evaluating $\sec^{-1}$, $\csc^{-1}$, or $\cot^{-1}$, follow Steps 4 and 5 in the section titled "Evaluating $\sin^{-1}$, $\cos^{-1}$, and $\tan^{-1}$."

Evaluating exponential and logarithmic functions

Evaluating exponential functions is fairly straightforward on a TI-89, but do you know how to find the value of *e*? Evaluating the natural log function is also straightforward, but do you know how to find $\log_{10}(2)$ or $\log_5(2)$? This section tells you how to accomplish these tasks.

Dealing with exponential functions

The exponential function e^x is above the [X] key in the left column of keys on the calculator. It appears in the same color as the [◆] key. So to evaluate this function on the Home screen, press [◆][X]. The function appears on the command line followed by a left parenthesis. Enter the value of the exponent, press [)] to close the parentheses, and then press [ENTER] to display the exact value of the function (as illustrated in the first line of the first picture in Figure 3-4) or press [◆][ENTER] to display an approximate value (as in the second line of this picture). For example, the keystrokes for approximating the value of *e* are [◆][X][1][)][◆][ENTER], as illustrated in the third line of the first picture in Figure 3-4.

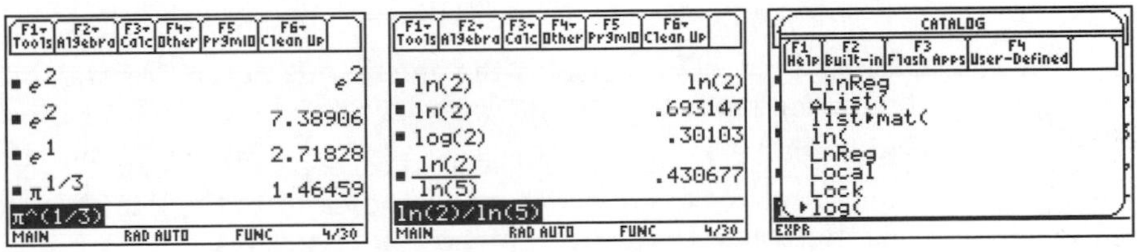

Figure 3-4: Evaluating exponential and logarithmic functions.

REMEMBER

If you press [ENTER] to evaluate an expression and get an exact answer when you really wanted an approximate answer, simply press [◆][ENTER] to tell the calculator to approximate that last answer.

You enter all other exponential functions by using the [^] key, which is located in the right column of keys on the calculator. For example, because π is located above the [^] key, the keystrokes for approximating the cubed root of

π are [2nd][^][^][(][1][÷][3][)][◆][ENTER], as illustrated in the last line of the first picture in Figure 3-4. On the other hand, if you want to display the approximate value of π, no exponent is necessary; simply press [2nd][^] to enter π on the command line and then press [◆][ENTER] to approximate its value.

Dealing with logarithmic functions

The calculator has commands for evaluating the natural log function ln(x) and the common log function $\log_{10}(x)$. Even though the calculator does not house commands for evaluating a logarithmic function to a base other than e or 10, you can always use the natural log function ln(x) and the conversion $\log_b(x) = \ln(x)/\ln(b)$ to evaluate such functions. In this section, I explain how to evaluate the natural and common logarithmic functions.

The natural log function ln(x) is above the [X] key in the left column of keys on the calculator. It appears in the same color as the [2nd] key. So to evaluate this function on the Home screen, press [2nd][X]. The function appears on the command line, followed by a left parenthesis. Enter the value of the argument, press [)] to close the parentheses, and then press [ENTER] to display the exact value of the function (as illustrated in the first line of the second picture in Figure 3-4) or press [◆][ENTER] to display an approximate value (as on the second line of this picture).

TIP

If you press [ENTER] to evaluate an expression and get an exact answer when you really wanted an approximate answer, press [◆][ENTER] to tell the calculator to approximate that last answer.

The natural log function ln(x) has base e. To evaluate the common log function $\log_{10}(x)$, follow these steps:

1. **If you aren't already on the Home screen, press [HOME].**
2. **Press [CATALOG][4] to display the commands in the CATALOG that begin with the letter *L*.**
3. **Press [2nd][↓] to display the second screen in this menu. Then repeatedly press [↓] to move the arrow indicator to the left of the log(entry, as illustrated in the third picture in Figure 3-4.**
4. **Press [ENTER] to place the log function on the command line.**
5. **Enter the argument of the function and then press [)] to close the parentheses.**
6. **Press [◆][ENTER] to approximate the value of the function, as illustrated in the third line of the second picture in Figure 3-4.**

TIP

To evaluate $\log_b(x)$, use the conversion $\log_b(x) = \ln(x)/\ln(b)$. This is illustrated in the last line in the second picture in Figure 3-4, where the value of $\log_5(2)$ is approximated.

Solving Equations and Inequalities

You use the **solve** command in the Algebra menu to solve equations and inequalities. You can even use it to solve small systems of two equations in two unknowns. For large systems of equations, such as a system of three equations in three unknowns, it is usually easier to use matrices to solve the system of equations. (See Chapter 17 for more on using matrices to solve systems of equations.) This section tells you how to use the **solve** command to solve equations, inequalities, and systems of equations.

Solving a single equation or inequality

The **solve** command is used to find the real solutions to an equation or inequality, and the **cSolve** command is used to find both the real and complex solutions. I explain using the **cSolve** command later in this chapter, in the section "Finding complex solutions." To use the **solve** command to find the real solutions to an equation or inequality, follow these steps:

1. **If you aren't already on the Home screen, press [HOME].**
2. **Press [F2][1] to invoke the solve command.**

   The **solve** command appears on the command line, followed by a left parenthesis.
3. **Enter the equation or inequality to be solved, as illustrated in the first two pictures in Figure 3-5.**

   You can use any letter or name for the variables and constants in the equation or inequality, as illustrated in Figure 3-5. The key for the equal sign is in the middle of the left column of keys. You enter the < symbol by pressing [2nd][0], and you enter > by pressing [2nd][.]. The other inequality symbols are housed in the MATH Test menu, which you access by pressing [2nd][5][8], as illustrated in the third picture in Figure 3-5. For example, to enter ≤ in an inequality, press [2nd][5][8][3].

   Use [×] (the times key) when entering the product of two letters that represent different variables or constants. If you don't, the calculator interprets the two juxtaposed letters as being a two-letter name for a single variable. For example, to the calculator, ax^2 is the square of the single variable having the two-letter name *ax*; whereas the calculator views $a \cdot x^2$ as the product of the variable *a* and the square of the variable *x*.

   The absolute value of a quantity is entered into an expression by pressing [2nd][5][▶][2] to select the **abs** command from the MATH Number menu, keying in the quantity, and then pressing [)] to enclose that quantity in the absolute value symbols, as illustrated on the command line in the first picture in Figure 3-5.

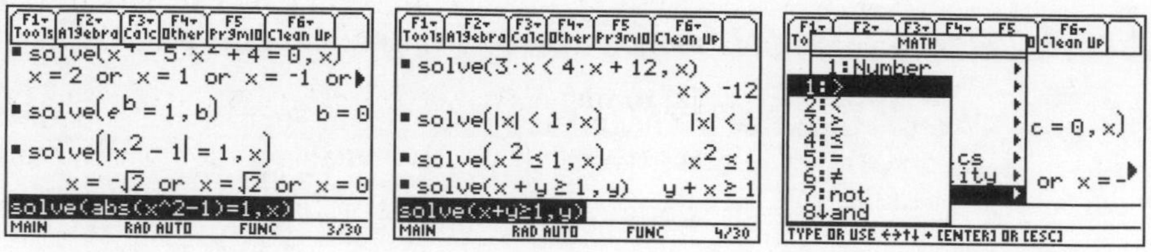
solve(x = 2 or x = 1 or x = -1 or
solve(e^b = 1, b) b = 0
solve(|x² - 1| = 1, x)
x = -√2 or x = √2 or x = 0
solve(abs(x^2-1)=1,x)
MAIN RAD AUTO FUNC 3/30
solve(3·x < 4·x + 12, x)
x > -12
solve(|x| < 1, x) |x| < 1
solve(x² ≤ 1, x) x² ≤ 1
solve(x + y ≥ 1, y) y + x ≥ 1
solve(x+y≥1,y)
MAIN RAD AUTO FUNC 4/30
MATH
1:Number
1:>
2:<
3:≥
4:≤
5:=
6:≠
7:not
8↓and
TYPE OR USE ←→↑↓ + [ENTER] OR [ESC]

Figure 3-5: Solving equations and inequalities.

4. **Press [,] and then enter the name of the variable you want to solve for, as illustrated in the first two pictures in Figure 3-5. Then press [)] to close the parentheses.**

5. **Press [ENTER] to solve the equation or inequality.**

   If the calculator is unable to display all solutions in one line, as in the second line in the first picture in Figure 3-5, use the ⊖ to highlight that line and then repeatedly press ⊙ to display the other solutions. To return to the command line, press [◆]⊖⊖.

The calculator isn't very good at solving inequalities, as illustrated in the second picture in Figure 3-5. The calculator has no problem with linear inequalities in one variable, such as in the first line in this picture, but, as the other entries in this picture indicate, its ability to solve other types of inequalities is usually quite disappointing. Sorry folks, but that's the nature of the beast!

TIP

If you used the **solve** command to solve a trigonometric equation and the answer contains symbols like @n1, @n2, and so on, see the section "Solving trigonometric equations" later in this chapter to discover the meaning of these symbols.

TIP

Have you forgotten the quadratic formula? No problem — just use the **solve** command to solve the equation $ax^2 + bx + c = 0$ for x.

Solving a system of equations

Solving large systems of equations on the Home screen is doable, but keying in all those equations is time consuming. If you often have to solve a system of three or more equations, I suggest that you look at Chapter 17 to find out how to use matrices to solve your large systems of equations.

To solve a system of equations on the Home screen, follow these steps:

1. **If you aren't already on the Home screen, press [HOME].**
2. **Press [F2][1] to invoke the solve command.**
3. **Enter the first equation in the system of equations.**

For details on entering the equation, see the explanation after Step 3 in the previous section titled "Solving a single equation or inequality."

4. **Press [2nd][5][8][8] to insert the word *and* from the MATH Test menu.**

   The MATH Test menu is displayed in the third picture in Figure 3-5.

5. **Enter the next equation in the system of equations.**
6. **If your system has more than two equations, repeat Steps 4 and 5 until you've entered all the equations in the system.**
7. **Press [,] to indicate that all equations in the system have been entered.**
8. **Enter the variables in the system as a list.**

   Lists are housed within braces and separated by commas, as illustrated on the command line in Figure 3-6. To enter your list of variables, follow these steps:

   a. **Press [2nd][(] to enter the first brace.**

   b. **Enter the first variable in your system of equations.**

   c. **Press [,] and enter the next variable.**

   d. **If your system has more than two variables, repeat the previous step until you've entered all the variables.**

   e. **Press [2nd][)] to close the braces.**

9. **Press [)] to close the parentheses.**
10. **Press [ENTER] to solve the system of equations, as illustrated in Figure 3-6.**

    If the calculator can't display all solutions in one line, as in the second line in the first picture in Figure 3-5, use the ▲ to highlight that line and then repeatedly press ► to display the other solutions. To return to the command line, press [◆]▼▼.

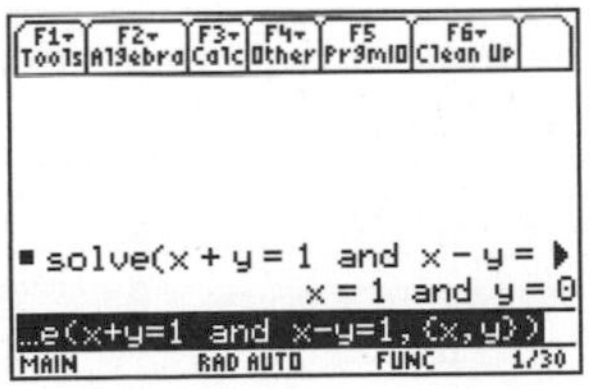

Figure 3-6: Solving a system of equations.

Finding particular solutions to equations

There are times when you might be interested in seeing only particular solutions to an equation or system of equations instead of seeing all solutions. For example, you might want only the positive solutions or only the solutions between zero and one. Being able to display only particular solutions instead of all solutions is quite handy when the equation or system of equations has so many solutions that you'd have to use the arrow keys to search amongst all solutions in order to find the ones you want. This is illustrated in the first picture in Figure 3-7, in which only a portion of the positive solutions is displayed on the screen. (I explain how to define a function, such as the function in this picture, earlier in this chapter in the section, "Creating user-defined functions.")

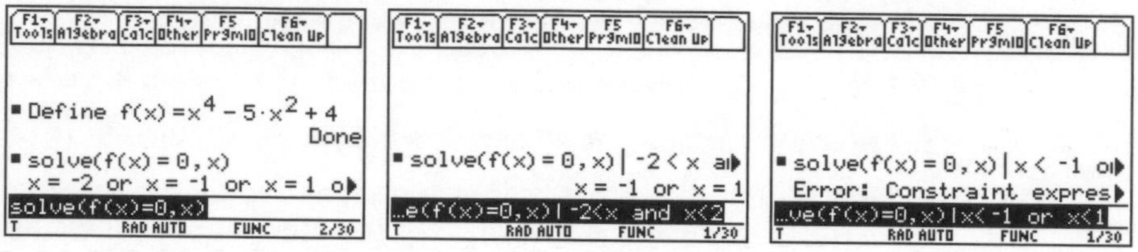

Figure 3-7: Finding particular solutions.

You find particular solutions to an equation or system of equations by using the **with** command to place restrictions on the solutions found by the calculator. These restrictions must be in the form $x \leq a$, $x \geq a$, or $a \leq x \leq b$, where x is the variable in the equation or system of equations and a and b are real numbers. The inequalities can be strict inequalities (that is, you can replace $\leq$ with $<$ or $\geq$ with $>$).

To find particular solutions to an equation or system of equations, follow these steps:

1. **If you're finding particular solutions to an equation, follow Steps 1 through 4 in the section "Solving a single equation or inequality." If you're finding particular solutions to a system of equations, follow Steps 1 through 9 in the section "Solving a system of equations."**
2. **Press [|] to insert the with command.**

   [|] is the fourth key from the bottom of the left column of keys.
3. **Enter the restrictions on the solutions that define the particular solutions you want to find.**

   REMEMBER

   These restrictions must be in the form $x \leq a$, $x \geq a$, or $a \leq x \leq b$, where x denotes the variable used in the equation or system of equations. The inequalities can be strict inequalities; for example, the strict inequality corresponding to $\leq$ is $<$.

   An expression of the form $a \leq x \leq b$ must be entered as "$a \leq x$ and $x \leq b$," as illustrated on the command line in the second picture in Figure 3-7, in which the inequalities are strict inequalities. The symbols for the inequalities and for the word *and* are housed in the MATH Test menu (as illustrated in the third picture in Figure 3-5). Press [2nd][5][8] to access this menu and then press the number of the desired option to insert that option on the command line. For example, to insert the word *and*, press [2nd][5][8][8].

You can enter the < symbol into an expression by simply pressing [2nd][0]; to enter > press [2nd][.].

It makes perfect sense to us to, for example, find those solutions to an equation that are less than –1 or greater than 1. But when using the **with** command to find particular solutions to an equation, the calculator doesn't accept the **or** command. If you use it, you get the "Constraint expression invalid" error message, as illustrated in the third picture in Figure 3-7. You can, of course, solve the equation twice — first using the restriction $x \le -1$ and then using $x \ge 1$ — but you can't solve them simultaneously on the calculator.

4. **Press [ENTER] to find the particular solutions, as illustrated in the second picture in Figure 3-7.**

Solving trigonometric equations

As illustrated in the first two pictures in Figure 3-8, you solve a trigonometric equation or a system of trigonometric equations the same way you solve an equation or a system of equations, which I explain earlier in this chapter in the sections "Solving a single equation or inequality" and "Solving a system of equations." But interpreting the answers the calculator gives for the solutions is a bit different, as seen on the second line in each of these pictures.

The calculator uses the notation @n1, @n2, and so on, to denote an arbitrary integer. For example, the notation 2 • @n • π in the second line of the first picture in Figure 3-8 is the calculator's way of saying "$2n\pi$ where n is an arbitrary integer." If the calculator needs to tell you that the arbitrary integer used in one expression isn't the same as that used in another expression, it places a different number after @n, as illustrated in the second picture in Figure 3-8.

If you're interested in finding only the solutions to a trigonometric equation that fall in a specified range, the preceding section, "Finding particular solutions to equations," tells you how to do this. For example, the last two entries in the first picture of Figure 3-8 illustrate the solution to the equation $\sin\theta = 1/2$ when $0 < \theta < \pi$.

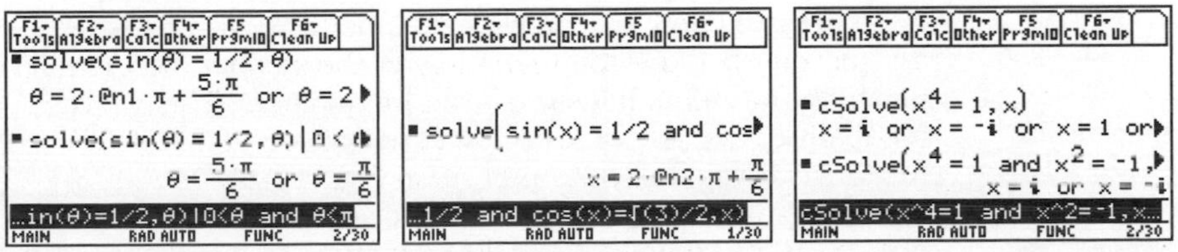

Figure 3-8: Solving trigonometric equations and finding complex solutions.

Finding complex solutions

Independent of the Mode setting of the calculator (which I explain in Chapter 1), you need a special command, **cSolve**, to find the real *and* complex solutions to an equation or system of equations.

To find the real and complex solutions to an equation, follow the steps in the section in this chapter titled "Solving a single equation or inequality," replacing Step 2 with "Press [F2]▲▲▶[1]" to invoke the **cSolve** command instead of the **solve** command, as illustrated in the first two lines in the third picture in Figure 3-8.

To find these real and complex solutions to a system of equations, follow the steps in the section "Solving a system of equations," replacing Step 2 with "Press [F2]▲▲▶[1]" to invoke the **cSolve** command instead of the **solve** command, as illustrated in the last two lines in the third picture in Figure 3-8.

Dealing with Polynomials

Do you need to find the quotient and remainder when dividing polynomials? Or maybe you need to find the partial fraction decomposition of a rational expression? Perhaps you simply need to add, subtract, multiply, or factor polynomials? Whatever the task is, this section tells you how to accomplish it.

Adding and subtracting rational expressions

The **comDenom** command combines rational expressions of polynomials into one rational expression, as illustrated in the first two pictures in Figure 3-9. To use this command on the Home screen, press [2nd][6], enter the combination of rational expressions, press [)] to close the parentheses, and then press [ENTER] to display the answer.

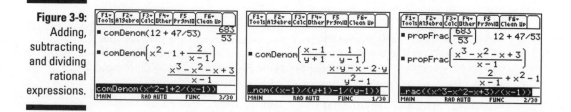

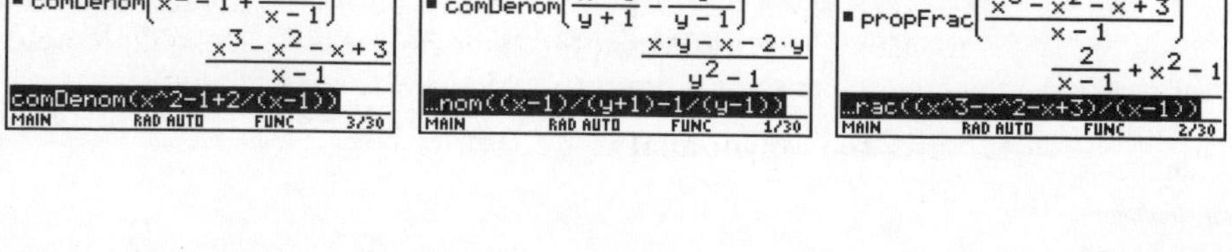

Figure 3-9: Adding, subtracting, and dividing rational expressions.

Dividing polynomials

The **propFrac** command is used to find the quotient and remainder when dividing polynomials. To use this command on the Home screen, press [F2][7], enter the division problem in the form dividend/divisor, press [)] to close the parentheses, and then press [ENTER] to display the answer. The answer is displayed in the form remainder/divisor + quotient, as illustrated in the third picture in Figure 3-9.

Multiplying polynomials

The **expand** command is used to expand a product of polynomials, as illustrated in the first entry in the first picture in Figure 3-10. To use this command on the Home screen, press [F2][3], enter the product of polynomials, press [)] to close the parentheses, and then press [ENTER] to expand the expression.

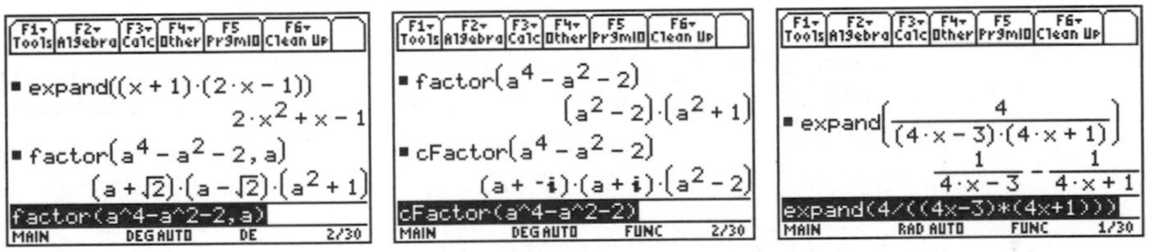

Figure 3-10: Multiplying and factoring polynomials; finding partial fraction decompositions.

Factoring polynomials

On the Home screen, you use the **factor** command to find the real factors of a polynomial, and you use the **cFactor** command to find both the real and the complex factors, as illustrated in the first two pictures in Figure 3-10. To use these commands on the Home screen, follow these steps:

1. **If you're interested only in the real factors of a polynomial, press [F2][2] to place the factor command on the command line. If you want to see all real and complex factors, press [F2][▲][▲][▶][2] to select the cFactor command from the Algebra Complex menu.**
2. **Enter the polynomial to be factored.**

3. **If you are not interested in seeing the irrational factors, skip this step; otherwise, press [,] and enter the variable used in expressing the polynomial.**

   The last entry in the first picture in Figure 3-10 illustrates how this step is used to find the irrational factors of a polynomial. The entries in the second picture in this figure show examples where this step was skipped so that the irrational factors would not be displayed.

4. **Press [)] to close the parentheses, and then press [ENTER] to factor the polynomial.**

Finding partial fraction decompositions

The **expand** command can be used to find the partial fraction decomposition of a rational expression having polynomials in the numerator and the denominator, as illustrated in the third picture in Figure 3-10. To use this command on the Home screen, press [F2][3], enter the rational expression, press [)] to close the parentheses, and then press [ENTER] to display the partial fraction decomposition of the expression.

Doing Trigonometry

This section tells you how to convert an angle measurement to degrees, radians, and DMS (degrees, minutes, and seconds). It also tells you how to get the calculator to simplify trigonometric expression. You even find out how to get the calculator to tell you those trigonometric formulas you've forgotten.

Converting angle units of measurement

The next section tells you how to enter an angle and its unit of measurement into the calculator, and the three sections after that tell you how to convert that measurement to radians, degrees, or DMS (degrees, minutes, and seconds).

Entering an angle and its unit of measurement

To enter an angle in DMS (degrees, minutes, and seconds), first enter the number of degrees and press [2nd][I] to insert the degree symbol. Then enter the number of minutes and press [2nd][=] to insert the minute symbol. Finally, enter the number of seconds and press [2nd][1] to insert the second symbol. This process is illustrated on the command line of the second picture in Figure 3-11.

Whether you enter a degree or radian angle in the calculator depends on the Angle mode (radians or degrees) of the calculator. This mode is displayed at the bottom of the screen. DEG on this bottom line indicates that the calculator is in Degree mode, as in the first picture in Figure 3-11. A calculator in Radian mode displays RAD at the bottom of the screen, as in the second picture in Figure 3-11. (You find out about setting the Angle mode in Chapter 1.)

If the unit of measurement of the angle you want to enter agrees with the Angle mode of the calculator, simply enter the numeric value of the angle. For example, you enter 3.72 degrees on a calculator in DEG mode as simply 3.72, as illustrated in the first entry in the first picture in Figure 3-11, where the calculator supplied the parentheses. On a calculator in RAD mode, you enter π/16 radians as π/16, as illustrated in the second entry in the second picture in Figure 3-11, where the calculator supplied the parentheses.

If the unit of measurement of the angle you want to enter *does not* agree with the Angle mode of the calculator, the numeric value of the angle must be followed by the appropriate symbol for the unit of measurement. You enter the degree symbol by pressing [2nd][|], and you enter the radian symbol by pressing [2nd][5][2][2]. If the numeric value of the angle contains an arithmetic operation, you must first enclose this expression in parentheses before entering the unit of measurement of the angle. For example, on a calculator in Degree mode, you enter π/16 radians by pressing [(][2nd][^][÷][1][6][)] to enclose the numeric value π/16 in parentheses and then press [2nd][5][2][2] to indicate that the angle is measured in radians, as illustrated in the second entry in the first picture in Figure 3-11, where the calculator then enclosed the result in parentheses. On a calculator in Radian mode, you enter 3.72 degrees by entering the numeric value 3.72 and then pressing [2nd][|] to indicate that the angle is in degrees, as illustrated in the first entry in the second picture in Figure 3-11, where the calculator then enclosed the result in parentheses.

Converting an angle to DMS (degrees, minutes, and seconds)

To convert an angle to degrees, minutes, and seconds (DMS), press [HOME] if you aren't already on the Home screen and then enter the angle to be converted. (I explain how to enter the angle in the preceding section.) Then press [2nd][5][2][8] to select the **DMS** command from the MATH Angle menu. Finally, press [ENTER] to convert the angle to degrees, minutes, and seconds, as illustrated in the first two entries in each picture in Figure 3-11.

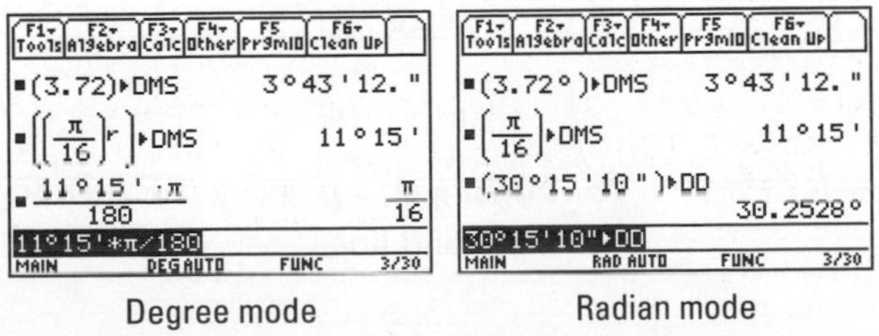

Degree mode Radian mode

Figure 3-11: Converting between degrees, radians, and DMS.

REMEMBER

When you're converting an angle to degrees, minutes, and seconds and the calculator is in Radian mode, the calculator might give you a result containing 60 seconds. Of course, we know that this should be expressed as 1 minute, but the calculator doesn't know that. For example, in Radian mode, the calculator converts 0.5 degrees to 29 minutes and 60 seconds; but in Degree mode, it converts it to 30 minutes. Why does it do this? Beats me!

Converting an angle to degrees

On the Home screen, follow these steps to convert to degrees an angle measured in DMS (degrees, minutes, and seconds) or in radians:

1. **Enter the angle to be converted.**

   You find out how to enter the angle earlier in this chapter, in the section "Entering an angle and its unit of measurement."

2. **If your calculator is in Degree mode, skip this step. If your calculator is in Radian mode, press [2nd][5][2][9] to select the DD command from the MATH Angle menu, as illustrated on the command line in the second picture in Figure 3-11.**

3. **Press [ENTER] to have the calculator find, if possible, the exact degree measure of the angle, or press [◆][ENTER] to approximate this value, as illustrated in the last entry of the second picture in Figure 3-11.**

Converting an angle to radians

On the Home screen, follow these steps to convert an angle to radians:

1. **Enter the angle to be converted.**

   I explain how to enter the angle earlier in this chapter in the section "Entering an angle and its unit of measurement."

2. **If your calculator is in Radian mode, skip this step. If your calculator is in Degree mode, multiply the entry from Step 1 by $\pi/180$, as illustrated on the command line in the first picture in Figure 3-11.**

   Because the calculator doesn't have a special command for converting degrees to radians when in Degree mode, you have no choice but to convert units by using the fact that 1 degree = $\pi/180$ radians.

3. **Press [ENTER] to have the calculator find, if possible, the exact radian measure of the angle, as illustrated in the last entry of the first picture in Figure 3-11, or press [◆][ENTER] to approximate this value.**

Dealing with trigonometric expressions

The calculator has two commands, **tExpand** and **tCollect**, that are somewhat useful when simplifying trigonometric expressions and even more useful if you've forgotten a trig formula, such as one of the half-angle formulas. You use the **tExpand** command with trigonometric functions whose arguments consists of sums, differences, or integer multiples of angles, such as those in Figure 3-12. It converts such a function into a combination of the sine and cosine functions, each being a function of a single angle. To use this command on the Home screen, press [F2]↑↑↑→[1] to place the **tExpand** command on the command line, enter the trigonometric function, press [)] to close the parentheses, and then press [ENTER] to expand the expression, as illustrated in Figure 3-12.

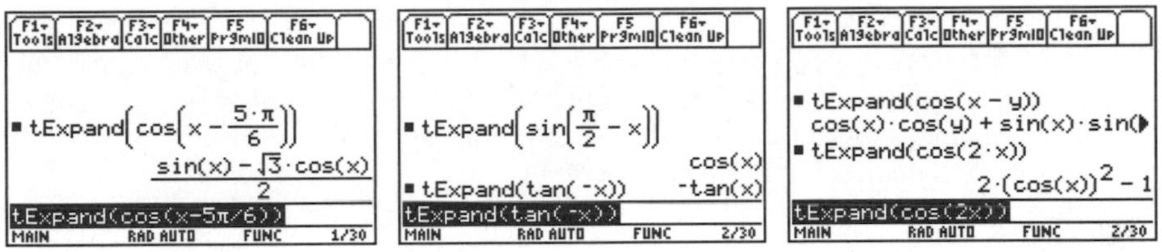

Figure 3-12: Expanding trigonometric functions.

You use the **tCollect** command with products and integer powers of trigonometric functions, such as those in Figure 3-13. It converts such expressions into a combination of the sine and cosine functions in which the argument of each of these functions is a sum, difference, or integer multiple of angles. To use this command on the Home screen, press [2nd]↑↑↑→[2] to place the **tCollect** command on the command line, enter the trigonometric expression, press [)] to close the parentheses, and then press [ENTER] to simplify the expression, as illustrated in Figure 3-13.

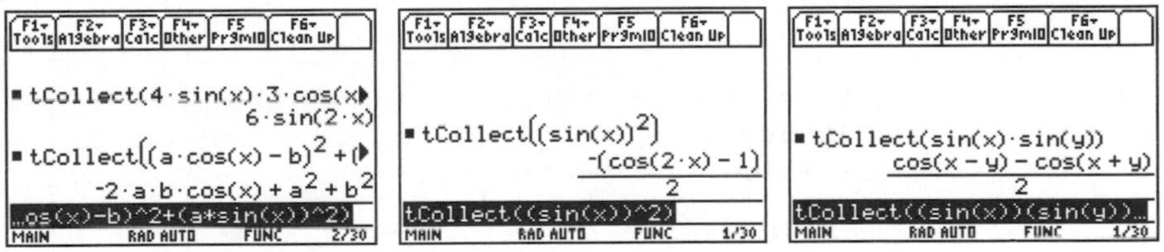

Figure 3-13: Simplifying products and powers of trigonometric functions.

TIP

Have you forgotten a trigonometric formula? No problem. As the last two pictures in Figure 3-12 show, you can use the **tExpand** command to give you a cofunction, reduction, sum and difference, or double-angle formula. And the last two pictures in Figure 3-13 show that you can use the **tCollect** command to display a power-reducing or product-to-sum formula. But be warned, all these formulas generated by the calculator are given in terms of the sine and cosine functions, so, for example, the double-angle formula for tangent won't be expressed on the calculator the way it is in textbooks. However, the formula given by the calculator is mathematically equivalent to the double-angle formula for tangent found in a textbook.

Dealing with an out-of-date OS

In an ideal world, all calculators have the current version of the operating system (OS). If you have the cable that links your calculator to a computer, see Chapter 21 for details on how to download the free upgrade of the operating system by using the free TI Connect software.

If your OS is out of date, your TI-89 might not, for example, have the MATH Trig menu. To check it out, press [2nd][5][▲]. If you see **A: Trig** on this menu, your calculator has the MATH Trig menu. If you see **A: Calculus**, your calculator doesn't contain the MATH Trig menu and your OS is out of date.

What do you do if your calculator's OS is out of date? Well, the best thing to do is to upgrade your OS. Even if you don't have the cable that links your calculator to a computer, you can always link your calculator to another calculator of the same kind that does have the current version of the OS. The cable that links these calculators came with the calculator, and Chapter 22 tells you how to upgrade your TI-89 by using an already upgraded TI-89.

And what do you do if you have no means of upgrading the OS on an older TI-89 that doesn't have the MATH Trig menu? Well, if you want to use the secant, cosecant, and cotangent functions and their inverses, you can use the following conversions:

- $\sec(x) = 1/\cos(x)$
- $\csc(x) = 1/\sin(x)$
- $\cot(x) = 1/\tan(x)$
- $\sec^{-1}(x) = \cos(1/x)$
- $\csc^{-1}(x) = \sin(1/x)$
- $\cot^{-1}(x) = \tan(1/x)$

Chapter 4

Dealing with Complex Numbers

In This Chapter

- Doing arithmetic with complex numbers
- Finding real and imaginary parts
- Finding conjugates and moduli

Do you know how to enter a complex number into the calculator? Do you know how to display solutions to arithmetic combinations of complex numbers in polar form? Do you know how to find the modulus of an arithmetic combination of complex numbers? If not, this chapter tells you how to accomplish these tasks.

Doing Arithmetic with Complex Numbers

Where's the *i* key? If you don't know, this section is for you. It tells you how to enter and display complex numbers and how to do arithmetic with complex numbers.

Displaying complex numbers

Before doing arithmetic with complex numbers, you should decide whether you want answers displayed in rectangular ($a + bi$) or polar ($re^{i\theta}$) form. The first two pictures in Figure 4-1 show answers displayed in rectangular form; the third picture displays answers in polar form.

When the calculator is in Real or Rectangular mode, the calculator displays complex numbers in rectangular form ($a + bi$); in Polar mode, you see them in polar form ($re^{i\theta}$). To change the Mode setting, press [MODE]▼▼▼▼▼▶ to display the Complex Format options in the Mode menu, press the number of the option you want, and then press [ENTER] to save the settings in the Mode menu.

TIP: See Chapter 11 for information on how to convert between the rectangular and polar forms of a complex number.

Entering and storing complex numbers

A complex number is entered in the calculator in the form $a + bi$. You enter the complex number i by pressing [2nd][CATALOG]. For example, you enter the complex number $1 + yi$ in the calculator by pressing [1][+][Y][2nd][CATALOG].

A complex number that you plan to use several times can be stored in the memory of the calculator. I explain how to do this in detail in Chapter 2. As an example, to store the complex number $1 + i$ in the variable named c, press [1][+][2nd][CATALOG] to enter the number, and then press [STO▸][ALPHA][)][ENTER] to store it in c. In the future, you can place this complex number in an expression simply by pressing [ALPHA][)] to enter the letter c.

REMEMBER

Entities stored under one-letter names can cause problems down the road if you forget you stored them. For example, if you store $1 + i$ in the variable c and then evaluate the expression $x + c$, the calculator replaces c with $1 + i$. To avoid such problems, when you finish using stored entities, press [2nd][F1][1][ENTER] to erase all entities stored in one-letter names. Pressing [2nd][F1][2][ENTER] instead of [2nd][F1][1][ENTER] erases the screen as well as the entities.

TIP

If you want to enter the conjugate, the real or imaginary part, or the modulus of a complex expression into your expression, the section in this chapter titled "Finding the conjugate, the real and imaginary parts, and the modulus" tells you how to do it. Simply follow the five steps in this section without pressing [ENTER] at the end of the last step.

Performing arithmetic operations

You perform arithmetic operations on complex numbers the same way you do them with real numbers, which I explain in detail in Chapter 2. Figure 4-1 illustrates various arithmetic operations performed on complex numbers. There is an exception: when attempting to find the nth roots of a complex number, the calculator finds only one of the n roots, as illustrated in the second picture in this figure.

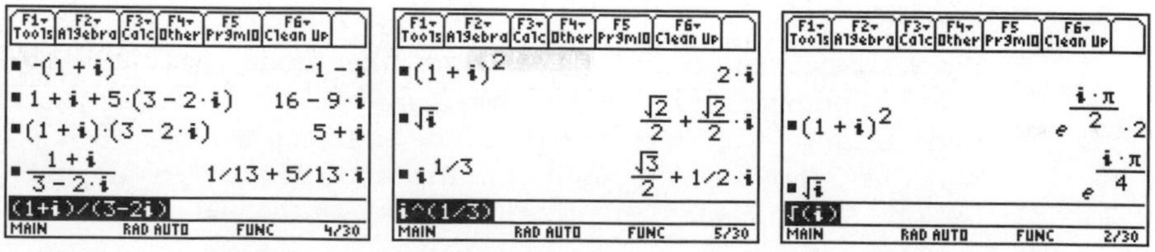

Figure 4-1: Performing arithmetic operations with complex numbers.

Answers in rectangular form

Polar form

Doing Algebra with Complex Numbers

This section tells you how to find the conjugate, the real and imaginary parts, and the modulus of complex expressions. It also tells you how to find complex factors of polynomials and how to find the complex solutions to an equation.

Finding the conjugate, the real and imaginary parts, and the modulus

To find the conjugate, real part, imaginary part, or modulus (absolute value) of an algebraic expression involving complex numbers, follow these steps:

1. **If you aren't already on the Home screen, press [HOME].**

   [HOME] is the fourth key from the top in the left column of keys.

2. **Press [2nd][5][5] to display the MATH Complex menu, as illustrated in the first picture in Figure 4-2.**

3. **Press [1] to find the conjugate, [2] to find the real part, [3] to find the imaginary part, or [5] to find the modulus (absolute value).**

   The calculator places the selected command on the command line and provides the first parenthesis that encloses the expression entered next.

4. **Enter the algebraic expression involving complex numbers.**

5. **Press [)] to close the parentheses of the command, and then press [ENTER] to execute the command, as illustrated in the second picture in Figure 4-2.**

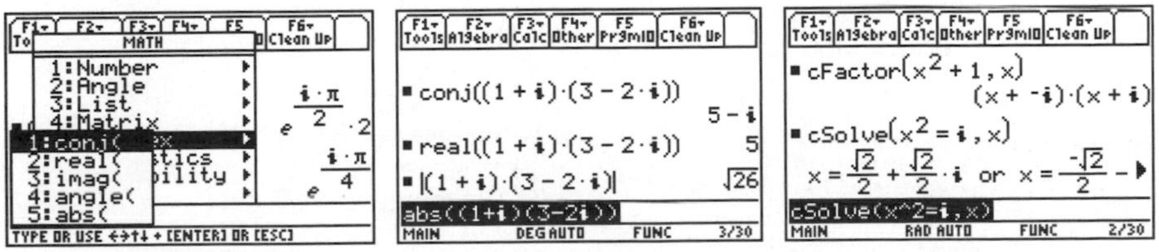

Figure 4-2: Performing algebraic operations with complex numbers.

Finding complex factors and solutions

Chapter 3 explains in detail how to find the complex factors of a polynomial and the complex solutions to an equation or system of equations, as illustrated in the third picture in Figure 4-2. The **cFactor** and **cSolve** commands are housed in the Algebra Complex menu, which you access on the Home screen by pressing [F2]⊖⊖⊙. You can find detailed information on how to use these commands in Chapter 3.

Part III
Graphing and Analyzing Functions

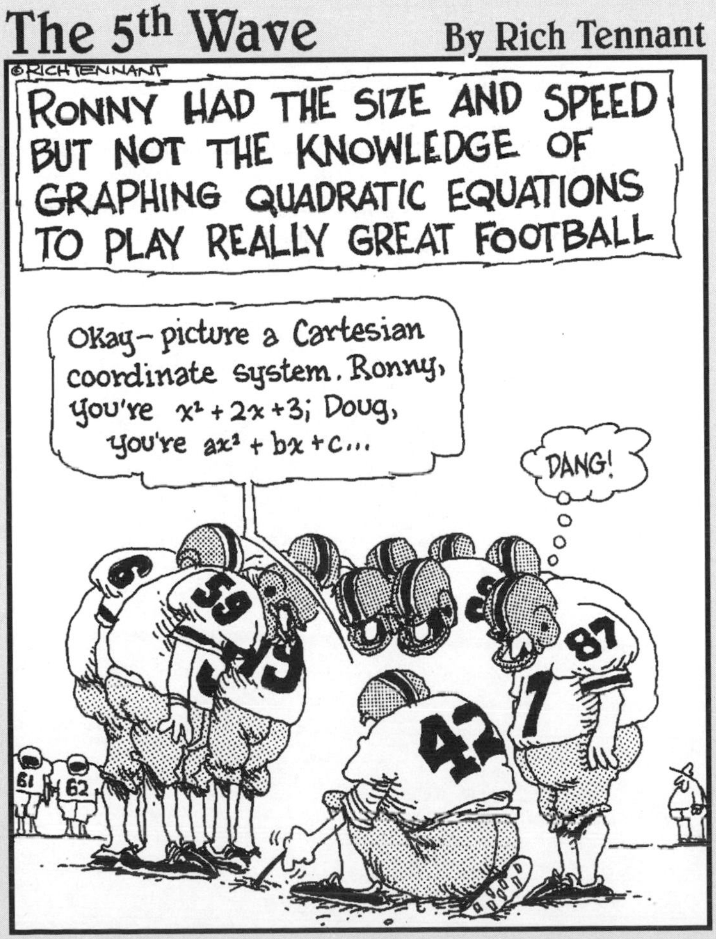
The 5th Wave
By Rich Tennant
©RICH TENNANT
RONNY HAD THE SIZE AND SPEED BUT NOT THE KNOWLEDGE OF GRAPHING QUADRATIC EQUATIONS TO PLAY REALLY GREAT FOOTBALL
Okay– picture a Cartesian coordinate system. Ronny, you're $x^2 + 2x + 3$; Doug, you're $ax^2 + bx + c$...
DANG!

In this part . . .

This part looks at graphing functions and then analyzing them by tracing the graph or by creating tables of functional values. You get pointers on how to find values associated with graphs, such as minimum and maximum points, points of intersection, and the slope of the curve.

Chapter 5

Graphing Functions of One Variable

In This Chapter

- Entering functions into the calculator
- Graphing functions
- Recognizing whether the graph is accurate
- Graphing piecewise-defined and trigonometric functions
- Viewing graphs and functions on the same screen
- Saving and recalling a graph and its settings in a Graph Database

The calculator has a variety of features that help you painlessly graph a function of one variable. The first step is to enter the function into the calculator. Then, to graph the function, you set the viewing window and press [♦][F3]. Or (better yet) you can use one of several Zoom commands to get the calculator to set the viewing window for you. Finally, after you graph the function, you can use Zoom commands to change the look of the graph. For example, you can zoom in or zoom out on a graph the same way that a zoom lens on a camera lets you zoom in or out on the subject of the picture you're about to take. This chapter tells you the ins and outs of graphing functions on the calculator.

Entering Functions

Before you can graph a function, you must enter it into the calculator. The calculator can save up to 99 functions, y_1 through y_{99}. To enter functions in the calculator, perform the following steps:

1. **If necessary, press [MODE] and put the calculator in Function mode, as shown in Figure 5-1.**

   If you see FUNC on the last line of the screen, you don't need to do anything in this step because the calculator is already in Function mode. If you don't see the word FUNC, press [MODE].

After pressing [MODE], the current Graph style in the Mode menu is highlighted and flashing. If this style is not the Function style, press ▶[1] to set the Graph style to Function. Then press [ENTER] to save the new Graph style. (For more about the other items on the Mode menu, shuffle over to Chapter 1.)

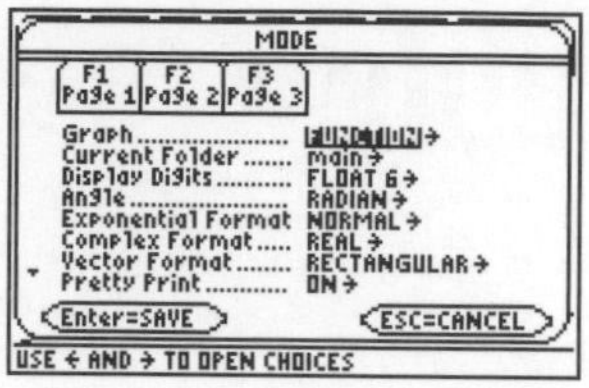

Figure 5-1: Setting Function mode.

2. **Press [◆][F1] to access the Y= editor.**
3. **Enter the definitions of your functions.**

   To erase an entry that appears after y_n, use the ▲▼ keys to place the cursor to the right of the equal sign and then press [CLEAR]. To clear all functions defined in the Y= editor, press [F1][8][ENTER].

   To enter or edit the definition of a function in y_n, use the ▲▼ keys to place the cursor to the right of the equal sign and press [ENTER] or [F3] to place the cursor on the command line. Then enter or edit the definition of the function and press [ENTER] when finished.

When you're defining functions, the only symbol the calculator allows for the independent variable is the letter x. Press [X] to enter this letter in the definition of your function. In Figure 5-2, I used this key to enter the functions y_1, y_2, y_4, y_5, y_6 and y_7.

As a timesaver, when entering functions in the Y= editor, you can reference another function in its definition provided that function is entered in the form $y_n(x)$. The first picture in Figure 5-2 shows function y_3 defined as $-y_2$, where you enter y_2 by pressing [Y][2].

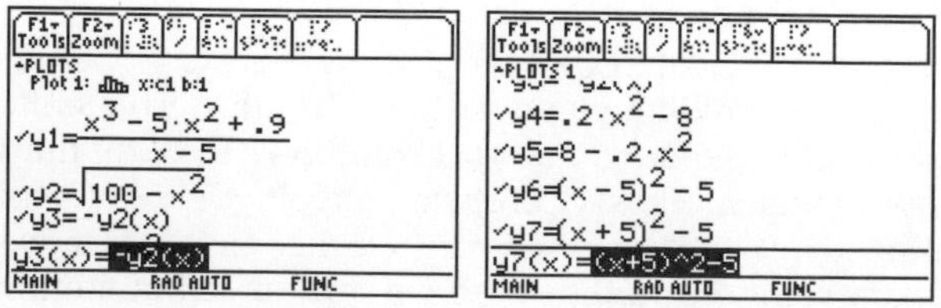

Figure 5-2: Examples of entering functions.

Graphing Functions

Here's where your calculator draws pretty pictures. After you enter the functions, as I describe in the preceding section, you can use the following steps to graph the functions:

1. **Turn off any Stat Plots that you don't want to appear in the graph of your functions.**

   The first line in the Y= editor tells you the graphing status of the Stat Plots. (I discuss Stat Plots in Chapter 20.) If, as in the first picture in Figure 5-2, no number appears after PLOTS on the very first line of the Y= editor, then no Stat Plots are selected to be graphed, even if a number appears on the second line after Plot. (If you haven't instructed the calculator to create any stat plots, or if you've deleted them, the word Plot doesn't appear.) If one or more numbers do appear after PLOTS, as in the second picture in Figure 5-2, those Stat Plots are selected to be graphed along with the graph of your functions.

   To prevent an individual Stat Plot from being graphed, use the ⊖⊖ keys to place the cursor on that Stat Plot and press [F4]. ([F4] toggles between deactivating and activating a Stat Plot.) To prevent all Stat Plots from being graphed, press [F5][5].

2. **Press [F1][9] to access the Graph Formats menu and select the formats that you want for your graph.**

   To do this, use the ⊖⊖ keys to place the cursor on the desired format and press ⊙ to display the options for that format. Then press the number of the option you want. When you finish setting the various formats for your graph, press [ENTER] to save your settings.

   An explanation of the Graph Formats menu options follows:

   - **Coordinates:** This format gives you a choice between having the coordinates of the location of the cursor displayed at the bottom of the graph screen in three ways. Select **RECT** to display (x, y) rectangular coordinates of the cursor or select **POLAR** to show the (r, θ) polar form. If, for some strange reason, you don't want to see the coordinates of the cursor, select **OFF**.

   - **Graph Order:** Select **SEQ** to graph functions one at a time or **SIMUL** to graph all functions at the same time. The SEQ option completes the graph of one function before starting the graph of the next function; the SIMUL option graphs the points for each function corresponding to a given *x*-value and then moves on to the next *x*-value to graph the next point for each function.

   - **Grid:** If you select **ON**, grid points appear in the graph at the intersections of the tick marks on the *x*- and *y*-axes (as illustrated in Figure 5-3). If you select **OFF**, no grid points appear in the graph.

- **Axes:** If you want to see the *x*- and *y*-axes on your graph, select **ON** (as illustrated in the first picture in Figure 5-3). If you don't want to see them, select **OFF** (as shown in the second picture in Figure 5-3).
- **Leading Cursor:** Select **ON** if you want to see the cursor as the function is being graphed (as shown in the second picture in Figure 5-3); otherwise, select **OFF** to hide the leading cursor.
- **Labels:** If you want the *x*- and *y*-axes to be labeled (as in the first picture in Figure 5-3), select **ON**. Because the location of the labels isn't ideal, selecting **OFF** is usually a wise choice.

Figure 5-3: A graph with grid points.

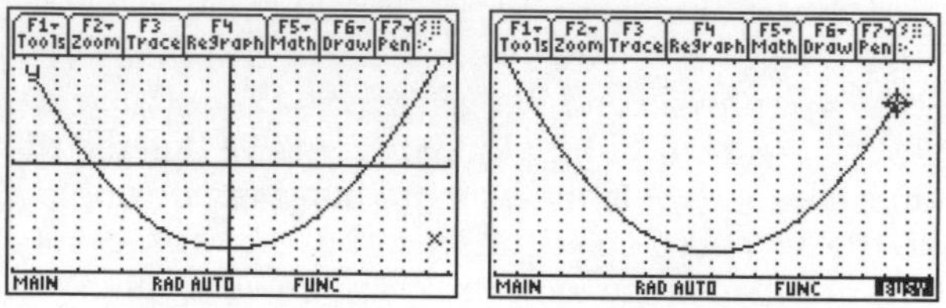

With labeled axes With axes hidden

3. **Press [◆][F2] to access the Window editor.**
4. **After each of the window variables, enter a numerical value that is appropriate for the functions you're graphing. Press [ENTER] or ⊝ to move to the next variable.**

   Figure 5-4 shows the Window editor when the calculator is in Function mode. The items in this menu determine the viewing window for your graph — in particular, how the *x*- and *y*-axes look on the screen. The following gives an explanation of the variables you must set in this editor:

   - **xmin and xmax:** These are, respectively, the smallest and largest values of *x* in view on the *x*-axis.

     TIP: If you don't know what values your graph will need for **xmin** and **xmax**, press [F2][6] to invoke the **ZoomStd** command. This command automatically graphs your functions in the Standard viewing window; the settings for this window appear in Figure 5-4. You can then, if necessary, use the other Zoom commands (see Chapter 6) to get a better picture of your graph.

Figure 5-4: The Window editor in Function mode.

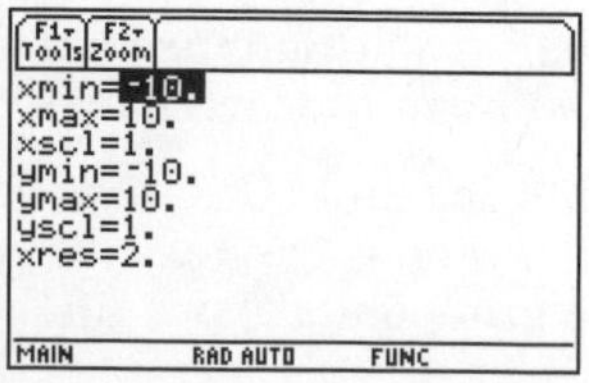

- **xscl:** This is the distance between tick marks on the *x*-axis. (Go easy on the tick marks; using too many makes the axis look like a railroad track. Twenty or fewer tick marks make for a nice-looking axis.)
- **ymin and ymax:** These are, respectively, the smallest and largest values of *y* that will be placed on the *y*-axis.

  If you have assigned values to **xmin** and **xmax** but don't know what values to assign to **ymin** and **ymax**, press [F2]▲▲▲[ENTER] to invoke the **ZoomFit** command. This command uses the **xmin** and **xmax** settings to determine the appropriate settings for **ymin** and **ymax** and then automatically draws the graph. It doesn't change the **yscl** setting. (You must return to the Window editor, if necessary, to adjust this setting yourself.)
- **yscl:** This is the distance between tick marks on the *y*-axis. (As with the *x*-axis, too many tick marks makes the axis look like a railroad track. Fifteen or fewer tick marks make a nice *y*-axis.)
- **xres:** This setting determines the resolution of the graph. It can be set to any of the integers 1 through 10. When **xres** is set equal to 1, the calculator evaluates the function at each of the 158 pixels on the *x*-axis and graphs the result. If **xres** is set equal to 10, the function is evaluated and graphed at every tenth pixel.

  xres is usually set equal to 2. If you're graphing a lot of functions, the calculator might take a while to graph them at this resolution, but if you change **xres** to a higher number, you might not get an accurate graph. You see this in Figure 5-5, which shows the function y_1 in Figure 5-2 graphed first using an **xres** of 1 and then using an **xres** of 10.

  If the calculator takes a long time to graph your functions and this causes you to regret setting **xres** equal to 1 or 2, press [ON] to terminate the graphing process. You can then go back to the Window editor and adjust the **xres** setting to a higher number.

5. **Press [2nd][F3] to graph the functions.**

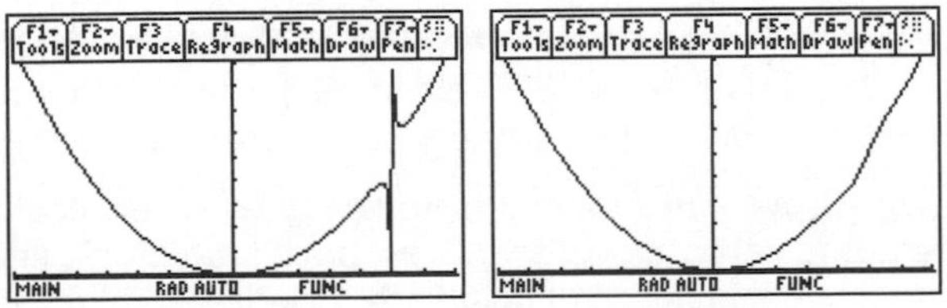

*x*res=1 *x*res=10

Figure 5-5: Graphing a function with a resolution of 1 and 10.

Graphing Several Functions

If you're graphing several functions at once, determining which graph each function is responsible for isn't easy. To help clear this up, the calculator allows you to identify the graphs of functions by setting a different graph style for each function. To do this, follow these steps:

1. **Press [◆][F1] to access the Y= editor.**
2. **Use the ⊙⊙ keys to place the cursor on the definition of the function.**
3. **Press [2nd][F1] to display the Style menu and then press the number of the desired style.**

   You have eight styles to choose from:

   - **Line, Dot, Square, and Thick graph styles:** Figure 5-6 shows these styles in order from top to bottom. As you can see in the first picture, where **xres** is 2, it's hard to distinguish between the Square and Thick line styles, but you can easily identify four different graphs. If you really want to distinguish between these two styles, set **xres** to at least 4, as in the second picture in Figure 5-6.

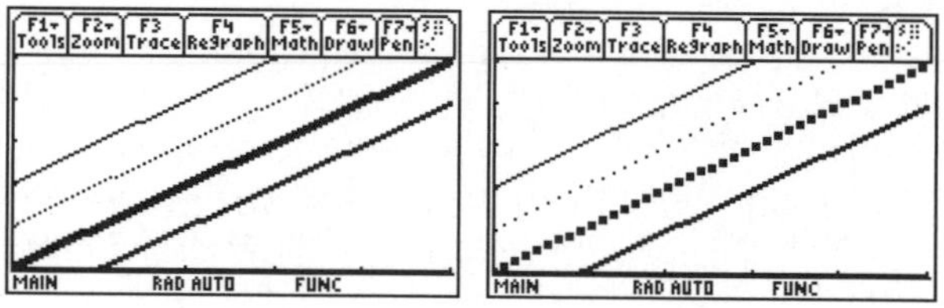

xres=2 xres=4

Figure 5-6: Line, Dot, Square, and Thick graph styles.

   - **Path and Animated styles:** The Path style uses a circle to indicate a point as it's being graphed (as illustrated in the first picture of Figure 5-7). When the graph is complete, the circle disappears and leaves the graph in Line style.

     The Animated style also uses a circle to indicate a point as it's being graphed, but when the graph is complete, no graph appears on the screen. For example, if this style is used, graphing $y = -x^2 + 9$ looks like a movie of the path of a ball thrown in the air.

   - **Shading Above and Below the curve styles:** The calculator has four shading patterns: vertical lines, horizontal lines, negatively sloping diagonal lines, and positively sloping diagonal lines. You don't get to select the shading pattern. If you're graphing only one function in this style, the calculator uses the vertical line pattern. If you're graphing two functions, the first is graphed in the vertical line pattern and the second in the horizontal line pattern. If you graph three functions in this style, the third appears in the negatively sloping diagonal lines pattern, and so on (as illustrated in the second picture in Figure 5-7).

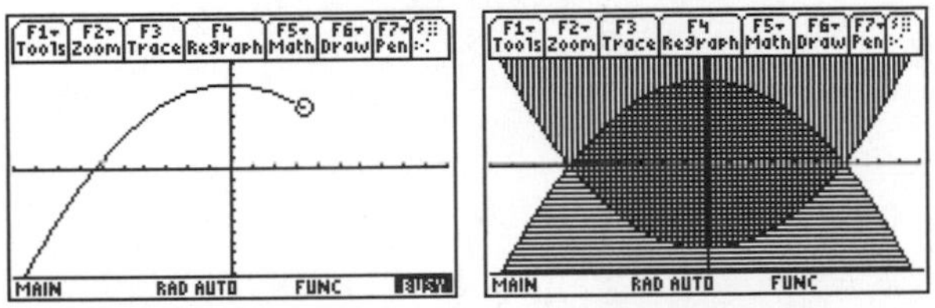

Path style

Shading style

Figure 5-7: Path and Shading styles.

If you don't want the calculator to graph a function in the Y= editor but you might need this function at a later date, remove the check mark to the left of the function. To do this from the Y= editor, use the ▲▼ keys to place the cursor on the function and press [F4]. ([F4] toggles between deactivating and activating a function.) To prevent all functions from being graphed, press [F5][3].

Is Your Graph Accurate?

The calculator can do only what you tell it to do — which doesn't always produce an accurate graph. The three main causes of inaccurate graphs are the following:

- **The graph is distorted by the size of the screen.**

  Because the calculator screen isn't square, circles don't look like circles unless the viewing window is properly set. How do you properly set the viewing window? No problem! Just graph the function, as I describe earlier in this chapter in the section "Graphing Functions," and then press [F2][5] to invoke the **ZoomSqr** command. **ZoomSqr** readjusts the window settings for you and graphs the function again in a viewing window in which circles look like circles. Figure 5-8 illustrates this. (The circle being drawn in each of these figures is the circle defined by y_2 and y_3, as shown earlier in Figure 5-2.)

  If your **ZoomSqr** graph leaves unwanted gaps around the x-axis, as in the second picture in Figure 5-8, change the **xres** setting in the Window menu to 1 to get the calculator to evaluate more points on the graph. Then press [♦][F3] to redraw the graph.

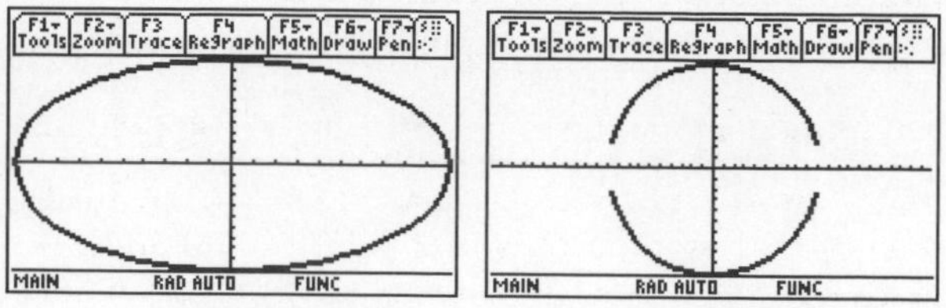

ZoomStd ZoomSqr

Figure 5-8: A circle graphed using ZoomStd and ZoomSqr.

- **The viewing window is too small or too big.**

  If you don't know what the graph should look like, after graphing it, you should zoom out to see more of the graph or zoom in to see a smaller portion of the graph. To do this, press [F2][3] to zoom out or press [F2][2] to zoom in.

  Then use the ◀▶▲▼ keys to move the cursor to the point from which you want to zoom out or in and press [ENTER]. It's just like using a camera. The point you move the cursor to is the focal point.

  After zooming in or out, you might have to adjust the window settings, as I describe earlier in this chapter in the section "Graphing Functions."

  As an example, the images in Figure 5-9 show the progression of graphing y_1 from Figure 5-2. I first graphed this function in the Standard viewing window. Then I zoomed out from the point (0, 10). And finally, I adjusted the window settings to get a better picture of the graph by removing the white space in the second picture in this figure to produce the third picture.

TIP

If you regret the change you made to a graph window and want to return to the window you had before making this change, press [F2]▲▲▶[1] to invoke the **ZoomPrev** command.

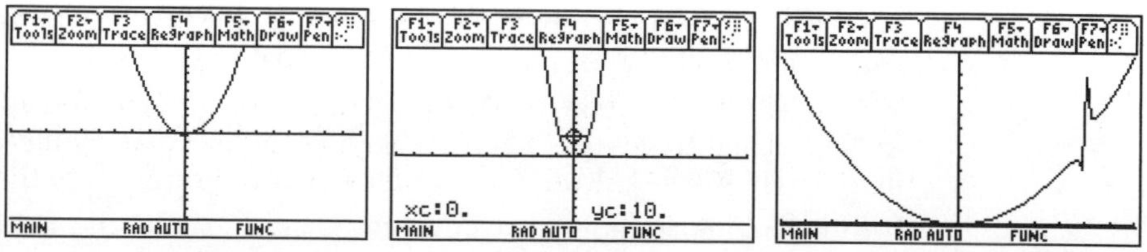

ZoomStd ZoomOut Adjusted window

Figure 5-9: A graph using ZoomStd, then ZoomOut, and then an adjusted viewing window.

- **Vertical asymptotes might not be recognizable.**

  In all graph styles except Dot, Square, and Animate, the calculator graphs one point, and then the next point, and connects those two points with a line segment. This sometimes causes vertical asymptotes to appear on the graph. The last graph in Figure 5-9 illustrates an example of when a vertical asymptote is present. Don't mistake this almost-vertical line for a part of the graph. It's not; it's just a vertical asymptote.

  In a different viewing window or with a high **xres** setting, the vertical asymptote might not even appear. This is illustrated earlier in Figure 5-5, where I graphed the same function in the same window but at different resolutions.

  TIP: If you want to ensure that vertical asymptotes don't appear on your graph, graph the function in the Dot or Square Graph style described in the previous section. This is illustrated in Figure 5-10.

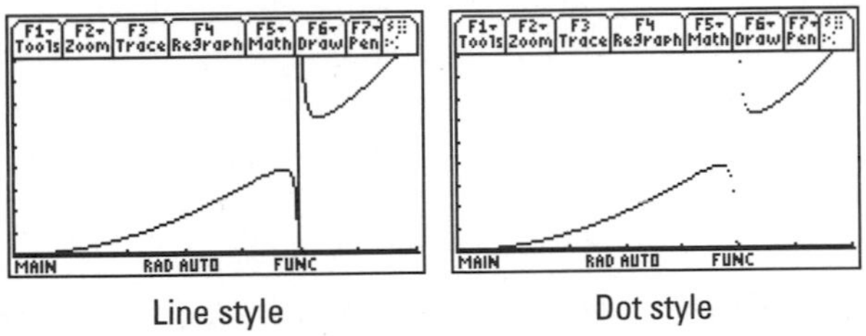

Figure 5-10: Showing and hiding asymptotes.

Piecewise-Defined Functions

Graphing a piecewise-defined function on the TI-89 is a bit like programming. If you're familiar with the When-Else programming format, you know what I'm talking about; if not, rest assured that I explain how you use this format to graph piecewise-defined functions. To graph a piecewise-defined function, perform the following steps:

1. **Press [◆][F1] to access the Y= editor and turn off any functions you don't want graphed.**

   You do this by using the ⊙⊙ keys to highlight the function and then pressing [F4] to toggle between displaying and not displaying a checkmark to the left of the function. The calculator graphs only the functions with checkmarks.

2. **Use the ⊙⊙ keys to place the cursor on the function in which you want to enter your piecewise-defined function and press [ENTER] to place the cursor on the command line.**

3. **Enter the definition of the piecewise-defined function.**

   A two-part piecewise-defined function, such as

   $$y = \begin{cases} y_1, & x < a \\ y_2, & x \geq a \end{cases}$$

   is defined by entering when($x < a$, y_1, y_2). This When-Else statement, when($x < a$, y_1, y_2), tells the calculator that when x is less than a, it should graph y_1; otherwise, when x is not less than a, it should graph y_2. This is illustrated on the command line in the first picture in Figure 5-11, where the When-Else statement defines the piecewise graph consisting of the line $y = -x$ when x is negative and the parabola $y = x^2$ when x is positive.

   A piecewise-defined function that has more than two parts is defined by using nested When-Else statements. For example, you use the nested statement when($x < a$, y_1,when($x < b$, y_2, y_3)) to define the function

   $$y = \begin{cases} y_1, & x < a \\ y_2, & a \leq x \leq a \\ y_3, & x > b \end{cases}$$

   Here's how you enter the word *when* and the inequalities:

   - **To enter the word *when*:** Press [CATALOG][.] to go to the section of the CATALOG housing words beginning with *W*. If necessary, use the ⊖⊖ keys to move the cursor to the **when(** item in the CATALOG. Press [ENTER] to insert **when(** in the definition of your function.
   - **To enter the inequalities:** Press [2nd][0] to enter less than (<) or [2nd][.] to enter greater than (>). When graphing, the calculator doesn't (for example) distinguish between less than and less than or equal to.

4. **Press [ENTER].**

   Notice that the way you enter the definition and the way the calculator displays the definition aren't the same, as illustrated in the first picture of Figure 5-11. You enter the definition the way a programmer would, and the calculator displays it the way a textbook would, except the calculator uses the word *else* to, for example, indicate the opposite of $x < 0$.

5. **Graph the piecewise-defined function.**

   "Graphing Functions," an earlier section of this chapter, explains how to graph functions. An example graph of a piecewise-defined function appears in the second picture in Figure 5-11.

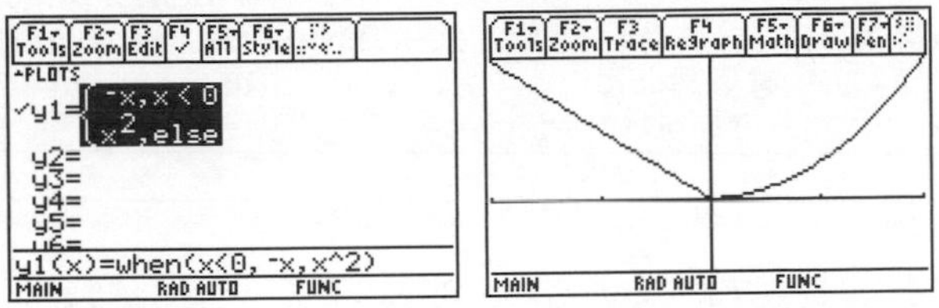

Figure 5-11: Defining and graphing a piecewise-defined function.

TIP

If one or more of the functions in your piecewise-defined function is a trigonometric function, make sure the calculator is in Radian and not Degree mode. Otherwise, your piecewise-defined function might look like a step function instead of the graph you were expecting. The next section tells you how to change the mode and how to graph trigonometric functions.

Graphing Trig Functions

The calculator has built-in features especially designed for graphing trigonometric functions. They produce graphs that look like graphs you see in textbooks, and when you trace these graphs, the *x*-coordinate of the tracing point is always given as a fractional multiple of π. To use these features when graphing trigonometric functions, follow these steps:

1. **Put the calculator in Function and Radian modes.**

   If you see the words RAD and FUNC displayed on the last line of your calculator (as in Figure 5-11), then you are already in the proper mode. If you don't see these words, press [MODE]. Then press ▶[1] to put the calculator in Function mode and ▼▼▼▶[1] to get Radian mode. Press [ENTER] to save your new Mode settings.

2. **Enter your trigonometric functions into the Y= editor.**

   Entering functions in the Y= editor is explained earlier in the chapter.

3. **Press [F2][7] to graph the function.**

   [F2][7] invokes the **ZoomTrig** command that graphs the function in a viewing window in which $-79\pi/24 \le x \le 79\pi/24$ and $-4 \le y \le 4$. It also sets the tick marks on the *x*-axis to multiples of $\pi/2$.

   When you trace a function graphed in a **ZoomTrig** window, the *x*-coordinate of the trace cursor will be a multiple of $\pi/24$, although the *x*-coordinate displayed at the bottom of the screen will be a decimal approximation of this value. (Tracing is explained in the next chapter.)

TIP: If you want to graph trigonometric functions in Degree mode, to get at least one period of a sine or cosine function, you must set **xmin** to 0 and **xmax** to 360. This is why I say that it is a lot easier to graph trig functions in Radian mode. I explain setting the Mode to degrees or radians in Chapter 1.

Viewing the Graph and the Y= Editor on the Same Screen

If you're planning to play around with the definition of a function you're graphing, it's quite handy to have both the Y= editor and the graph on the same screen. That way you can edit the definition of your function and see the effect your editing has on your graph.

To do this, press [MODE][F2]▶[2] to split the screen horizontally, press ▼▶[4] to put the graph at the top of the screen, and then press ▼▶[2] to put the Y= editor at the bottom of the screen, as illustrated in the first picture in Figure 5-12. Finally, press [ENTER] to save the changes you just made. You now see a screen similar to that in the second picture in Figure 5-12.

The active screen is the one with the bold border, as illustrated in the second picture in Figure 5-12, in which the Graph screen is active. You toggle between these two screens by pressing [2nd][APPS].

There's no rule that says you have to have the graph at the top of the screen; reverse the order if you so desire. And for that matter, maybe you'd rather use the Home screen with your graph. If so, go ahead and select it as one of the applications appearing in your split screen. You can even select the **LEFT-RIGHT** mode to split the screen vertically, but this mode doesn't leave much room for editing functions or doing calculations on the Home screen. The vertically split screen is best used for displaying a graph and its table of values, a topic addressed in the next chapter.

TIP: If you're in split screen mode and want to use an application other than the two appearing in your split screen, just press the keys that activate that application and it will replace the active window in your split screen. For example, if the Graph screen is active and you want to do a calculation on the Home screen, press [HOME] to replace the Graph screen with the Home screen. When you are finished with the Home screen, press [◆][F3] to make the Graph screen active again.

TIP: To return to FULL screen mode without making changes in the Mode menu, just press [2nd][ESC]. This automatically sets the Split Screen mode on Page 2 of the Mode menu to FULL.

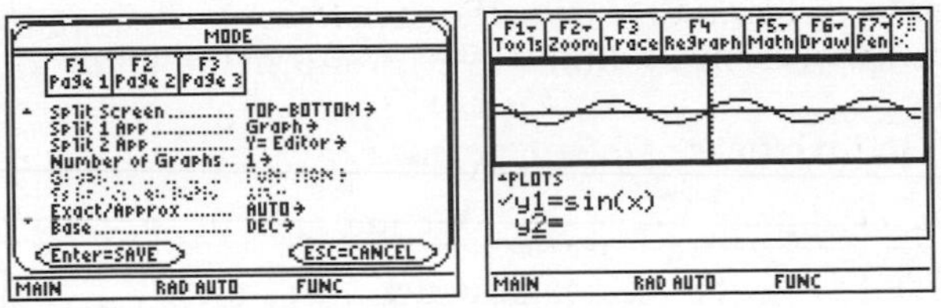

Figure 5-12: Function and graph on the same screen.

Saving and Recalling a Graph

If you save your graph as a Graph Database, when you recall the graph at a later time, the graph remains interactive. This means that you can, for example, trace the graph and resize the viewing window because a Graph Database also saves the Graph mode, Window, Format, and Y= editor settings. It does not, however, always save the Split Screen settings (TOP-BOTTOM and LEFT-RIGHT) entered on the second page of the Mode menu. This section explains how to save, delete, and recall a graph in a Graph Database.

To save a Graph Database from the Graph screen (♦F3), the Y= editor (♦F1), or the Windows editor (♦F2), perform the following steps:

1. **Press F1 2 to invoke the Save Copy As command.**
2. **If GBD does not appear on the first line under Type, press ▶1.**
3. **Press ▼▶ and select the folder in which you want to store the Graph Database.**

   If you haven't created a special folder, store it in the default (main) folder.
4. **Press ▼ and enter a name for your Graph Database and press ENTER to save the name.**

   The calculator is automatically in Alpha mode, so you can enter a name without first pressing the ALPHA key. But be warned, the calculator will accept no more than an eight-letter name. If you key in more than eight letters, it just disregards those extra letters.
5. **Press ENTER to save your Database.**

To delete a Graph Database from your calculator, perform the following steps:

1. **Press 2nd − to access the VAR-LINK menu.**
2. **Arrow down to highlight the name of the Database.**
3. **Press F1 ENTER ENTER to delete the Database.**
4. **Press ESC to exit this menu and return to the previous screen.**

To recall a saved Graph Database from the Graph screen ([◆][F3]), the Y= editor ([◆][F1]), or the Windows editor ([◆][F2]), perform the following steps:

1. **Press [F1][1] to invoke the Open command.**
2. **If GBD doesn't appear on the first line under Type, press ▶[1].**
3. **Press ▼▶ and select the folder in which you stored the Graph Database.**
4. **Press ▼▶ and select the name of your database and press [ENTER].**

   If you see only an arrow but no filename after Variable in the third line of the Open menu, there are no saved Graph Databases on your calculator.

When you recall a Graph Database, the Mode, Window, Format, and Y= editor settings in your calculator change to those saved in the Graph Database. If you don't want to lose the settings you have in the calculator, save them in another Graph Database before recalling your saved Graph Database.

Chapter 6

Exploring Functions

In This Chapter

- Using Zoom commands
- Tracing the graph of a function
- Constructing tables of functional values
- Creating and clearing user-defined tables
- Viewing graphs and tables on the same screen

The calculator has three very useful features that help you explore the graph of a function: zooming, tracing, and creating tables of functional values. Zooming allows you to quickly adjust the viewing window for the graph so that you can get a better idea of the nature of the graph. Tracing shows you the coordinates of the points that make up the graph. And creating a table — well, I'm sure you already know what that shows you. This chapter explains how to use each of these features.

Using Zoom Commands

After you've entered your functions in the Y= editor (as I describe in Chapter 5), you can use Zoom commands to graph or to adjust the view of your previously drawn graph. From the Graph screen or Y= editor, press [F2] to see the ten Zoom commands that you can use to graph or to adjust the view of an already drawn graph — they're labeled 1 through 9 and A. In the following four sections, I explain how and when to use these Zoom commands.

Zoom commands that help you to graph or regraph a function

- **ZoomStd:** This command graphs your function in a preset viewing window where $-10 \le x \le 10$ and $-10 \le y \le 10$. You access it by pressing [F2][6].

This is a nice Zoom command to use when you haven't the slightest idea what size viewing window to use for your function. After graphing the function using **ZoomStd**, you can, if necessary, use the **ZoomIn** and **ZoomOut** commands to get a better idea of the nature of the graph. I discuss how to use **ZoomIn** and **ZoomOut** later in this section.

- **ZoomDec:** This command graphs your function in a preset viewing window where $-7.9 \le x \le 7.9$ and $-3.8 \le y \le 3.8$. You access it by pressing [F2][4].

  When you trace a function graphed in a **ZoomDec** window, the *x*-coordinate of the trace cursor is a multiple of 0.1. (I explain tracing later in the chapter.)

- **ZoomTrig:** This command, which is most useful when graphing trigonometric functions, graphs your function in a preset viewing window where $-79\pi/24 \le x \le 79\pi/24$ and $-4 \le y \le 4$ if your calculator is in Radian mode or where $-592.5 \le x \le 592.5$ and $-4 \le y \le 4$ if your calculator is in Degree mode. It also sets the tick marks on the *x*-axis to multiples of $\pi/2$ in Radian mode or to multiples of 90 in Degree mode. You access **ZoomTrig** by pressing [F2][7].

  When you trace a function graphed in a **ZoomTrig** window, the *x*-coordinate of the trace cursor will be a multiple of $\pi/24$ when your calculator is in Radian mode or a multiple of 15 degrees when it's in Degree mode. Tracing is explained later in this chapter, in the section "Tracing a Graph."

  To use the **ZoomTrig** command, enter your function into the calculator (as I describe in Chapter 5) and then press [F2][7]. The graph automatically appears.

Zoom commands that help you find an appropriate viewing window

- **ZoomFit:** This is my favorite Zoom command. If you know how you want to set the *x*-axis, **ZoomFit** automatically figures out the appropriate settings for the *y*-axis.

  To use **ZoomFit**, press [◆][F2] and enter the values you want for **xmin**, **xmax**, and **xscl**. Then press [F2][▲][▲][▲][ENTER] (or press [F2][ALPHA][=]) to get **ZoomFit** to figure out the *y*-settings and graph your function. **ZoomFit** doesn't figure out an appropriate setting for **yscl**, so if the *y*-axis looks a little cluttered, you might want to go back to the Window editor and adjust this value. You can find out more about the Window editor in Chapter 5.

- **ZoomData:** If you're graphing functions, this command is useless. But if you're graphing Stat Plots (as I explain in Chapter 20), this command finds the appropriate viewing window for your plots. See Chapter 20 for information on how this works.

Zoom commands that readjust the viewing window of your graph

- **ZoomSqr:** Because the calculator screen isn't perfectly square, graphed circles won't look like real circles unless you properly set the viewing window by using **ZoomSqr**. **ZoomSqr** readjusts the existing Window settings for you and then regraphs the function in a viewing window in which circles look like circles.

  To use **ZoomSqr**, graph the function as I describe in Chapter 5 and then press [F2][5]. The graph automatically appears.

- **ZoomInt:** This command is quite useful when you want the trace cursor to trace your functions using integer values for the *x*-coordinate, such as when graphing a function that defines a sequence. (I explain tracing in the next section.) **ZoomInt** allows you to select a new center for your window, and then it regraphs the function in a viewing window in which the trace cursor displays integer values for the *x*-coordinate.

  To use **ZoomInt**, graph the function as I describe in Chapter 5 and then press [F2][8]. Use the ◀▶▲▼ keys to move the cursor to the spot on the screen that will become the center of the new screen. Then press [ENTER]. The graph is redrawn centered at the cursor location.

Zoom commands that zoom in or out from your graph

- **ZoomIn and ZoomOut:** After the graph is drawn (as I describe in Chapter 5), these commands allow you to zoom in on a portion of the graph or to zoom out from the graph. They work very much like a zoom lens on a camera. To use these commands, follow these steps:

  1. **Press [F2][2] to zoom in or press [F2][3] to zoom out.**
  2. **Use the ◀▶▲▼ keys to move the cursor to the spot on the screen from which you want to zoom in or zoom out.**

     This spot becomes the center of the new viewing window.

  3. **Press [ENTER].**

     The calculator redraws the graph centered at the cursor location.

- **ZoomBox:** This command allows you to define a new viewing window for a portion of your graph by enclosing it in a box, as illustrated in Figure 6-1. The box becomes the new viewing window.

To construct the box, follow these steps:

1. **Press [F2][1] and use the ◀▶▲▼ keys to move the cursor to the spot where you want one corner of the box to be located.**
2. **Press [ENTER] to anchor that corner of the box.**
3. **Use the ◀▶▲▼ keys to construct the rest of the box.**

   When you press these keys, the calculator draws the sides of the box.

   TIP: When you use **ZoomBox**, if you don't like the size of the box, you can use any of the ◀▶▲▼ keys to resize the box. If you don't like the location of the corner you anchored in Step 2, press [ESC] and start over with Step 1. If you're drawing a rather large box, to move the cursor a large distance, press [2nd] and then press an appropriate Arrow key.

4. **Press [ENTER] when you finish drawing the box.**

   The calculator then redraws the graph in the window defined by your box (as shown in Figure 6-1).

   REMEMBER: When you use **ZoomBox**, [ENTER] is pressed only two times. The first time you press it is to anchor a corner of the zoom box. The next time you press [ENTER] is when you're finished drawing the box, and you're ready to have the calculator redraw the graph.

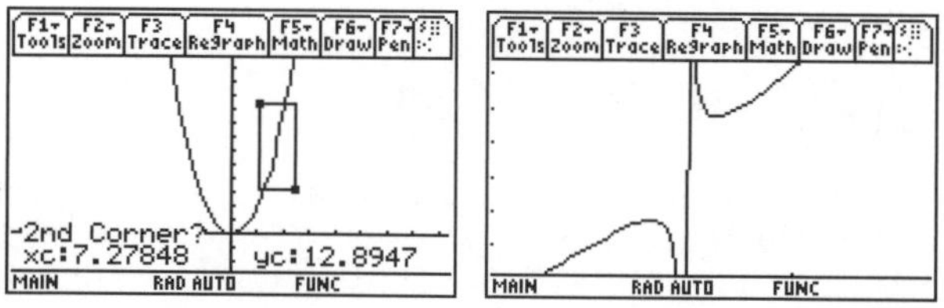

The box

Redrawn graph

Figure 6-1: Constructing a zoom box and redrawing the graph.

Texas Instruments' first graphing calculator didn't have all the nice Zoom commands possessed by its current line of graphing calculators. If you owned one of these early calculators, you had to set the Window to get the calculator to do what the current Zoom commands automatically do for you. For example, if you wanted the *x*-coordinate of the cursor to move in increments of 0.1, you set **xmin** to –4.7 and **xmax** to +4.7 because the screen of the calculator was 94 pixels wide. And each time you wanted to use this window you had to reenter it in the Window editor because there was no way to save the settings for later use. You don't have these inconveniences with Texas Instruments' current graphing calculators.

Undoing a zoom

If you used a Zoom command to redraw a graph and then want to undo what that command did to the graph, follow these steps:

1. **Press [F2]↑↑→ to access the Zoom Memory menu.**
2. **Press [1] to select ZoomPrev.**

   The graph is redrawn as it appeared in the previous viewing window.

If you have a favorite or often-used viewing window, be sure to save it in the calculator and then recall it for later use. To save a viewing window, first construct the window in the Window editor (as I explain in Chapter 5). Then press [F2]↑↑→[2] to save the window settings. When you want to graph a function in the Y= editor by using the saved viewing window, all you need to do is press [F2]↑↑→[3] and the graph appears in your saved window. You can save only one viewing window at a time. If you save a new window, it overwrites the previously saved window.

If you like creating your own viewing windows, you might find it helpful to know that the screen on the TI-89 is 158 pixels wide and 76 pixels long.

Tracing a Graph

After you have graphed your function, as described in the previous chapter, you can press [F3] (Trace) and then use → and ← to more closely investigate the points that are on the graph of the function.

If you use only the ←→↑↓ keys instead of [F3] (Trace) to locate a point on a graph, all you will get is an *approximation* of the location of that point. You rarely get an actual point on the graph. So always use [F3] (Trace) to identify points on a graph.

The following list describes what you see, or don't see, as you trace a graph:

- **The number of the function:** The number of the function in the Y= editor that you're tracing is displayed in the top-right corner of the screen. For example, if you see the number 1, you're tracing the first graph (y_1) in the Y= editor.

  If you've graphed more than one function and you would like to trace a different function, press ↓ or ↑. Each time you press ↓, the cursor jumps to the next higher-numbered function in the graph. If the cursor is already on the highest-numbered function, pressing ▼ sends the cursor back to the lowest-number function in the graph. Pressing ▲ cycles the cursor through the graphed functions in the opposite order.

- **The values of *x* and *y*:** At the bottom of the screen, you see either the coordinates that define the cursor location or nothing at all. If you do see the coordinates, they appear in either rectangular (*x*, *y*) or polar (*r*, θ) form with, for example, the rectangular *x*-coordinate denoted by "xc" or the polar *r*-coordinate denoted by "rc."

  If you don't like what you see at the bottom of the screen, press [F1]▲[ENTER]► and then press the number of the option you want. (The OFF option displays nothing at the bottom of the screen.) Finally, press [ENTER] to save your new selection, and then press [F3] to start tracing your graph.

  When you press [F3] (Trace), the cursor is placed on the lowest-numbered graph at the point having an *x*-coordinate that is approximately midway between **xmin** and **xmax**. If the *y*-coordinate of the cursor location isn't between **ymin** and **ymax**, the cursor doesn't appear on the screen. The sidebar, "Panning in Function mode," tells you how to correct this situation.

  Each time you press ►, the cursor moves right to the next plotted point on the graph, and you see the coordinates of that point at the bottom of the screen (provided Coordinates in the Graph Formats menu is not set to OFF). If you press ◄, the cursor moves left to the previously plotted point. And if you press ▼ or ▲ to trace a different function, the tracing of that function starts at the point on the graph that has the *x*-coordinate displayed on-screen before you pressed this key.

  When tracing a graph, you can press [2nd]► or [2nd]◄ to get the cursor to move a greater distance between plotted points.

  When tracing a graph, if the value of the *x*-coordinate appears at the bottom of the screen but the value of the *y*-coordinate is blank, this means that the function is undefined at the value of the *x*-coordinate.

Press [CLEAR] to terminate tracing the graph. This also removes the number of the function and the coordinates of the cursor from the screen. But if you press any Arrow key, the coordinates of the cursor location appear at the bottom of the screen (provided Coordinates in the Graph Formats menu is not set to OFF).

When you're using Trace, if you want to start tracing your function at a specific value of the independent variable *x*, just key in that value and then press [ENTER]. (The value you assign to *x* must be between **xmin** and **xmax**; if it's not, you get an error message.) After you press [ENTER], the trace cursor moves to the point on the graph with the *x*-coordinate you just entered. If that point isn't on the portion of the graph appearing on the screen, the sidebar, "Panning in Function mode," tells you how to get the cursor and the graph in the same viewing window.

TIP

Panning in Function mode

When you're tracing a function and the cursor hits the top or bottom of the screen, you still see the coordinates of the cursor location displayed at the bottom of the screen (provided Coordinates in the Graph Formats menu is not set to OFF), but you won't see the cursor itself on the screen because the viewing window is too small. Press [ENTER] to get the calculator to adjust the viewing window to a viewing window that is centered about the cursor location. Then press [F3] to continue tracing the graph. If the function you were tracing isn't displayed at the top of the screen, repeatedly press ⊙ until it is. The trace cursor then appears in the middle of the screen, and you can use ⊙ and ⊙ to continue tracing the graph.

When you're tracing a function and the cursor hits the left or right side of the screen, the calculator automatically pans left or right. It also appropriately adjusts the values assigned to **xmin** and **xmax** in the Window editor, but it doesn't change the values of **ymin** and **ymax**. So you might not see the cursor on the screen. If this happens, follow the directions in the previous paragraph to see both the function and the cursor on-screen.

TIP

If the display of the coordinates of the cursor is interfering with your view of the graph when you're tracing the graph, create more space at the bottom of the screen by pressing [♦][F2] and decreasing the value of **ymin**. Then press [♦][F3] to redisplay and press [F3] to continue tracing the graph. The coordinates of the trace cursor will be in the same location as they were before you edited the **ymin** setting in the Window editor.

Displaying Functions in a Table

After you enter the functions in the Y= editor (as I describe in Chapter 5), you can have the calculator create one of three different types of tables of functional values:

- An automatically generated table displaying the plotted points that appear in the calculator's graphs of the functions
- An automatically generated table that displays the points according to an initial *x*-value and *x*-increment that you specify
- A user-defined table in which you enter the *x*-values of your choice and the calculator determines the corresponding *y*-values

I discuss these tables in the sections that follow.

Creating automatically generated tables

To create a table for which you specify the initial *x*-value and the value by which *x* is incremented, perform the following steps. After the last step, I tell you how to create an automatically generated table displaying the points graphed by the calculator.

1. **Place a checkmark to the left of the functions in the Y= editor that you want to appear in the table; leave the other functions unchecked.**

   Only the functions in the Y= editor that have checkmarks to the left of their names will appear in the table. To check or uncheck a function, press ◆F1, use the ▲▼ keys to place the cursor on the function, and then press F4 to toggle between checking and unchecking the function.

2. **Press ◆F4 to access the Table Setup editor (shown in Figure 6-2).**

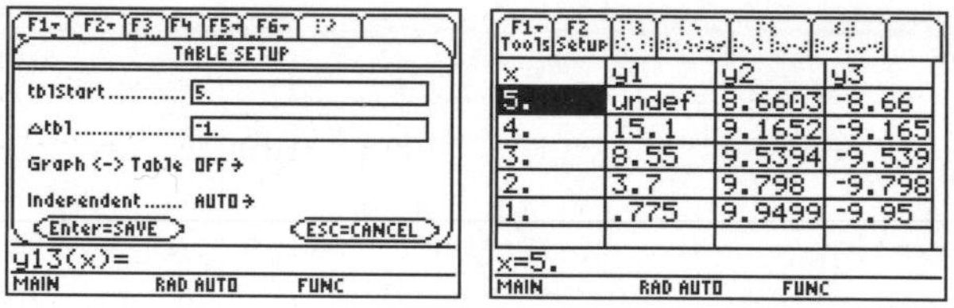

Figure 6-2: A table generated by specifying the initial x-value and the x-increment.

3. **Enter a number in tblStart and press ▼. Then enter a number in Δtbl and press ▼.**

   If **Independent** is the only item you see in the Table Setup editor, and if the cursor is on ASK, press ►1 to change this setting to AUTO and then press ▲▲▲ to enter numbers in **tblStart** and **Δtbl**.

   tblStart is the first value of the independent variable *x* to appear in the table, and **Δtbl** is the increment for the independent variable *x*. The calculator constructs the table by adding the value you assign to **Δtbl** to the previous *x*-value and calculating the corresponding *y*-values of the function or functions. In Figure 6-2, **tblStart** is assigned the value 5 and **Δtbl** the value –1 to produce the decreasing values in the *x*-column of the table.

   Use the number keys to enter values in **tblStart** and **Δtbl**. Press ▼ after making each entry.

4. **Set Graph <–> Table to OFF, press ▽, and then set Independent to AUTO (as shown in Figure 6-2).**

   You have two choices for the **Graph <–> Table** setting (OFF and ON) and two for the **Independent** setting (AUTO and ASK). When **Independent** is set to **ASK**, no matter what other settings appear in the Table Setup menu, a user-generated table is created where you manually place values in the *x*-column of the table and the calculator computes the corresponding *y*-values. (I explain dealing with user-generated tables in the next section.)

   When **Independent** is set to **AUTO** and **Graph <–> Table** is set to **ON**, a table of the points plotted in the calculator's graphs of the functions is created. However, if **Graph <–> Table** is set to **OFF**, the calculator uses the values in **tblStart** and **Δtbl** to generate the table.

   To change the **Graph <–> Table** or **Independent** setting, use △▽ to place the cursor on the existing setting, press ▷ to display the options for that setting, and press the number of the option you want.

5. **Press [ENTER] to save the settings in the Table Setup editor.**

6. **Press [◆][F5] to display the table.**

   The table appears on the screen, as illustrated in Figure 6-2. You find out about navigating the table later in this chapter in the section "Displaying hidden rows and columns."

   REMEMBER

   The word *undef* in a table means that that particular function is not defined at the specified *x*-value. For example, in Figure 6-2, function y_1 is undefined when $x = 5$.

To create a table of the plotted points appearing in the calculator's graphs of the functions, follow Steps 1 and 2 in the preceding step list. If **Independent** is the only item you see in the Table Setup editor, press ▷[1] to change this setting to **AUTO** and then press △▷[2][ENTER] to set **Graph <–> Table** to **ON** and to save your settings. If you see all four items in this editor, press ▽▽▷[2]▽▷[1][ENTER] to set **Graph <–> Table** to **ON**, **Independent** to **AUTO** and to save your settings. (With these settings, the calculator ignores the other values in the Table Setup menu.) If necessary, press [◆][F5] to display the table, as shown in Figure 6-3.

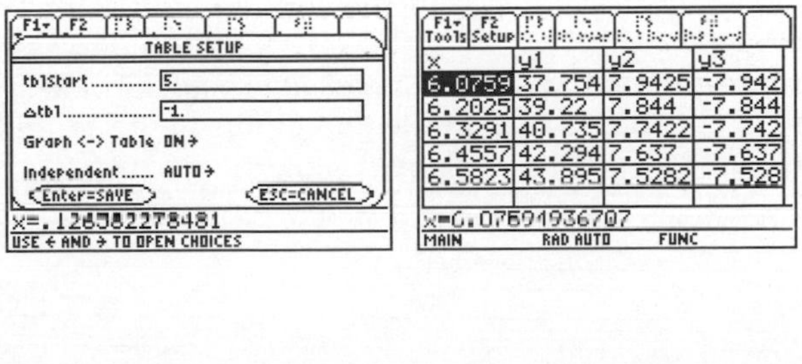

Figure 6-3: Generating a table of plotted points.

Creating a user-defined table

To create a user-defined table in which you enter the *x*-values and the calculator figures out the corresponding *y*-values, follow these steps:

1. **Place a checkmark to the left of the functions in the Y= editor that you want to appear in the table; leave the other functions unchecked.**

   Only those functions in the Y= editor that have checkmarks to the left of their names will appear in the table. To check or uncheck a function, press [◆][F1], use the ▲▼ keys to place the cursor on the function, and then press [F4] to toggle between checking and unchecking the function.

2. **Press [◆][F4] to access the Table Setup editor.**

3. **Set Independent to ASK and press [ENTER].**

   Independent is the last item in the menu. If it is already set to ASK, press [ENTER] and go to the next step. If it is set to AUTO, press ▼▼▼▶[2] to set it to ASK and then press [ENTER] to save your settings.

4. **If the calculator doesn't automatically display the table for you, press [◆][F5] to display it. If the table contains entries, press [F1][8][ENTER] to clear the table.**

   You see a table without entries, as in the first picture in Figure 6-4.

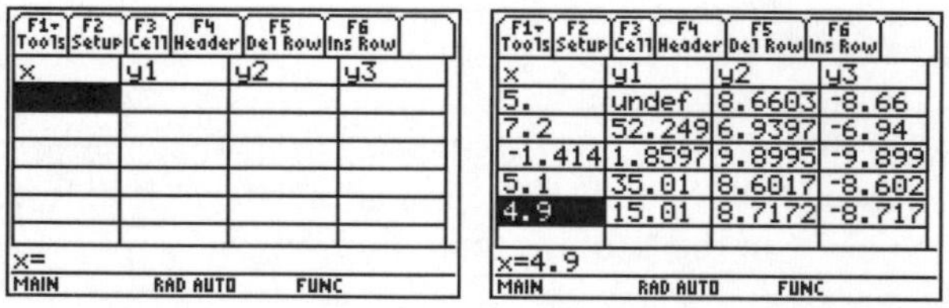

Figure 6-4: Generating a user-defined table.

5. **Enter values for the independent variable *x* in the first column.**

   You must enter values consecutively; you cannot skip any cells in the first column of the table. To make the first entry, press [ENTER], use the number keys to enter a value, and then press [ENTER] after you finish. To make entries after the first entry, press ▼[ENTER], enter a value, and then press [ENTER].

   After you make each entry, the calculator automatically calculates the corresponding *y*-values of the checked functions in the Y= editor, as shown in the second picture of Figure 6-4.

Displaying hidden rows and columns

To display more rows in an automatically generated table, repeatedly press ▲ or ▼. If you have constructed a table for more than three functions, repeatedly press ► or ◄ to navigate through the columns of the table.

Each time the calculator redisplays a table with a different set of rows, it also automatically resets **tblStart** to the value of *x* appearing in the first row of the newly displayed table. To return the table to its original state, press [F2] to access the Table Setup editor and then change the value that the calculator assigned to **tblStart**. Press [ENTER] to save the value you entered, and then press [ENTER] again to return to the table.

TIP

While displaying a table of functional values, you can edit the definition of a function without going back to the Y= editor. To do this, use the ►◄ keys to place the cursor in the column of the function you want to redefine and press [F4]. This places the definition of the function on the command line so that you can edit the definition. (I cover editing expressions in Chapter 1.) When you finish your editing, press [ENTER]. The calculator automatically updates the table and the edited definition of the function in the Y= editor.

Viewing the Table and the Graph on the Same Screen

After you have graphed your functions and generated a table of functional values, you can view the graph and the table on the same screen. To do so, follow these steps:

1. **Press [MODE][F2]►[3] to vertically divide the screen.**
2. **Press ▼►[5] to place the table in the left screen. Then press ▼►[4] to place the graph in the right screen.**

   If you prefer to have the graph on the left and the table on the right, just reverse the order of keystrokes in this step.
3. **Press [ENTER] to save your Mode settings.**

   You now see a screen similar to that in the first picture in Figure 6-5. Notice that the right screen doesn't contain the graph of the function. This graph appears in the next step.
4. **Press [2nd][APPS] to toggle between the left and right screens.**

   The right screen is no longer empty, as in the second picture in Figure 6-5.

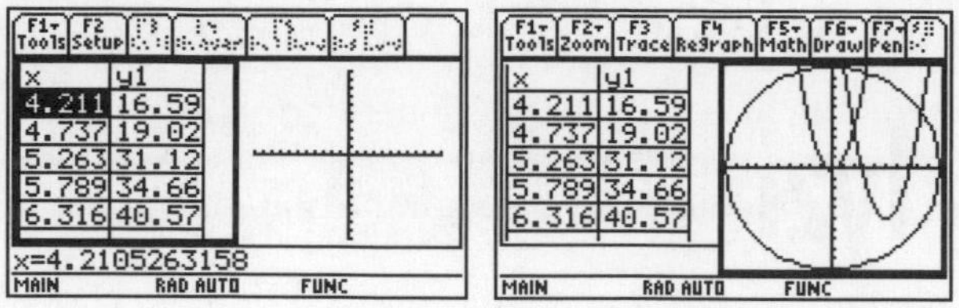

Figure 6-5: Viewing a graph and a table on the same screen.

On a split screen, the screen with the bold border is the active screen, and the commands for the application in that screen appear at the top of the screen. For example, in the second picture in Figure 6-5, the Graph screen is active. So if you press [F3] while in this application, you can trace the functions graphed on that screen. (See "Tracing a Graph," earlier in this chapter.)

TIP: To return to FULL screen mode without making changes in the Mode menu, just press [2nd][ESC]. This automatically sets the Split Screen mode on Page 2 of the Mode menu to FULL.

Chapter 7

Analyzing Functions

In This Chapter

- Evaluating a function and finding its *x*-intercepts
- Finding the maximum and minimum values of a function
- Finding the point of intersection of two functions
- Finding the slope and equation of a tangent to the graph of a function
- Locating inflection points
- Finding arc length

After graphing a function (as I describe in Chapter 5), you can use the options on the F5 Graph Math menu to find the value of the function at a specified value of *x*, to find the zeros (*x*-intercepts) of the function, and to find the maximum and minimum values of the function. You can even find the derivative of the function at a specified value of *x*, or you can evaluate a definite integral of the function. The calculator will even find the inflection points on your graph and the equation of a tangent to the graph.

These options enable you to find, for example, the slope and equation of the tangent to the graph of the function at a specified value of *x* or to find the area between the graph and the *x*-axis. And if you have graphed two or more functions, the calculator will tell you where they intersect. It will even find the arc length of a curve. This chapter shows you how to do all these things.

WARNING!

Is the answer approximate or exact?

The routines used by the calculator to find all the neat Graph Math menu options that you find in this chapter are numerical routines. This means that they find really good *approximations* of what you want, but they don't always find the *exact* value. So if you're a student, think twice about the appropriateness of using one of these calculations when answering a test question. As an example, the last figure in this chapter shows that the circumference of a semicircle of radius 5 is 15.708. That's an *approximate* answer; the *exact* answer is 5π.

Finding the Value of a Function

When you trace the graph of a function, the trace cursor doesn't hit every point on the graph. So tracing isn't a reliable way of finding the value of a function unless you tell the calculator the value of the independent variable *x*. (You discover how to do this in Chapter 6 in the section on tracing a graph.) The [F5] Graph Math menu, however, contains a command that evaluates a function at a specified *x*-value. To access and use this command, perform the following steps:

1. **Graph the functions in a viewing window that contains the specified value of *x*.**

   Graphing functions and setting the viewing window are explained in Chapter 5. To get a viewing window containing the specified value of *x*, that value must be between **xmin** and **xmax** in the viewing window.

2. **If necessary, set the Graph Formats menu to display coordinates in rectangular (*x*, *y*) form.**

   To do this, press [F1][▲][ENTER][▶][1][ENTER].

3. **Press [F5][1] to invoke the Value command in the Graph Math menu.**

4. **Enter the specified value of *x*.**

   When using the **Value** command to evaluate a function at a specified value of *x*, that value you specify must be an *x*-value that appears on the *x*-axis of the displayed graph — that is, it must be between **xmin** and **xmax** in the Window editor (see Chapter 5). If it isn't, you get an error message.

   Use the keypad to enter the value of *x*; if you make a mistake when entering your number, press [CLEAR] and reenter the number.

5. **Press [ENTER].**

   After you press [ENTER], the number of the first checked function in the Y= editor appears at the top of the screen, the cursor appears on the graph of that function at the specified value of *x*, and the coordinates of the cursor appear at the bottom of the screen. This is illustrated in the first graph in Figure 7-1.

6. **Repeatedly press [▲] to see the value of the other graphed functions at your specified value of *x*.**

   Each time you press [▲], the number of the function being evaluated appears at the top of the screen and the coordinates of the cursor location appear at the bottom of the screen. This is illustrated in the second graph in Figure 7-1.

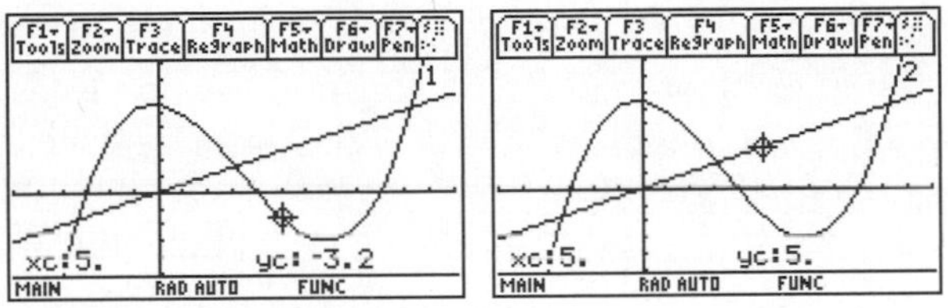

Figure 7-1: Evaluating two functions at a specified value of x.

Enter *x*; press [ENTER]   Press [▲]

After using the **Value** command to evaluate your functions at one value of x, you can evaluate your functions at another value of x by keying in the new value and pressing [ENTER]. Press any function key (such as [ENTER] or [►]) *after* evaluating a function to deactivate the **Value** command.

TIP: If you plan to evaluate functions at several specified values of x, consider constructing a user-defined table of functional values (as I explain in Chapter 6).

Finding the Zeros of a Function

The *zeros* of the function $y = f(x)$ are the solutions to the equation $f(x) = 0$. Because $y = 0$ at these solutions, these zeros (solutions) are really just the x-coordinates of the x-intercepts of the graph of $y = f(x)$. (An x-intercept is a point where the graph crosses or touches the x-axis.) To find a zero of a function, perform the following steps:

1. **Graph the function in a viewing window that contains the zeros of the function.**

   I explain graphing a function and finding an appropriate viewing window in Chapter 5. To get a viewing window containing a zero of the function, that zero must be between **xmin** and **xmax** in the viewing window, and the x-intercept at that zero must be visible on the graph.

2. **If necessary, set the Graph Formats menu to display coordinates in rectangular (x, y) form.**

   To do this, press [F1][▲][ENTER][►][1][ENTER].

3. **Press [F5][2] to invoke the Zero command in the Graph Math menu.**

   TIP: If the calculator instructions at the bottom of the screen interfere with your view of the graph, create more space at the bottom of the screen by pressing [◆][F2] and decreasing the value of **ymin**. Then press [◆][F3] to redraw the graph and press [F5][2] to invoke the **Zero** command.

4. **If necessary, repeatedly press [▲] until the number of the appropriate function appears at the top of the screen.**

5. **Set the lower bound for the zero you want to find and press [ENTER].**

   The lower bound for the zero is simply *any* number that is less than the zero. To set the lower bound, use the ◀ and ▶ keys to place the cursor on the graph a little to the left of the zero (as illustrated in the first picture in Figure 7-2), and then press [ENTER]. A Left Bound indicator appears at the top of the screen (as illustrated by the triangle in the second picture in Figure 7-2) and you are prompted to enter an upper bound.

   TIP

   If you prefer, you can set the bound by keying in a number and then pressing [ENTER].

6. **Set the upper bound for the zero and press [ENTER].**

   As with the lower bound, the upper bound is *any* number that is a bit larger than the zero. To set the upper bound, use the ◀ and ▶ keys to place the cursor on the graph a little to the right of the zero (as illustrated in the second picture in Figure 7-2) and then press [ENTER]. The value of the zero automatically appears on the bottom of the screen, as shown in the second picture in Figure 7-2.

WARNING!

If more than one zero exists between the lower and upper bounds you specify when using the **Zero** command, the calculator finds *only* the smallest zero and *does not* warn you that more zeros might exist. To find the other zero, if such a zero exists, use the **Zero** command again, but this time set the lower bound to a number slightly larger than the first zero found by the calculator and set the upper bound to the same number you used when finding the first zero. For example, if the first zero found by the calculator is 0.1, set the lower bound to 0.1000001. After using the **Zero** command a second time, the calculator either displays the value of another zero or it displays an error message telling you that there are no more zeros in the interval defined by the new lower and upper bounds. After getting the error message, press [ESC] or x̄ to make it go away.

TIP

The calculator uses scientific notation to denote really large or small numbers. For example, –0.00000001 is displayed on the calculator as –1.E–8, and 0.000000005 is displayed as 5.E–9.

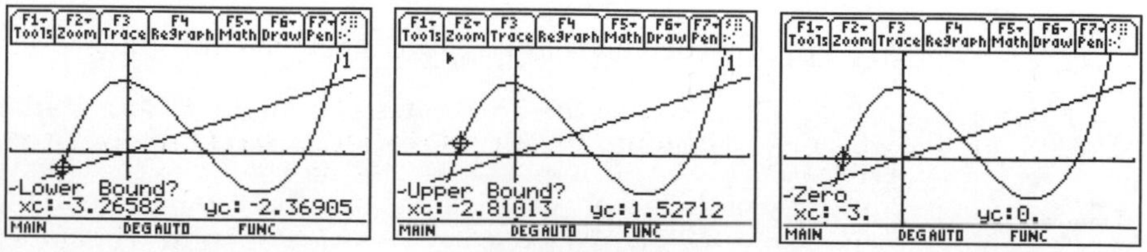

Figure 7-2: Steps for finding a zero of a function.

Locate the lower bound | Press [ENTER], then locate the upper bound | Press [ENTER]

Finding Min and Max

Finding the maximum or minimum point on a graph has many useful applications. For example, the maximum point on the graph of a profit function not only tells you the maximum profit (the *y*-coordinate), it also tells you how many items (the *x*-coordinate) the company must manufacture and sell to achieve this profit. To find the minimum or maximum value of a function, perform the following steps:

1. **Graph the function in a viewing window that displays the maximum and minimum points you want to find.**

   See Chapter 5 for details on graphing a function and finding an appropriate viewing window.

2. **If necessary, set the Graph Formats menu to display coordinates in rectangular (*x*, *y*) form.**

   To do this, press [F1][▲][ENTER][▶][1][ENTER].

3. **Press [F5][3] to find a minimum or press [F5][4] to find a maximum.**

   Do the calculator instructions at the bottom of the screen interfere with your view of the graph? If so, create more space at the bottom of the screen by pressing [◆][F2] and decreasing the value of **ymin**. When you're done, press [◆][F3] to redraw the graph and then press [F5][3] or [F5][4] to find a minimum or maximum, respectively.

4. **If necessary, repeatedly press [▲] until the number of the appropriate function appears at the top of the screen.**

5. **Set the lower bound for the minimum or maximum point and press [ENTER].**

   To do so, use the [◀] and [▶] keys to place the cursor on the graph a little to the left of the point (as illustrated in the first picture in Figure 7-3 for the maximum point) and then press [ENTER]. A Left Bound indicator appears at the top of the screen (as illustrated by the triangle in the second picture in Figure 7-3), and you are prompted to enter an upper bound.

   TIP: If you prefer, you can set the bound by keying in a number and then pressing [ENTER].

6. **Set the upper bound for the minimum or maximum point and press [ENTER].**

   To do so, use the [◀] and [▶] keys to place the cursor on the graph a little to the right of the point (as illustrated in the second picture in Figure 7-3), and then press [ENTER]. The coordinates of the minimum or maximum point automatically appear at the bottom of the screen, as shown in the third picture in Figure 7-3.

Figure 7-3: Steps for finding the maximum value of a function.

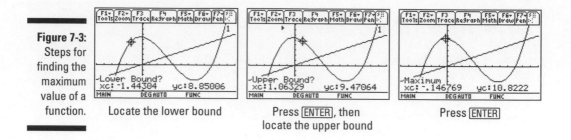

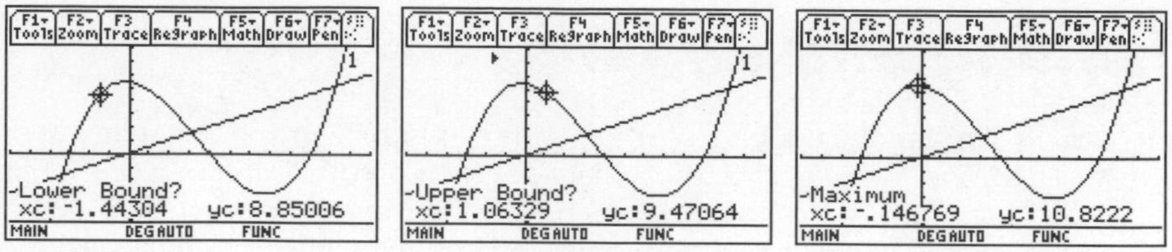

Locate the lower bound | Press [ENTER], then locate the upper bound | Press [ENTER]

Finding Points of Intersection

Using the ◀▶▲▼ keys to locate the point of intersection of two graphs gives you an *approximation* of that point, but this method rarely gives you the actual point of intersection. To accurately find the coordinates of the point where two functions intersect, perform the following steps:

1. **Graph the functions in a viewing window that contains the point of intersection of the functions.**

   For information on graphing a function and finding an appropriate viewing window, see Chapter 5.

2. **If necessary, set the Graph Formats menu to display coordinates in rectangular (*x*, *y*) form.**

   To do this, press [F1]▲[ENTER]▶[1][ENTER].

3. **Press [F5][5] to select the Intersection option.**

4. **Select the first curve.**

   If the number of one of the intersecting functions doesn't appear at the top of the screen, repeatedly press ▲ until it does, as illustrated in the first picture in Figure 7-4. When the cursor is on one of the intersecting functions, press [ENTER] to select it. The calculator places a small + sign on the selected curve, as illustrated in the second picture in Figure 7-4.

5. **Select the second curve.**

   If the calculator doesn't automatically display the name of the second intersecting function at the top of the screen, repeatedly press ▲ until it does, as illustrated in the second picture Figure 7-4. When the cursor is on the second intersecting function, press [ENTER] to select it. The calculator places a small + sign on the curve you just selected.

6. **Set the lower bound for the point of intersection and press [ENTER].**

   To do so, use the ◀ and ▶ keys to place the cursor on the graph a little to the left of the point of intersection, as illustrated in the first picture in Figure 7-5, and then press [ENTER]. A Left Bound indicator appears at the top of the screen (as illustrated by the triangle in the second picture in Figure 7-5), and you are prompted to enter an upper bound.

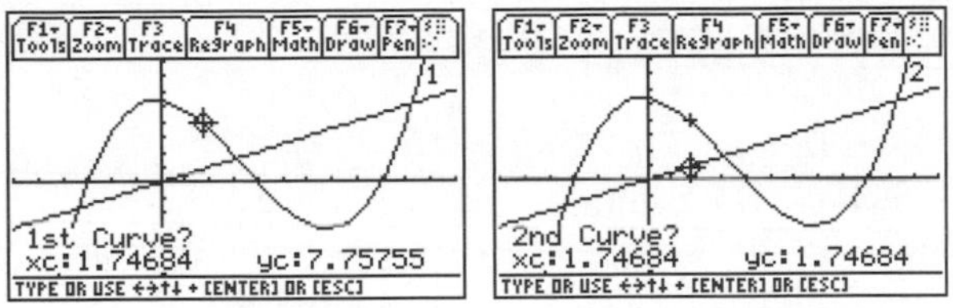

Select 1st curve

Select 2nd curve

Figure 7-4: The initial steps for finding a point of intersection.

TIP

If you prefer, you can set the bound by keying in a number and then pressing [ENTER].

7. **Set the upper bound for the point of intersection and press [ENTER].**

   To do so, use the ◀ and ▶ keys to place the cursor on the graph a little to the right of the point, as illustrated in the second picture in Figure 7-5, and then press [ENTER]. The coordinates of the point of intersection automatically appear at the bottom of the screen, as shown in the third picture in Figure 7-5.

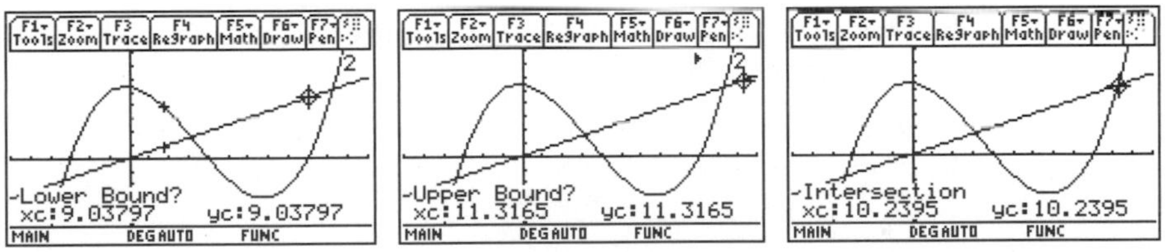

Locate the lower bound

Press [ENTER], then locate the upper bound

Press [ENTER]

Figure 7-5: The final steps for finding a point of intersection.

WARNING!

If more than one point of intersection exists between the lower and upper bounds that you specify when using the **Intersection** command, the calculator finds *only* the point of intersection that has the smallest *x*-coordinate and *does not* warn you that more points of intersection might exist. To find another point of intersection, if such a point exists, use the **Intersection** command again, but this time set the lower bound to a number slightly larger than the *x*-coordinate of the first point of intersection found by the calculator, and set the upper bound to the same number you used when finding the first point of intersection. For example, if the *x*-coordinate of the first point of intersection found by the calculator is 0.1, set the lower bound to 0.1000001. After using the **Intersection** command a second time, the calculator either displays the value of another point of intersection or it displays an error message telling you that there are no more points of intersection in the interval defined by the new lower and upper bounds. After getting the error message, press [ESC] or [ENTER] to make it go away.

Finding the Slope of a Curve

From the calculator's Home screen you can find the derivative of a function. For example, you can get the calculator to tell you that the derivative of x^2 is $2x$. (I tell you how to do this Chapter 12.) But when you're in a Graph screen, the calculator uses a numerical routine to evaluate the derivative at a specified value of x. This numerical value of the derivative is the slope of the tangent to the graph of the function at the specified x-value. It's also called the *slope* of the curve. To find the slope (derivative) of a function at a specified value of x, perform the following steps:

1. **Graph the function in a viewing window that displays the point at which you want to find the slope.**

   Graphing a function and finding an appropriate viewing window are explained in Chapter 5.

2. **If necessary, set the Graph Formats menu to display coordinates in rectangular (x, y) form.**

   To do this, press [F1][▲][ENTER][▶][1][ENTER].

3. **Press [F5][6][1] to select the dy/dx option.**

   If the instructions that appear at the bottom of the screen interfere with your view of the graph, create more space at the bottom of the screen by pressing [♦][F2] and decreasing the value of **ymin**. When you're finished, press [♦][F3] to redraw the graph, and then press [F5][6][1] to reselect the **dy/dx** option.

4. **If necessary, repeatedly press [▲] until the number of the appropriate function appears at the top of the screen.**

5. **Enter the x-coordinate of the point where you want to find the slope and then press [ENTER].**

   To do so, use the keypad to enter the value of x. As you use the keypad, the x-coordinate of the cursor location is replaced by the number you key in. (See the first picture in Figure 7-6.) If you make a mistake when entering your number, press [CLEAR] and reenter the number.

   After pressing [ENTER], the cursor moves to the point on the curve having the specified x-coordinate and the slope of the curve at that point is displayed at the bottom of the screen, as in the second picture in Figure 7-6.

   TIP: If you're interested only in finding the slope of the function in a general area of the function instead of at a specific value of x, don't enter a value for x. Instead, just use the [◀] and [▶] keys to move the cursor to the desired location on the graph of the function and press [ENTER].

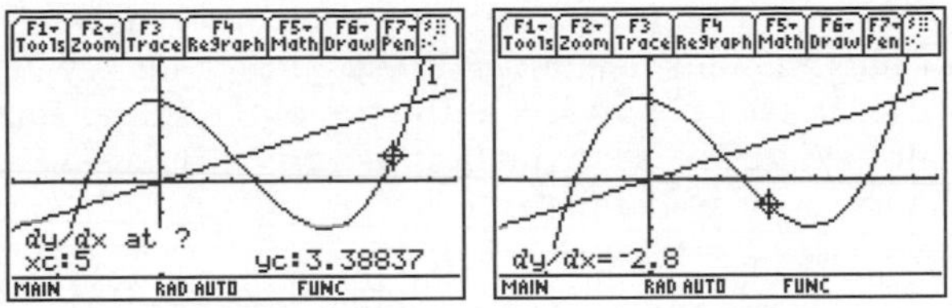

Enter x Press [ENTER]

Figure 7-6: Steps for finding the slope at a specified value of x.

Evaluating a Definite Integral

If $f(x)$ is positive (above the x-axis) for $a \le x \le b$, the definite integral

$$\int_a^b f(x)\,dx$$

gives the area between the curve and the x-axis for $a \le x \le b$; if $f(x)$ is negative, the definite integral equals negative the area between the curve and the x-axis. If $f(x)$ is both above and below the x-axis, the definite integral equals the area above the x-axis minus the area below the x-axis (as in the second picture in Figure 7-7). To evaluate the definite integral, perform the following steps:

1. **Graph the function $f(x)$ in a viewing window that contains the lower limit a and the upper limit b.**

   See Chapter 5 for details on how to Graph a function and find an appropriate viewing window. To get a viewing window containing both the upper and lower limits, these limits must be between **xmin** and **xmax** in the viewing window.

2. **If necessary, set the Graph Formats menu to display coordinates in rectangular (x, y) form.**

   To do this, press [F1][▲][ENTER][▶][1][ENTER].

3. **Press [F5][7] to select the $\int_a^b f(x)\,dx$ option in the Graph Math menu.**

   If the calculator instructions at the bottom of the screen interfere with your view of the graph, create more space at the bottom of the screen by pressing [♦][F2] and decreasing the value of **ymin**. Then press [♦][F3] to redraw the graph and then repeat Step 3.

4. **If necessary, repeatedly press [▲] until the number of the appropriate function appears at the top of the screen.**

5. **Enter a value for the lower limit a and press [ENTER].**

   Use the keypad to enter the value. As you use the keypad, the x-coordinate of the cursor location is replaced by the number you key in, as illustrated in the first picture in Figure 7-7, in which –1 is entered for the lower limit. If you make a mistake when entering your number, press [CLEAR] and reenter the number. Press [ENTER] after keying in a value for a. A triangular symbol appears at the top of the screen, marking the lower limit of the integral, as in the first picture in Figure 7-7. You are prompted to enter the upper limit.

TIP: You can also set the lower limit by using the ◀ and ▶ keys to move the cursor to the appropriate point on the graph and then setting this limit by pressing [ENTER].

6. **Enter the value of the upper limit *b* and press [ENTER].**

   After pressing [ENTER], the value of the definite integral appears at the bottom of the screen, and the area between the curve and the *x*-axis, for $a \le x \le b$, is shaded, as illustrated in the third picture in Figure 7-7.

Figure 7-7: Steps for evaluating a definite integral.

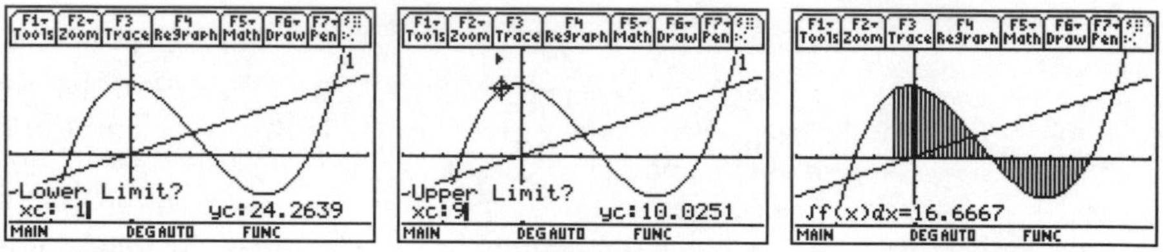

Enter the lower limit

Press [ENTER], then enter the upper limit

Press [ENTER]

REMEMBER: The shading of the graph produced by using the $\int f(x)\,dx$ option on the Graph Math menu doesn't automatically vanish when you use another Graph Math menu option. To erase the shading, press [F4] to redraw the graph without the shading.

Finding Inflection Points

An *inflection point* is a point on the graph where the graph changes concavity. For example, the point in the second picture in Figure 7-8 is the point at which the function in this picture changes from concave down to concave up. To find an inflection point, follow these steps:

1. **Graph the function in a viewing window that displays the inflection point you want to find.**

   You can find details on graphing a function and finding an appropriate viewing window in Chapter 5.

The inflection points on the graph are the points where a wiggle in the graph changes concavity. To have the calculator find these inflection points, you must be able to eyeball the approximate location of where a wiggle in the graph changes from concave up to concave down or vice versa. If your graph has two closely spaced wiggles, it is very difficult to spot the location where each inflection point begins. If this happens in your graph, adjust the Window settings for **xmin** and **xmax** so that you can get a better view of the approximate location where the wiggles change concavity. I explain changing Window settings in Chapter 5.

2. **If necessary, set the Graph Formats menu to display coordinates in rectangular (*x*, *y*) form.**

   To do this, press [F1][▲][ENTER][▶][1][ENTER].

3. **Press [F5][8] to invoke the Inflection command in the Graph Math menu.**

   Are the calculator instructions at the bottom of the screen interfering with your view of the graph? If so, create more space at the bottom of the screen by pressing [♦][F2] and decreasing the value of **ymin**. Then press [♦][F3] to redraw the graph and press [F5][8] to reinstate the **Inflection** command.

4. **If necessary, repeatedly press [▲] until the number of the appropriate function appears at the top of the screen.**

5. **Set the lower bound for the inflection point and press [ENTER].**

   To do so, use the [◀] and [▶] keys to place the cursor on the graph a bit to the left of where you suspect the inflection point is located. Be generous; eyeballing inflection points isn't easy. If in doubt, move the cursor a bit more to the left. When you're sure the inflection point is definitely to the right of the cursor, press [ENTER] to set the lower bound. A Left Bound indicator appears at the top of the screen (as illustrated by the triangle in the first graph of Figure 7-8), and you are prompted to enter an upper bound.

   If you prefer, you can set the bound by keying in a number and then pressing [ENTER].

6. **Set the upper bound for the inflection point and press [ENTER].**

   To do so, use the [◀] and [▶] keys to place the cursor on the graph a little to the right of the inflection point and then press [ENTER]. The coordinates of the inflection point automatically appear at the bottom of the screen, and the cursor is placed on this point, as shown in the second graph in Figure 7-8.

Figure 7-8: Steps for finding an inflection point.

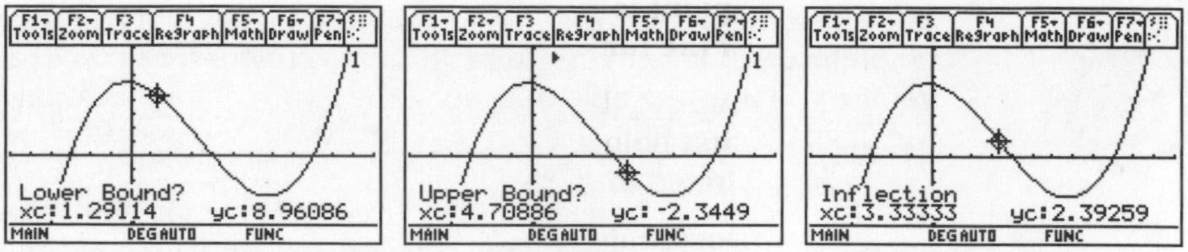

Locate the lower bound — Press [ENTER], then locate the upper bound — Press [ENTER]

Finding the Distance between Two Points

The calculator has a neat feature that draws a line between two points (on either the same or different graphs) and calculates the distance between them. To use this feature, follow these steps:

1. **Graph the functions in a viewing window that displays the two points between which you want to find the distance.**

   Graphing a function and finding an appropriate viewing window are explained in Chapter 5.

2. **If necessary, set the Graph Formats menu to display coordinates in rectangular (*x*, *y*) form.**

   To do this, press [F1][▲][ENTER][▶][1][ENTER].

3. **Press [F5][9] to select the Distance command from the Graph Math menu.**

   Are the instructions at the bottom of the screen interfering with your view of the graph? If so, create more space at the bottom of the screen by pressing [◆][F2] and decreasing the value of **ymin**. When you're finished, press [◆][F3] to redraw the graph and then press [F5][9] to reinstate the **Distance** command.

4. **If necessary, repeatedly press [▲] until the number of the function containing the first point appears.**

5. **Enter the *x* coordinate of the first point and press [ENTER].**

   Use the keypad to enter the *x*-coordinate, as illustrated in the first picture in Figure 7-9. After you press [ENTER], the calculator might display an *x*-coordinate that is close to, but not equal to, the one you entered. If this happens, don't worry about it — the *x*-coordinate you entered is stored in the memory of the calculator, and the calculator uses this stored value to compute the distance.

6. **If the second point is on a different curve, repeatedly press ⊖ until the number of the function containing the second point appears.**

   The cursor jumps to the point on the other curve that is directly above or below the first point, and the calculator prompts you for the second point, as illustrated in the second picture in Figure 7-9.

7. **Enter the *x*-coordinate of the second point, as illustrated in the second picture in Figure 7-9, and then press [ENTER].**

   After you press [ENTER], the calculator connects the two points with a line and displays the distance between these points, as illustrated in the third picture in Figure 7-9.

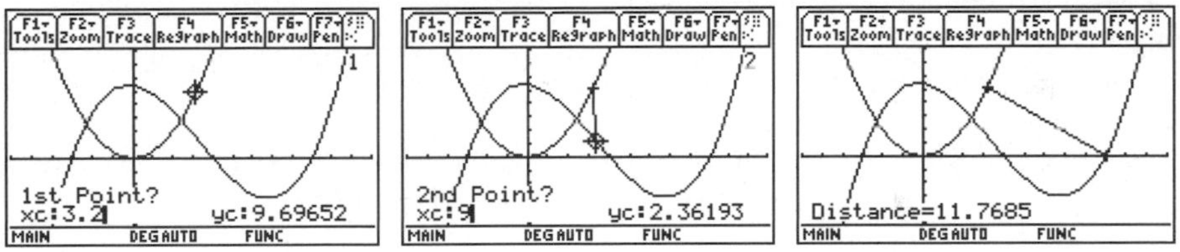

Figure 7-9: Steps for finding the distance between two points.

Finding the Equation of a Tangent Line

Not only can the calculator draw a tangent to a curve at a specified point, but it can also tell you the equation of that tangent. To get the calculator to do this, follow these steps:

1. **Graph the function in a viewing window that displays the point of tangency.**

   I tell you how to graph a function and find an appropriate viewing window in Chapter 5.

2. **If necessary, set the Graph Formats menu to display coordinates in rectangular (*x*, *y*) form.**

   To do this, press [F1]⊖[ENTER]⊙[1][ENTER].

3. **Press [F5]⊖⊖⊖[ENTER] to select the Tangent command from the Graph Math menu.**

   If the calculator instructions at the bottom of the screen interfere with your view of the graph, create more space at the bottom of the screen by pressing [♦][F2] and decreasing the value of **ymin**. Then press [♦][F3] to redraw the graph and press [F5]⊖⊖⊖[ENTER] to reinstate the **Tangent** command.

4. **If necessary, repeatedly press ▲ until the number of the function containing the point of tangency appears.**
5. **Enter the *x*-coordinate of the point of tangency.**

   Use the keypad to enter the *x*-coordinate, as illustrated in the first picture in Figure 7-10. If you make a mistake, press [CLEAR] and reenter the number.
6. **Press [ENTER] to construct the tangent.**

   The tangent appears on the graph with the cursor located at the point of tangency. The equation of the tangent appears at the bottom of the screen, as illustrated in the second picture in Figure 7-10.

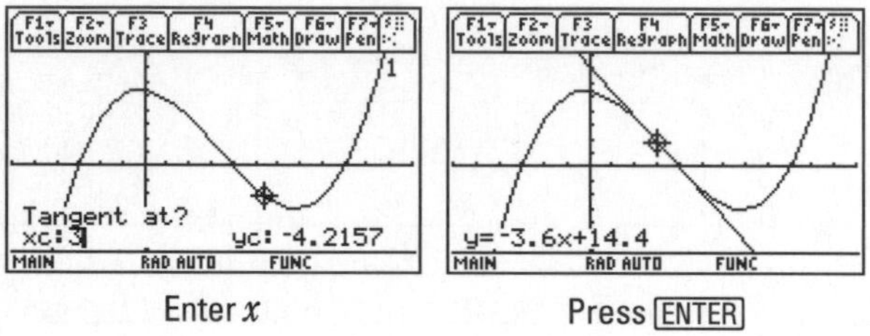

Figure 7-10: Steps for finding the equation of a tangent.

REMEMBER: A tangent drawn by the **Tangent** command in the Graph Math menu doesn't automatically vanish when you use another Graph Math menu option or when you trace the graph. To erase the tangent, press [F4] to redraw the graph without the tangent line.

Evaluating Arc Length

The **Arc** command in the Graph Math menu can find the length of a section (arc) of a graphed curve when you specify the beginning and ending point of the section. To get the calculator to find this length, follow these steps:

1. **Graph the function in a viewing window displaying the part of the curve you want to measure.**

   I explain graphing a function and finding an appropriate viewing window in Chapter 5.
2. **If necessary, set the Graph Formats menu to display coordinates in rectangular (*x*, *y*) form.**

   To do this, press [F1][▲][ENTER][▶][1][ENTER].

3. **Press [F5]↑↑[ENTER] to select the Arc command from the Graph Math menu.**

   Do the calculator instructions at the bottom of the screen interfere with your view of the graph? If so, create more space at the bottom of the screen by pressing [◆][F2] and decreasing the value of **ymin**. When you're finished, press [◆][F3] to redraw the graph and then press [F5]↑↑[ENTER] to reinstate the **Arc** command.

4. **If necessary, repeatedly press ↑ until the number of the appropriate function appears at the top of the screen.**

5. **Enter the *x*-coordinate of one of the points that define the arc and then press [ENTER].**

   Use the keypad to enter the *x*-coordinate, as illustrated in the first picture in Figure 7-11. As you enter a number, the *x*-coordinate of the cursor location is replaced by your number. If you make a mistake, press [CLEAR] and reenter the number.

   When you press [ENTER] after entering the *x*-coordinate, the cursor moves to a point on the graph that has approximately the same *x*-coordinate you entered, and the calculator prompts you to enter the second point, as illustrated in the second picture in Figure 7-11.

6. **Key in the *x*-coordinate of the second point defining the arc, as illustrated in the second picture in Figure 7-11, and then press [ENTER].**

   The calculator displays the length of the arc at the bottom of the screen, as illustrated in the third picture in Figure 7-11.

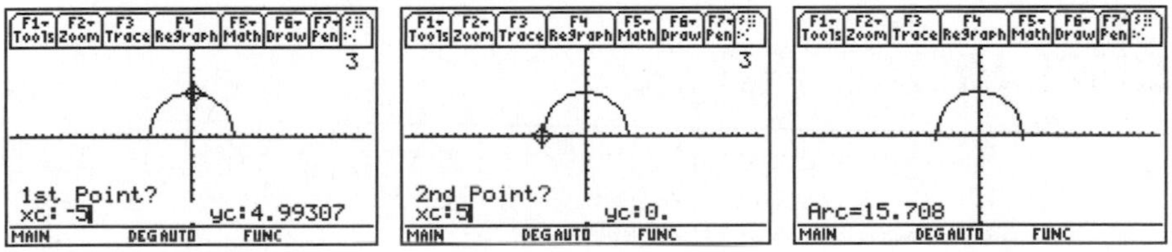

Enter first point — Enter second point — Press [ENTER]

Figure 7-11: Steps for finding arc length.

Part IV

Working with Sequences, Parametric Equations, and Polar Equations

The 5th Wave
By Rich Tennant
Beyond Euclidean and Cartesian geometry, there is Ed Dubrowski geometry which proves that the volume of a sphere changes in proportion to the amount of food at an All-U-Can-Eat buffet.
©RICHTENNANT

In this part . . .

This part shows you how to graph sequences, parametric equations, and polar equations. You get a look at how to trace graphs, create tables, and save your graphs for future use. I also show you how to convert between rectangular and polar coordinates.

Chapter 8

Graphing Sequences

In This Chapter

- Entering sequences into the calculator
- Graphing sequences
- Graphing web and custom plots

A *sequence* is simply an ordered list of terms or numbers. The most famous sequence is perhaps the Fibonacci sequence, {0, 1, 1, 2, 3, 5, 8, 13, . . .}, where the first two terms are given and each following term is found by adding the previous two terms. The formula used to generate a sequence is referred to by various names, including recurrence relation, recursive function, and iterative function. Texas Instruments calls these formulas *sequence functions,* so I use this term as well.

Entering a Sequence

The calculator can accommodate up to 99 sequences. To avoid notational confusion, I denote these sequences by u_1 through u_{99}; the calculator uses the notation *u*1 through *u*99. For the sake of simplicity, I tell you how to enter the sequence $u_k(n)$. However, what I state in this chapter and in Chapter 9 for $u_k(n)$ also applies to the sequences $u_1(n)$, $u_2(n)$, and so on. To enter a sequence into the calculator, follow these steps:

1. **Press [MODE][▶][4] to put the calculator in SEQUENCE mode (as shown in Figure 8-1) and then press [ENTER] to save the Mode settings.**

   TIP: If your sequence deals with money, set Display Digits in the Mode menu to FIX 2 (as shown in Figure 8-1) to make the calculator round all numbers to two decimal places. To do this, press [MODE] if you aren't already in the Mode menu and then press [▼][▼][▶][3]. Finally, press [ENTER] to save the Mode settings.

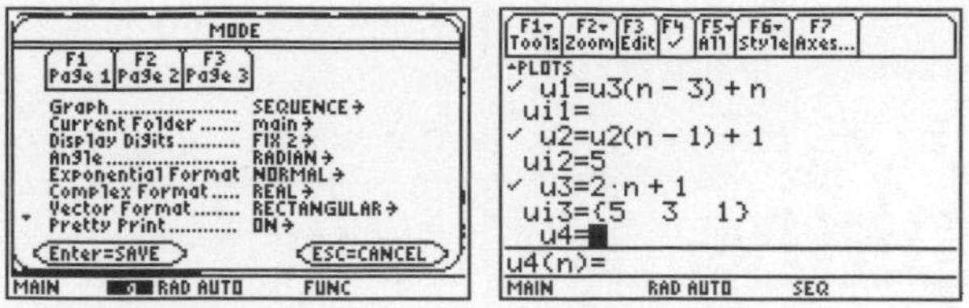

Figure 8-1: Defining sequences.

Set mode   Enter sequences

2. **Press [◆][F1], enter the definition of u_k, and then press [ENTER] to store the definition in the Y= editor.**

   To do this, use the ⊖ and ⊖ keys to place the cursor on u_k and then press [ENTER] to place the cursor on the command line, where you enter the definition. Any previously entered definition is overwritten when you type the new definition.

   The letters *u* and *n* are entered in a sequence definition by pressing [ALPHA][+] and [ALPHA][6], respectively, and you enter the subscript by pressing the appropriate number key. For example, to enter the term $u_1(n-1)$ press [ALPHA][+][1][(][ALPHA][6][−][1][)]. After entering the definition, press [ENTER] to store it in the Y= editor. Examples of definitions of sequences appear in the second picture in Figure 8-1.

   The definition of a sequence can reference any previous terms of itself or of another sequence, but it cannot reference a current term. For example, $u_2(n) = u_1(n-1) + 5$ is a valid sequence definition, but defining $u_2(n) = u_1(n) + 5$ results in an error message.

3. **If you're graphing more than one sequence, give your sequence a unique graphing style so you can identify it in a graph of multiple sequences.**

   In Sequence mode, you have four style choices: Line, Dot, Square, and Thick (line). To set the style of sequence u_k, use the ⊖⊖ keys to highlight the definition of u_k and press [2nd][F1] to display the style menu. Then press the number of the desired style.

   In some windows, the Square style looks like a super thick line, thicker than the Thick style.

4. **Enter the initial value or values of the sequence in ui_k and then press [ENTER] to save the setting in the Y= editor.**

   ui_k is left blank, set equal to the first term in the sequence $u_k(n)$, or set equal to several terms in $u_k(n)$. It depends on how *all* the sequences, u_1 through u_{99}, are defined in the Y= editor.

The following tells you what value to assign to ui_k and how it's entered in the calculator:

- ui_k can be left blank if no other sequence in the Y= editor, including u_k, references a previous term of u_k in its definition. If ui_k has previously been assigned a value, you can leave it there because the calculator doesn't need (and won't use) it. But if you really want to leave it blank, use the ⊖⊖ keys to place the cursor on ui_k and press [CLEAR] to get rid of that value.
- ui_k is set equal to the first term in the sequence $u_k(n)$ if any of the sequences defined in the Y= editor use $u_k(n-1)$ in their definitions but none use any other previous term of u_k, such as $u_k(n-2)$ or $u_k(n-4)$. To set ui_k equal to the first term in the sequence $u_k(n)$, use the ⊖⊖ keys to place the cursor on ui_k and press [ENTER] to place the cursor on the command line. Use the keypad to enter the initial value and then press [ENTER] to store that value in the Y= editor. As you enter this number, the calculator automatically erases any previous value assigned to ui_k.
- ui_k is set equal to the first two terms in the sequence $u_k(n)$ if any of the sequences in the Y= editor use $u_k(n-2)$ in their definitions but none use any other previous term of u_k, such as $u_k(n-3)$ or $u_k(n-5)$. Similarly, ui_k is set equal to the first three terms in the sequence $u_k(n)$ if any of the sequences in the Y= editor use $u_k(n-3)$ in their definitions but none use any other previous term of u_k, such as $u_k(n-4)$ or $u_k\ (n-6)$; and so on. And to complicate matters, *these initial values must be entered in reverse order*, separated by commas, and surrounded by curly braces. For example, the format for entering three initial conditions is $\{u_k(3), u_k(2), u_k(1)\}$. To enter multiple initial values, use the ⊖⊖ keys to place the cursor on ui_k and press [ENTER] to place the cursor on the command line. Then enter the initial conditions in the specified format; the curly braces are entered into the calculator by pressing [2nd][(] and [2nd][)]. When you're finished entering multiple initial conditions, press [ENTER] to store them in the Y= editor. Although the calculator requires that you enter multiple initial values separated by commas, the calculator removes these commas after you press [ENTER]. An example appears at the bottom of the second picture in Figure 8-1 (shown previously).

In the example in the second picture in Figure 8-1, ui_1 is left blank because none of the defined sequences u_1, u_2, or u_3 use any term of u_1 in their definitions. ui_2 is set equal to the first term in $u_2(n)$ because sequence u_2 used $u_2(n-1)$ in its definition. In this example, ui_2 is assigned the value of 5. Because u_1 uses $u_3(n-3)$ in its definition, ui_3 is set equal to the first three terms in u_3, listed in reverse order.

TECHNICAL STUFF

A comment on setting initial values

The second picture in Figure 8-1 illustrates a situation in which you might want to — but don't have to — deviate from the requirements for setting initial values. In this example, the calculator doesn't know the value of the first three terms of u_1, nor can it evaluate them. After all, $u_1(1) = u_3(-2)$, $u_1(2) = u_3(-1)$, and $u_1(3) = u_3(0)$, but the calculator is told to evaluate u_3 only for $n \geq 1$ (as indicated by the **nmin** = 1 setting in the first picture in Figure 8-2).

Luckily, the calculator doesn't have to know these terms. It just calculates $u_1(n)$ for $n \geq 4$. When you graph $u_1(n)$, or produce a table of the values for $u_1(n)$, the first three terms of $u_1(n)$ don't appear, but the rest of the sequence does. This is illustrated in Figure 8-2 (elsewhere in this chapter), where the top graph doesn't start at the same place as the other graphs. If you can't live with the omission of these terms in the graph or in the table, set ui_1 equal to the first three terms in $u_1(n)$, listed in reverse order.

In this example, $u_1(n) = u_3(n-3) + n = [2(n-3) + 1] + n = 3n - 5$. So evaluating the first three terms of $u_1(n)$ isn't that difficult. However, for more complicated sequences, it might not be worth your time to figure out the values of these terms. After all, didn't you get the calculator so you wouldn't have to do scads of math by hand? In situations like this, I recommend just living with the omission of the first three terms in the graph and the table.

Graphing Sequences

After you have entered the sequence functions into the calculator, as I describe in the preceding section, you can use the following steps to graph the sequences:

1. **If necessary, press [◆][F1] to enter the Y= editor.**
2. **Press [2nd][F2] and set the Axes for the graph to TIME, WEB, or CUSTOM.**

   The following list explains how your graph is displayed in each format and how you set the format.

   - **TIME:** The Time format is the most common format for graphing sequences because it graphs the sequences as a function of the independent variable *n* by graphing the points $(n, u_k(n))$ for each sequence. To set the Axes format to the Time format, press [◆][F1] if you are not already in the Y= editor and press [2nd][F2] if you haven't already displayed the Axes menu. Then press [▶][1][ENTER] to set the axes to the **TIME** format.
   - **WEB:** This format produces a *web plot* (also known as a *cobweb plot* because of its shape) for a sequence $u_k(n)$. Use it when you want to see whether $u_k(n)$ converges to an equilibrium point or just veers off into space. It graphs the points $(u_k(n-1), u_k(n))$ and the

line $y = x$. Using the [F3] (Trace) and ▶ keys then shows you whether the sequence is converging or diverging. I explain this tracing process in Chapter 9.

In Web format, the calculator places two restrictions on how the sequence $u_k(n)$ is defined. First, it requires that $u_k(n - 1)$ appear as a variable in the definition of $u_k(n)$; second, it requires that $u_k(n - 1)$ be the only variable used in this definition. Defining $u_k(n)$ as $u_k(n) = u_k(n - 1) + u_k(n - 2)$ or as $u_k(n) = u_{k+1}(n - 1)$ results in an error message. And because $u_k(n) = u_k(n - 1) + n$ uses the variable n in its definition, it too results in an error message when it's used in the Web format.

To set the Axes format to WEB, press [◆][F1] if you aren't already in the Y= editor, and press [2nd][F2] if you haven't already displayed the Axes menu. Then press ▶[2] to select the **WEB** option. If you want to use Trace to build the web plot yourself, press ▼▶[1]; otherwise, if you want the calculator to automatically build the web plot for you, press ▼▶[2]. Finally, press [ENTER] to save your settings.

- **CUSTOM:** Use this format to create phase plots when you want to see how one sequence affects another sequence. The Custom format graphs the points (u(n), v(n)) where you specify which sequences u(n) and v(n) in the Y= editor are used as the x- and y-coordinates.

  To set the Axes format to the Custom format, press [◆][F1] if you aren't already in the Y= editor and press [2nd][F2] if you haven't already displayed the Axes menu. Then press ▶[3] to select the **CUSTOM** option. Then press ▼▶, use the ▲▼ keys to highlight the sequence you want graphed on the x-axis, and press [ENTER]. Repeat this process to select the function you want graphed on the y-axis. Finally, press [ENTER] to save your settings.

3. **If necessary, turn off any sequences and Stat Plots that you don't want to appear in your graph.**

   If you don't know what I mean here, press [◆][F1] to display the Y= editor. If a checkmark appears to the left of a sequence in the Y= editor, that sequence will be graphed. To prevent this from happening, use the ▲▼ keys to highlight the definition of the sequence and then press [F4] to remove the checkmark, thus preventing the sequence from being graphed. At a later time, when you do want to graph the sequence, repeat this process to reinstate the checkmark.

   When graphing a web plot, only the sequence used to create the plot should have a checkmark next to it in the Y= editor. If another sequence is checked and its definition doesn't abide by the restrictions placed on sequences that can be used to produce web plots (as specified in Step 2) you get an error message. If the other sequence does abide by these restrictions, it will graph, but its existence usually clutters the screen.

The first line in the Y= editor tells you the graphing status of the Stat Plots. (See Chapter 20 for more on Stat Plots.) If no number appears after PLOTS on the very first line of the Y= editor, as in the second picture in Figure 8-1, no Stat Plots are selected to be graphed. If one or more numbers do appear, those Stat Plots will be graphed along with the graph of your sequences.

To prevent an individual Stat Plot from being graphed, use the ▲▼ keys to place the cursor on that Stat Plot and press [F4]. ([F4] toggles between deactivating and activating a Stat Plot.) To prevent all Stat Plots from being graphed, press [F5][5].

WARNING!

When you're graphing sequences, Stat Plots that are active when you don't really want them graphed cause problems that could result in an error message when you attempt to graph your sequences. So if you aren't planning to graph a Stat Plot along with your sequences, make sure all Stat Plots are turned off by pressing [F5][5].

4. **If necessary, set the format for the graph by pressing [F1]▲[ENTER] and changing the selections in the Graph Formats menu.**

   I explain the Graph Formats menu in Chapter 5. It allows you to have coordinates of the cursor location displayed in rectangular (x, y) form, polar (r, θ) form, or not displayed at all. So if you want to know the coordinates of the cursor location, my recommendation is to display the coordinates in either rectangular or polar form.

   The **Grid** option is your choice, either OFF or ON. And as far as the other options go, my recommendation is to set **Axes** ON, **Leading Cursor** OFF, and **Labels** OFF. Why? See Chapter 5.

5. **Press [♦][F2] to access the Window editor.**

6. **After each of the first four window variables, enter a numerical value that is appropriate for the sequences you're graphing, and press [ENTER] after entering each number.**

   Figure 8-2 pictures the Window editor when the calculator is in Sequence mode. The following list explains the variables you must set in this editor:

   - **nmin:** This setting contains the first value of the independent variable n for each of your sequences. For example, if you set **nmin** = 5, the first number in the each sequence is $u_k(5)$. You usually want to set **nmin** equal to 1 so your sequences look like $\{u_k(1), u_k(2), u_k(3), \ldots\}$. But if your sequences represent an experiment that starts at "time zero," you want to set **nmin** equal to 0. This way your sequences look like $\{u_k(0), u_k(1), u_k(2), \ldots\}$.

nmin must be set equal to 0 or to a positive integer. The calculator isn't equipped to handle negative values for **nmin**. Setting **nmin** equal to something other than 0 or a positive integer results in an error message.

- **nmax:** This setting contains the largest value of the independent variable *n* that you want the calculator to use to evaluate your sequences. For example, if **nmin** = 0 and **nmax** = 100, the calculator will evaluate the first 101 terms in each sequence.

  nmax must be set equal to a positive integer that is greater than **nmin**. Be reasonable when entering a value for **nmax**. If you're graphing three sequences and **nmax** = 1,000, it's going to take the calculator at least a minute to get the job done.

  If it's taking a long time for the calculator to graph your sequences and you start to regret that rather large value you placed in **nmax**, press [ON] to terminate the graphing process. You can then go back to the Window editor and adjust the **nmax** setting.

- **plotStrt:** This setting tells the calculator where you want to start graphing in each sequence. For example, if **plotStrt** = 20, the calculator starts the graph with the 20th term in each sequence.

  This isn't always intuitive. If **nmin** = 1, the 20th term in the sequence $u_k(n)$ is $u_k(20)$. But if **nmin** = 0, the 20th term in this sequence is $u_k(19)$.

  plotStrt must be set equal to a positive integer. Usually it's set equal to 1 so that the graph starts at the beginning of a sequence.

- **plotStep:** This setting tells the calculator whether you want to skip graphing any terms in each sequence. A setting of 1 tells the calculator to graph every term in each sequence; a setting of 2 tells it to graph every other term; a setting of 3 tells it to graph every third term; and so on. **plotStep** must be set equal to a positive integer. Usually it's set equal to 1 so that the graph shows all terms in your sequences.

You use the remaining items in the Window editor to set a viewing window for the graph (a procedure that I explain in detail in Chapter 5). If you know the dimensions of the viewing window required for your graph, go ahead and assign numerical values to the remaining variables in the Window editor and advance to Step 10. If you don't know the minimum and maximum *y* values required for your graph, the next step tells you how to get the calculator to find them for you.

7. **If your calculator is in Web format, enter numerical values for xmin and xmax.**

   If you're graphing in the Time or Custom format, skip this step. If the calculator is in Web format, set **xmin** equal to the value in **nmin** and **xmax** equal to the value in **nmax**. After producing the graph (in Step 8), you can adjust the window to your liking (in Step 9).

8. **Press [F2][▲][▲][▲][ENTER] to get ZoomFit to graph your sequences.**

   After you have assigned values to **nmin**, **nmax**, **plotStrt,** and **plotStep** (and **xmin** and **xmax** when using the Web format), **ZoomFit** determines appropriate values for **xmin**, **xmax**, **ymin**, and **ymax** and graphs your sequences. **ZoomFit** graphs the sequences in the smallest possible viewing window, however, and it won't assign appropriate values to **xscl** and **yscl**. (See the second picture in Figure 8-2.)

   If you like the way the graph looks, you can skip the remaining steps. If you'd like to spruce up the graph, Step 9 gives you some pointers.

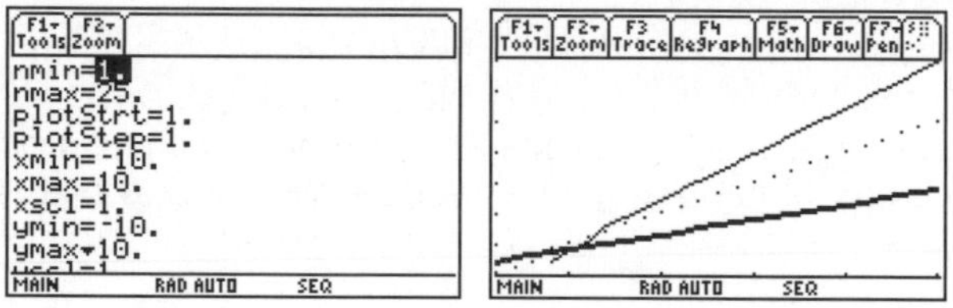

Set window Use ZoomFit

Figure 8-2: Graphing sequences.

9. **Press [◆][F2] and adjust the values assigned to the *x* and *y* settings, and press [ENTER] after entering each new number.**

   Here's how to readjust the viewing window after using **ZoomFit:**

   - **xmin and xmax:** If you want to see the *y*-axis on your graph, set **xmin** equal to 0. If you're interested in seeing a more detailed graph of the beginning of the sequence, decrease the value of **xmax**.
   - **xscl:** Set this equal to a value that doesn't make the *x*-axis look like railroad tracks — that is, an axis with far too many tick marks. Twenty or fewer tick marks makes for a nice-looking axis.
   - **ymin and ymax:** If you don't want the graph to hit the top and bottom of the screen, decrease the value assigned to **ymin** and increase the value assigned to **ymax**. If you want to see the *x*-axis on the graph, assign values to **ymin** and **ymax** so that zero is strictly between these two values.
   - **yscl:** Set this equal to a value that doesn't make the *x*-axis look like railroad tracks (too many tick marks). Fifteen or fewer tick marks is a nice number for the *y*-axis.

10. **Press [◆][F3] to regraph the sequences.**

    If you get an error message when graphing sequences and then select the **GOTO** option, the calculator sends you to the definition of the "problem" sequence that you placed in the Y= editor. More often than not, however, the true location of the problem is not in the way you defined the sequence (as the calculator leads you to believe) but in the way you defined the initial value ui_k or in the settings in the Axes menu. For example, if $u_3(n) = u_1(n-2) + u_2(n-1)$ and the calculator places the cursor after $u_1(n-2)$ when you select the **GOTO** option, then the problem is most likely caused by not setting ui_1 equal to the first two terms in $u_1(n)$. (The previous section explains how to set ui_k.) Or the problem could be caused by the Axes being set to Web format when you were expecting to graph in Time format. (You change the Axes format in Step 2.)

Saving a Sequence Graph

After you've graphed your sequences, you can save the graph and its entire Window, Y= editor, Mode, and Format settings in a Graph Database. When you recall the Database at a later time, you get more than just a picture of the graph. The calculator also restores the Window, Y= editor, Mode, and Format settings to those stored in the Database. Thus you can, for example, trace the recalled graph.

You can find the procedures for saving and recalling a Graph Database in Chapter 5.

Chapter 9

Exploring Sequences

In This Chapter

- Using Zoom commands with sequences
- Tracing the graph of a sequence
- Constructing a table of sequence values
- Evaluating sequences

The calculator has three very useful features that help you explore the graph of a sequence: zooming, tracing, and creating tables. Zooming allows you to quickly adjust the viewing window for the graph so that you can get a better idea of the nature of the graph of the sequence. Tracing shows you the coordinates of the points that make up the graph. And creating a table — well, I'm sure you already know what that shows you. This chapter explains how to use each of these features.

Exploring Sequence Graphs

When you explore the graph of a sequence, you're usually interested in seeing where the sequence goes. Does it level off and converge to a number? Does it veer off into space? What does it look like from the 50th term on? What's the value of the 100th term? The calculator's Zoom and Trace features help you answer such questions.

Using Zoom commands in Sequence mode

After graphing your sequence, as I explain in Chapter 8, press [F2] to access the Zoom menu. On this menu you see all the Zoom commands that are available for graphing functions. Not all of them are useful when graphing sequences, however. Here are the ones that I use when graphing sequences (see Chapter 8 for details on how to use these commands):

- **ZoomFit:** This command finds a viewing window for a specified portion of the sequence. How to use **ZoomFit** is explained in Chapter 8 in the section on graphing sequences.
- **ZoomBox:** After the calculator draws the graph, this command allows you to define a new viewing window for a portion of your graph by enclosing it in a box, as illustrated in Figure 9-1. The box becomes the new viewing window. I explain how to use this command in Chapter 6 in the section on zoom commands that zoom in or out from your graph.
- **Zoom In and Zoom Out:** After the calculator draws the graph, these commands allow you to zoom in on a portion of the graph or to zoom out from the graph. They work very much like a zoom lens on a camera. (The section on zoom commands that zoom in or out in Chapter 6 gives detailed instructions on how to use these commands.)

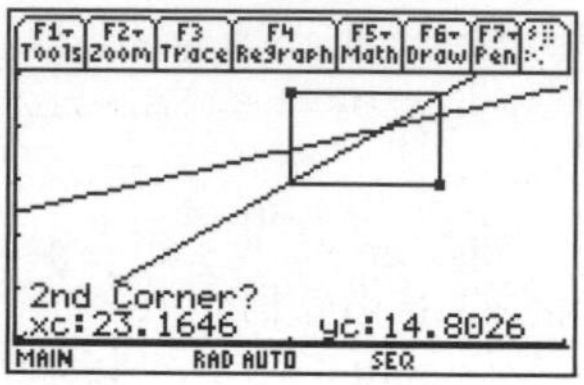

Figure 9-1: Using ZoomBox.

Tracing a sequence

After you have graphed your sequences (as I describe in Chapter 8), you can have the cursor display the coordinates of the points on the graph by tracing the graph. To do this, press [F3] and then use ▶ to trace the graph. As you investigate a sequence by tracing it, here's what you see:

- **The number of the sequence:** In the Time and Web formats (which I explain in Chapter 8), the number of the sequence in the Y= editor that you're tracing is displayed in the top-right corner of the screen. For example, if you see the number 1, you're tracing the first sequence (u_1) in the Y= editor. If you've graphed more than one sequence and you would like to trace a different sequence, repeatedly press ▲ until the number of the desired sequence appears.

  In the Custom format (which you also find out about in Chapter 8), no sequence number appears on the screen because there is only one graph.
- **The value of the independent variable *n*:** At the bottom of the screen you see a value for **nc**. This is the value of the independent variable *n* at the cursor location. Each time you press ▶, the cursor moves to the next plotted point in the graph and updates the value of **nc**. If you press ◀, the

cursor moves left to the previously plotted point in the sequence (provided you aren't at the beginning of the graph). And if you press ⊖ to trace a different sequence, the tracing of that sequence starts at the same value of **nc** that was displayed on-screen before you pressed this key.

If no value for **nc** appears on the screen, press [F1]⊖x̄⊙ and then press [1] to display coordinates in rectangular form or press [2] to display them on polar form. When you're finished, press x̄ to save your settings and then press [F3] to trace the graph. The value of **nc** now appears on the screen.

When you're using Trace, if you want to start tracing your sequence at a specific value of the independent variable n, just key in that value and press [ENTER]. (The value you assign to n cannot be greater than **nmax**; if it is, you get an error message.) After you press [ENTER], the trace cursor moves to the point on the graph corresponding to that value of n. But the calculator does not change the viewing window. So if the value you assigned to n is greater than **xmax**, you won't see the cursor; it's on the part of the graph outside the viewing window. The sidebar "Panning in Sequence mode" tells you how to get the cursor and the graph in the same viewing window.

- **The coordinates of the cursor:** On the last line of the screen you see the coordinates of the cursor location. The relationship between these coordinates and the sequence depends on how you set the Axes format for the sequence. (You discover how to set the Axes format in Chapter 8.) Here's what you see in the various formats:

  - **TIME format:** When coordinates are displayed in rectangular form, the x-coordinate of the cursor location is the independent variable n and the y-coordinate is the corresponding value of the sequence at n.

  - **WEB format:** In this format, the trace cursor starts on the x-axis at the first term in the sequence $u_k(n)$. When you press [π], the cursor moves vertically to the graph of the points $(u_k(n), u_k(n + 1))$. And when you press ⊙ again, it moves horizontally to the line $y = x$. Each time you press ⊙ after that, the cursor repeats this alternating vertical and horizontal movement. In addition, a vertical or horizontal line connects all points traced by the cursor. This is illustrated in Figure 9-2.

    In Web format, if **nmin** = 1, the points traced by the trace cursor are $(u_k(1), 0)$, $(u_k(1), u_k(2))$, $(u_k(2), u_k(2))$, $(u_k(2), u_k(3))$, $(u_k(3), u_k(3))$, $(u_k(3), u_k(4))$, and so on. You use this format when you want to see whether a sequence converges or diverges. Figure 9-2 shows an example of using the Web format and Trace to investigate the convergence of a sequence.

  - **CUSTOM format:** When you defined this format (see Chapter 8) you specified which sequences u_k and u_l in the Y= editor were to be used as the x- and y-coordinates. Because this format graphs the points $(u_k(n), u_l(n))$, xc = u_k(nc) and yc = u_l(nc).

TIP

Panning in Sequence mode

When you're tracing a sequence and the cursor hits an edge of the screen, if **nc** is less than **nmax** and you continue to press ▶, the coordinates of the cursor are displayed at the bottom of the screen. However, because the calculator doesn't automatically adjust the viewing window, you don't see the cursor on the graph. To rectify this situation, make a mental note of the value of **nc** (or jot it down) and then press [ENTER]. The calculator then redraws the graph centered at the location of the cursor at the time you pressed [ENTER]. Then press [F3] to continue tracing the graph.

Unfortunately, after the graph is redrawn, the trace cursor is placed at the beginning of the first sequence appearing in the Y= editor. To get the trace cursor back on the part of the graph displayed in the new viewing window, key in the value of **nc** that you made note of, and then press [ENTER]. If you weren't tracing the first sequence appearing in the Y= editor, use ▲ to place the cursor on the graph you *were* tracing. The trace cursor then appears in the middle of the screen and you can use ▶ or ◀ to continue tracing the sequence. When **nc** equals **nmax**, pressing ▶ has no effect because the cursor has reached the end of the sequence and can go no farther.

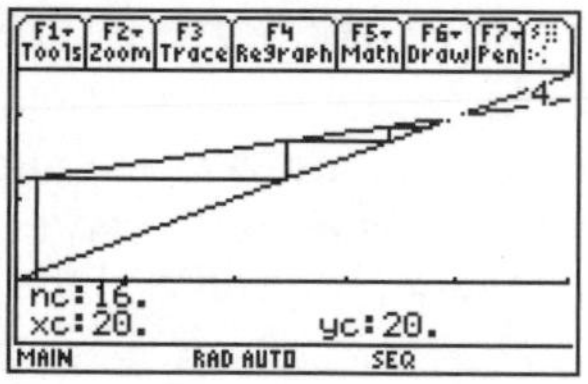

Figure 9-2: Tracing a convergent sequence in Web format.

Press [ESC] to terminate tracing the graph. This also removes the number of the sequence and the coordinates of the cursor from the screen.

TIP

If the number of the sequence and the value of the independent variable *n* are interfering with your view of the graph when you use Trace, increase the height of the window by pressing [◆][2nd] and decreasing the value of **ymin** and increasing the value of **ymax**.

Displaying Sequences in a Table

After you've entered the sequences into the calculator (as I describe in Chapter 8), you can have the calculator create one of two different types of tables displaying the terms in your sequence; one is automatically generated by the calculator and the other, the user-defined table, allows you to manually place a value of the independent variable in the table and have the calculator determine the value of the corresponding term in the sequence.

Creating automatically generated tables

You have some flexibility in the way the calculator displays sequence terms in an automatically generated table. You can have it display every term starting with the first term or starting with any term of your choice. And instead of having the table display every term, you can have it display every other term or every tenth term or any interval between terms that you want. To create an automatically generated table, follow these steps:

1. **Place a checkmark to the left of the sequences in the Y= editor that you want to appear in the table; leave the other sequences unchecked.**

   Only the sequences in the Y= editor that have checkmarks to the left of their names will appear in the table. To check or uncheck a sequence, press ◆F1 to access the Y= editor, use the ▲▼ keys to place the cursor on the sequence, and then press F4 to toggle between checking or unchecking the sequence.

2. **Press ◆F4 to access the Table Setup editor (shown in Figure 9-3).**

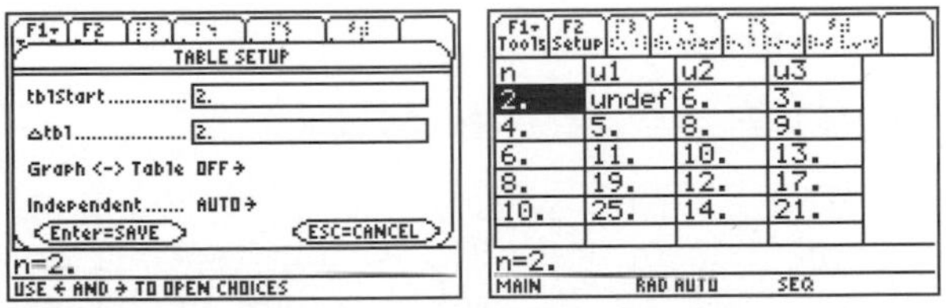

Figure 9-3: Generating a table of sequence values.

3. **Enter a number in tblStart and press ▼; then enter a number in Δtbl and press ▼.**

   If **Independent** is the only item you see in the Table Setup editor and if the cursor is on ASK, press ▶1 to change this setting to AUTO and then press ▲▲▲ to enter numbers in **tblStart** and **Δtbl**.

   tblStart is the first value of the independent variable *n* to appear in the first column of the table. It must be an integer greater than or equal to **nmin**. **Δtbl** is the increment for the independent variable *n*. The calculator constructs the table by adding the value you assign to **Δtbl** to the previous *n*-value and calculating the corresponding values of the terms in the sequences displayed in the table. Usually **Δtbl** is assigned the value 1, so the table displays each term of the sequence. If you want to see every other term in the sequence, assign it the value 2. To see every tenth term, assign it the value 10, and so on.

In Figure 9-3, **tblStart** is assigned the value 2 and **Δtbl** is assigned the value 2 to display every other term in the sequences starting with the second term.

Use the number keys to enter values in **tblStart** and **Δtbl**. Press ▼ after making each entry.

4. **If necessary, set Graph <-> Table to OFF (as shown in Figure 9-3).**

   You have two choices for the **Graph <-> Table** setting: OFF and ON. When **Graph <-> Table** is set to ON, a table of the points plotted in the calculator's graphs of the sequences is created. However, if **Graph <-> Table** is set to OFF, the calculator uses the values in **tblStart** and **Δtbl** to generate the table.

   To change the **Graph <-> Table** setting, use ▲▼ to place the cursor on the existing setting, press ▶ to display the options for that setting, and press the number of the option you want.

5. **Press [ENTER] to save the settings in the Table Setup editor.**
6. **If necessary, press [♦][F5] to display the table.**

   The table appears on the screen, as illustrated in Figure 9-3. You find out about navigating the table later in this chapter.

   The word *undef* in a table means that a particular sequence is not defined at the specified *n*-value. For example, in Figure 9-3, sequence u_1 is undefined when $n = 2$.

Creating a user-defined table

User-defined tables are quite handy when you want to know the values of terms in a sequence that don't necessarily appear in a specific order. For example, you might want to know the value of $u_k(100)$. And based on that value, you might decide that you need to know the value of $u_k(50)$. The value of $u_k(50)$ might then inspire you to investigate the value of $u_k(75)$, and so on. To create a user-defined table in which you enter the *n*-values and the calculator figures out the corresponding sequence values, follow these steps:

1. **Place a checkmark to the left of each sequence in the Y= editor that you want to appear in the table; leave the other sequences unchecked.**

   Only the sequences in the Y= editor that have checkmarks to the left of their names will appear in the table. To check or uncheck a sequence, press [♦][F1], use the ▲▼ keys to place the cursor on the sequence, and then press [F4] to toggle between checking and unchecking the sequence.

2. **Press [♦][F4] to access the Table Setup editor.**

3. **Set Independent to ASK and press [ENTER].**

   Independent is the last item in the menu. If it is already set to ASK, press [ENTER] and go to the next step. If **Independent** is set to AUTO, press ▼▼▼▶[2] to set it to ASK and then press [ENTER].

4. **If necessary, press [♦][F5] to display the table, and if the table contains entries, press [F1][8][ENTER] to clear the table.**

   You see a table without entries, as in the first picture in Figure 9-4.

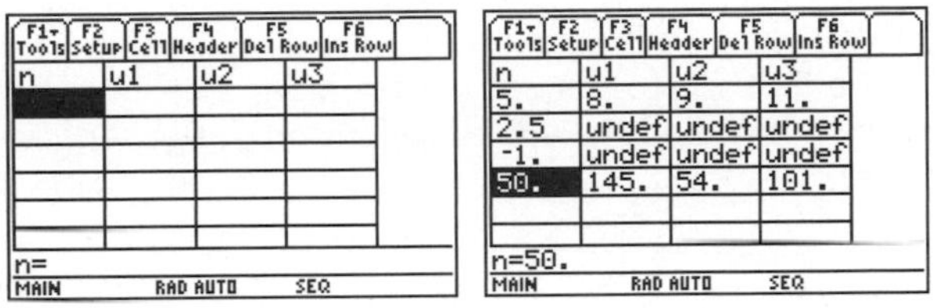

Figure 9-4: Generating a user-defined table.

5. **Enter values for the independent variable *n* in the first column.**

   You must enter values consecutively so no cell in the column is skipped — but you don't have to enter these values in numerical order. To make the first entry, press [ENTER], use the number keys to enter a value, and then press [ENTER] when you're finished. To make entries after the first entry, press ▼[ENTER], enter a value, and then press [ENTER].

   After you make each entry, the calculator automatically calculates the corresponding terms of the checked sequences in the Y= editor, as shown in the second picture in Figure 9-4.

As you can see in the second row of the table in the second picture in Figure 9-4, the calculator allows you to enter non-integer values for the independent variable *n*. But because sequences are defined only for integer values of *n*, the calculator will naturally tell you that the sequence is undefined. It will also tell you that the sequence is undefined if you enter values of *n* that are less than **nmin**, as indicated in the third row of this table. However, the calculator is quite comfortable with telling you the value of a term when *n* is larger than **nmax**, as you see on the fourth row when **nmax** is set to 25.

Displaying hidden rows and columns

To display more rows in an automatically generated table, repeatedly press ▲ or ▼. If you have constructed a table for more than three sequences, repeatedly press ◀ or ▶ to navigate through the columns of the table.

Each time the calculator redisplays a table with a different set of rows, it also automatically resets **tblStart** to the value of *n* appearing in the first row of the newly displayed table. To return the table to its original state, press [F2] to access the Table Setup editor and then change the value that the calculator assigned to **tblStart**. Press [ENTER] to save the value you entered and then press [ENTER] again to return to the table.

TIP

While the calculator displays a table of sequence terms, you can edit the definition of a sequence without going back to the Y= editor. To do this, use the ◀▶ keys to place the cursor in the column of the sequence you want to redefine and press [F4]. This places the definition of the sequence on the command line so that you can edit the definition. (I explain editing expressions in Chapter 1.) After you finish your editing, press [ENTER]. The calculator automatically updates the table and the edited definition of the sequence in the Y= editor.

Viewing the table and the graph on the same screen

After you graph your sequences and generate a table of sequence terms, you can view the graph and the table on the same screen. To do so, follow these steps:

1. **Press [MODE][F2]▶[3] to vertically divide the screen.**
2. **Press ▼▶[5] to place the table in the left screen and then press ▼▶[4] to place the graph in the right screen.**

   If you prefer to have the graph on the left and the table on the right, just reverse the order of keystrokes in this step by first pressing ▼▶[5] and then pressing ▼▶[4].

3. **Press [ENTER] to save your Mode settings.**

   You now see a screen similar to that in the first picture in Figure 9-5. Notice that the right screen may be empty. It appears in the next step.

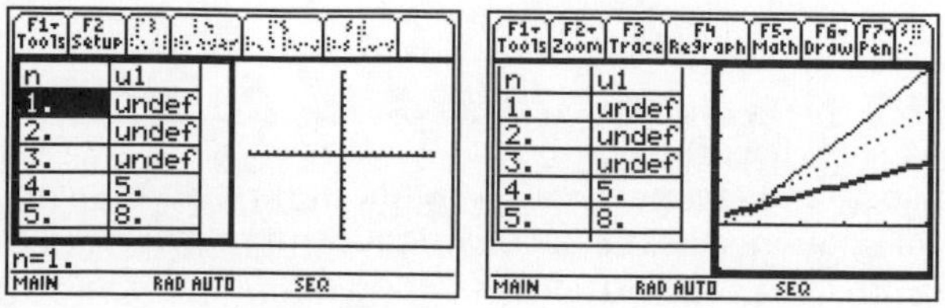

Figure 9-5: Viewing a graph and a table on the same screen.

4. **Press [2nd][APPS] to toggle between the left and right screens.**

   The right screen is no longer empty, as in the second picture in Figure 9-5.

In a split screen, the screen with the bold border is the active screen, and the commands for the application in that screen appear at the top of the screen. In the second picture in Figure 9-5, the Graph screen is active. So, for example, if you press [F3] while in this application, you can trace the sequences graphed in that screen. (Tracing a graph is explained earlier in this chapter.) If you press [2nd][APPS] to make the table screen active and then press [▶], the table displays the terms in the next checked sequence in the Y= editor.

TIP

To return to FULL screen mode without making changes in the Mode menu, just press [2nd][ESC]. This automatically sets the **Split Screen** mode on Page 2 of the Mode menu to **FULL**.

Evaluating Sequences

The [F5] Graph Math menu contains a command that evaluates a sequence at any specified value of the independent variable *n*. To access and use this command, perform the following steps:

1. **Graph the sequences in a viewing window that contains the specified value of *n*.**

   I explain graphing sequences and setting the viewing window in Chapter 8. To get a viewing window containing the specified value of *n*, that value must be between **nmin** and **nmax** in the viewing window.

2. **If necessary, set the Graph Formats menu to display coordinates in rectangular (*x*, *y*) form.**

   To do this, press [F1][▲][ENTER][▶][1][ENTER].

3. **Press [F5][1] to invoke the Value command in the Graph Math menu.**

4. **Enter the specified value of *n*.**

   Use the keypad to enter the value of *n*. The number you enter overwrites the value previously assigned to the value of **nc** that appears on the screen.

5. **Press [ENTER].**

   After you press [ENTER], the number of the first checked sequence in the Y= editor appears at the top of the screen, the cursor appears on the graph of that sequence at the specified value of *n*, and the coordinates of the cursor appear at the bottom of the screen. This is illustrated in the first picture in Figure 9-6.

6. **Repeatedly press ▲ to see the value of the other graphed sequences at your specified value of *n*.**

   Each time you press ▲, the number of the sequence being evaluated appears at the top of the screen, and the coordinates of the cursor location appear at the bottom of the screen. This is illustrated in the second picture in Figure 9-6, which displays the value of the third sequence in the Y= editor.

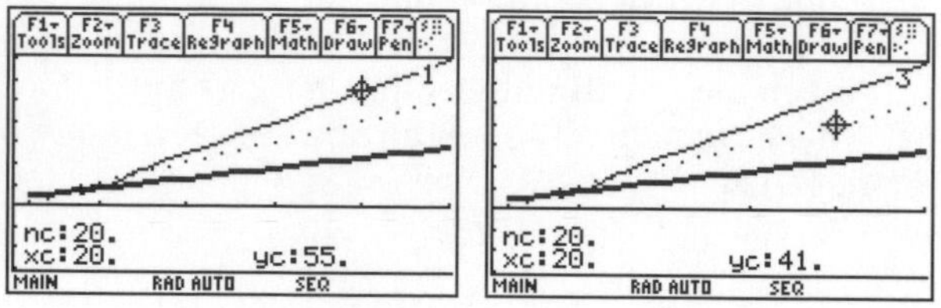

Enter *n*; press ENTER

Press ▲

Figure 9-6: Steps for evaluating sequences at a specified value of *n*.

After using the **Value** command to evaluate your sequences at one value of *n*, you can evaluate your sequences at another value of *n* by keying in the new value and pressing ENTER. Pressing any function key (such as ENTER or ▶) *after* evaluating a sequence deactivates the **Value** command.

TIP

If you plan to evaluate sequences at several specified values of *n*, consider constructing a user-defined table of sequence terms (as I explain in this chapter).

Chapter 10

Parametric Equations

In This Chapter

- Entering and graphing parametric equations
- Tracing the graph of a pair of parametric equations
- Constructing a table of values of parametric equations
- Evaluating derivatives of parametric equations
- Finding the equation of a tangent to a parametric curve

As a particle moves along a curve, you usually want to know two things about that particle: Where is it? How long did it take to get there?

The answer to the first question is pretty straightforward: The x- and y-coordinates of the particle tell you where it's at. To get an answer to the second question, you can bring a third variable — time — into the picture. To answer both questions simultaneously, you express the x and y variables as functions of t, the time variable.

The motion of the particle is thus described by the pair of parametric equations $x = f(t)$ and $y = g(t)$. In these equations, t is called the *parameter,* hence the equation type: *parametric.*

In addition to motion, such equations also describe graphs that aren't the graphs of functions. An example of such a graph appears later in this chapter, in Figure 10-3.

Entering Parametric Equations

The calculator can accommodate up to 99 pairs of parametric equations. To avoid notational confusion, I denote these pairs of equations by xt_1 and yt_1 through xt_{99} and yt_{99}; the calculator uses the notation *xt*1 and *yt*1 through *xt*99 and *yt*99. For the sake of simplicity, I tell you how to enter the pair xt_k and yt_k.

But what is stated in the following for xt_k and yt_k also applies to the pairs xt_1 and yt_1, xt_2 and yt_2, and so on. To enter a pair of parametric equations into the calculator, follow these steps:

1. **Press [MODE][▶][2] to put the calculator in Parametric mode (as shown in Figure 10-1) and then press [ENTER] to save the Mode settings.**

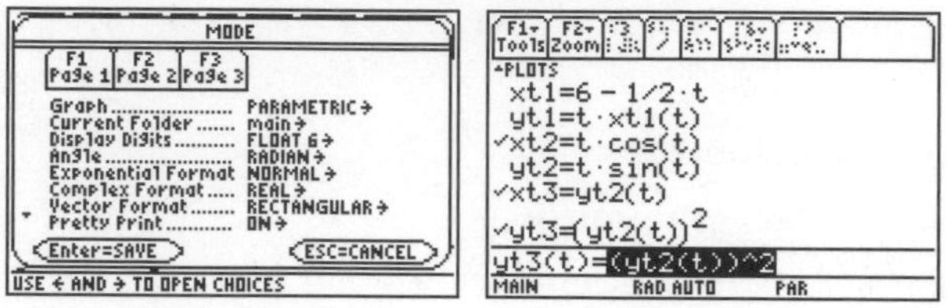

Set mode

Enter equations

Figure 10-1: Entering parametric equations.

2. **Press [◆][F1] and enter the definitions of xt_k and yt_k and then press [ENTER] to store each definition in the Y= editor.**

   To do this, use the ⊖ and ⊖ keys to place the cursor on xt_k and then press [ENTER] to place the cursor on the command line, where you enter the definition. Any previously appearing definition is overwritten when you key in the new definition. Press [ENTER] to store the definition in the calculator and then repeat this process to store the definition of yt_k in the Y= editor. Examples of definitions of parametric equations appear in Figure 10-1.

   When you're defining parametric equations, the only symbol the calculator allows you to use for the independent variable is the letter *t*. Press [T] to enter this letter in your definitions. In Figure 10-1, I used this key to define xt_1, yt_1, xt_2, and yt_3.

   As a timesaver, when entering parametric equations in the Y= editor, you can reference another equation in its definition provided that equation is entered in the form $xt_k(t)$ or $yt_k(t)$ where k is the number of the equation. The second picture in Figure 10-1 shows equation xt_3 defined as yt_2. You enter the definition in xt_3 by pressing [Y][T][2][(][T][)].

3. **If you're graphing more than one pair of parametric equations, give each pair a unique graphing style so you can identify it in a graph of multiple pairs of parametric equations.**

   In Parametric mode, you have six style choices: Line, Dot, Square, Thick, Animate, and Path. The Path style uses a circle to indicate a point as it's being graphed, and when the graph is complete, the circle disappears and leaves the graph in Line style. The Animate style also uses a circle to indicate a point as it's being graphed, but when the graph is complete, no graph appears on-screen.

To set the style for a pair of equations xt_k and yt_k, use the ▲▼ keys to highlight the definition of one of the equations in the pair and press [2nd][F1] to display the style menu. Then press the number of the desired style. Setting the style for one equation in the pair of equations automatically sets the style for both equations in the pair.

You must enter parametric equations in pairs. If you enter just one equation in the pair, say xt_k but not yt_k, then when you go to graph these parametric equations, you won't see a graph nor will you get an error message warning you that there's a problem.

Graphing Parametric Equations

After you enter the parametric equations into the calculator, you can use the following steps to graph the equations:

1. **If necessary, turn off any parametric equations and Stat Plots that you don't want to appear in your graph.**

   If you don't know what I mean by this, press [♦][F1] to display the Y= editor. If a checkmark appears to the left of one or both equations in a pair of parametric equations in the Y= editor, that pair of equations will be graphed. To prevent this from happening, use the ▲▼ keys to highlight the definition of the equation, and then press [F4] to remove the checkmark. At a later time, when you do want to graph that pair of equations, repeat this process to reinstate the checkmark. In the second picture in Figure 10-1, the second and third pairs of equations have at least one equation in the pair checked, so they will appear when graphed; the first pair has no equation checked, so it won't be graphed.

   The first line in the Y= editor tells you the graphing status of the Stat Plots. (I cover Stat Plots in Chapter 20.) If no number appears after **PLOTS** on the very first line of the Y= editor (as in the second picture in Figure 10-1), no Stat Plots are selected to be graphed. If one or more numbers do appear, those Stat Plots will be graphed along with the graph of your parametric equations.

   To prevent an individual Stat Plot from being graphed, use the ▲▼ keys to place the cursor on that Stat Plot and press [F4]. ([F4] toggles between deactivating and activating a Stat Plot.) To prevent all Stat Plots from being graphed, press [F5][5].

   When you're graphing parametric equations, Stat Plots that are active when you don't really want them graphed cause problems that could result in an error message when you attempt to graph your parametric equations. So if you aren't planning to graph a Stat Plot along with your equations, make sure all Stat Plots are turned off.

2. **If necessary, set the format for the graph by pressing F1 ⊙ ENTER and changing the selections in the Graph Formats menu.**

   I explain the Graph Formats menu in Chapter 5. It allows you to have Coordinates displayed in rectangular (x, y) form, polar (r, θ) form, or not displayed at all. Because I see no point in not knowing the coordinates of the cursor location, my recommendation is to display the coordinates in either rectangular or polar form.

   The **Grid** option is your choice, either OFF or ON. And as far as the other options go, my recommendation is to set **Axes** ON, **Leading Cursor** OFF, and **Labels** OFF.

3. **Press ◆ F2 to access the Window editor.**

4. **After each of the first three window variables, enter a numerical value that is appropriate for the equations you're graphing. Press ENTER after entering each number.**

   Figure 10-2 shows the Window editor when the calculator is in Parametric mode. The following list explains the variables you must set in this editor:

   - **tmin:** This setting contains the first value of the independent variable (parameter) *t* that the calculator will use to evaluate all parametric equations in the Y= editor. It can be set equal to any real number. When *t* denotes time or angle measurements, it is usually set to 0.

   - **tmax:** This setting contains the largest value of the independent variable (parameter) *t* that you want the calculator to use to evaluate all parametric equations in the Y= editor. It can be set equal to any real number. When *t* denotes angle measurements, it is usually set to 2π.

   - **tstep:** This setting tells the calculator how to increment the independent variable *t* as it evaluates the parametric equations in the Y= editor and graphs the corresponding points. It is usually set equal to a positive real number. But you can set **tstep** equal to a negative number provided that **tmin** is larger than **tmax**.

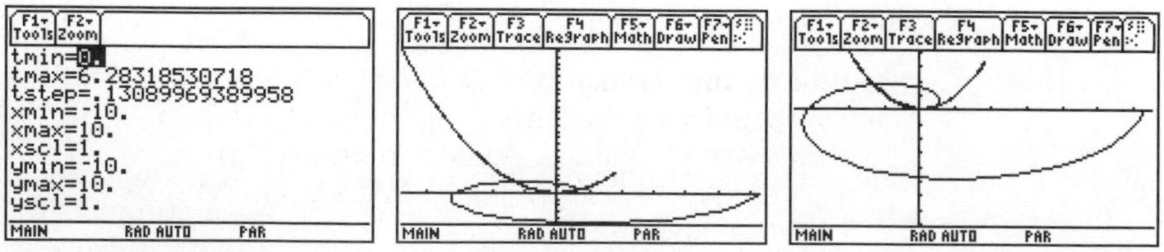

Set window | Use ZoomFit | Adjust window

Figure 10-2: Graphing parametric equations.

You want **tstep** to be a small number, like 0.1 or $\pi/24$, but you don't want it to be too small, like 0.001. If **tstep** is too small, it will take a few minutes for the calculator to produce the graph; if it's really, really small (like 0.0001) you get an error message. And if **tstep** is a large number like 1, the calculator might not graph enough points to show you the true shape of the curve.

If it's taking a long time for the calculator to graph your parametric equations, and this causes you to regret that very small number you placed in **tstep**, press [ON] to terminate the graphing process. You can then go back to the Window editor and adjust the **tstep** setting.

When graphing parametric equations that use trigonometric functions in their definitions, it's best to set **tmin**, **tmax**, and **tstep** to multiples of π. In Figure 10-2, **tmax** is set equal to 2π, and **tstep** is set equal to $\pi/24$. You enter π into the calculator by pressing [2nd][^].

You use the remaining items in the Window editor to set a viewing window for the graph (a procedure I explain in detail in Chapter 5). If you know the dimensions of the viewing window required for your graph, go ahead and assign numerical values to the remaining variables in the Window editor and advance to Step 6. If you don't know the minimum and maximum *x* and *y* values required for your graph, the next step tells you how to get the calculator to find them for you.

5. **Press [F2]▲▲▲[ENTER] to get ZoomFit to graph your equations.**

   After you've assigned values to **tmin**, **tmax**, and **tstep**, **ZoomFit** determines appropriate values for **xmin**, **xmax**, **ymin**, and **ymax**, and then graphs your parametric equations. Note, however, that **ZoomFit** graphs parametric equations in the smallest possible viewing window, and it doesn't assign new values to **xscl** and **yscl**. This is illustrated in the second picture in Figure 10-2.

   If you like the way the graph looks, you can skip the remaining steps. If you'd like to spruce up the graph, Step 6 gives you some pointers. The third picture in Figure 10-2 displays a spruced-up graph.

6. **Press [◆][F2] and adjust the values assigned to the x and y settings. Press [ENTER] after entering each new number.**

   Here's how to readjust the viewing window after using **ZoomFit**:

   - **xmin and xmax:** If you don't want the graph to hit the left and right sides of the screen, decrease the value assigned to **xmin** and increase the value assigned to **xmax**. If you want to see the *y*-axis on the graph, assign values to **xmin** and **xmax** so that zero is between these two values.

   - **xscl:** Set this equal to a value that doesn't make the *x*-axis look like railroad tracks — that is, an axis with far too many tick marks. Twenty or fewer tick marks makes a nice-looking axis.

- **ymin and ymax:** If you don't want the graph to hit the top and bottom of the screen, decrease the value assigned to **ymin** and increase the value assigned to **ymax**. If you want to see the *x*-axis on the graph, assign values to **ymin** and **ymax** so that zero is between these two values.
- **yscl:** Set this equal to a value that doesn't make the *y*-axis look like railroad tracks. Fifteen or fewer tick marks is a nice number for the *y*-axis.

7. **Press [◆][F3] to regraph the parametric equations.**

After you graph your parametric equations, you can save the graph and its entire Window, Y= editor, Mode, and Format settings in a Graph Database. When you recall the Database at a later time, you get more than just a picture of the graph. The calculator also restores the Window, Y= editor, Mode, and Format settings to those stored in the Database. Thus you can, for example, trace the recalled graph.

To find out about the procedures for saving and recalling a graph database, see Chapter 5.

Exploring Parametric Graphs

The calculator has three very useful features that help you explore the graphs of parametric equations: zooming, tracing, and creating tables. Zooming allows you to quickly adjust the viewing window for the graph so that you can get a better idea of the nature of the graph. Tracing shows you the coordinates of the points that make up the graph. Creating a table — well, I'm sure you already know what that shows you. These sections explain how to zoom and trace; the section "Displaying Equations in a Table" tells you how to create tables.

Using Zoom commands

After you've graphed your parametric equations (as I explain earlier in this chapter), press [F2] to access the Zoom menu. On this menu, you see all the Zoom commands that are available for graphing functions. Not all of them are useful when graphing parametric equations, however. Here are the ones that I use when graphing parametric equations:

- **ZoomFit:** This command finds a viewing window for a specified portion of the graph. I cover how to use **ZoomFit** in the section, "Graphing Parametric Equations," earlier in this chapter.

- **ZoomSqr:** Because the calculator screen is rectangular instead of square, circles will look like ellipses if the viewing window isn't properly set. **ZoomSqr** adjusts the settings in the Window editor so that circles look like circles. To use **ZoomSqr**, first graph your parametric equations and then press [F2][5]. The graph is then redrawn in a viewing window that makes circles look like circles. Figure 10-3 illustrates the effect that **ZoomSqr** has on the spiral $x = t\cos(t)$, $y = t\sin(t)$.
- **ZoomBox:** After the graph is drawn (as I describe earlier in this chapter), this command allows you to define a new viewing window for a portion of your graph by enclosing it in a box, as illustrated in Figure 10-4. The box becomes the new viewing window. To construct the box:

  1. **Press [F2][1].**
  2. **Define a corner of the box.**

     To do so, use the ◀▶▲▼ keys to move the cursor to the spot where you want one corner of the box to be located and then press [ENTER]. The calculator marks that corner of the box with a small square.
  3. **Construct the box.**

     To do so, press the ◀▶▲▼ keys. As you press these keys, the calculator draws the sides of the box.

     TIP: When you use **ZoomBox**, if you don't like the size of the box, you can use any of the ◀▶▲▼ keys to resize the box. If you don't like the location of the corner you anchored in Step 2, press [ESC] and start over with Step 1. If you're drawing a rather large box, to move the cursor a large distance, press [2nd] and then press an appropriate arrow key.
  4. **When you're finished drawing the box, press [ENTER].**

     The calculator redraws the graph in the viewing window specified by the box.

     REMEMBER: When you use **ZoomBox**, you press [ENTER] only two times. The first time you press it is to anchor a corner of the zoom box. The next time you press [ENTER] is when you're finished drawing the box and you're ready to have the calculator redraw the graph.

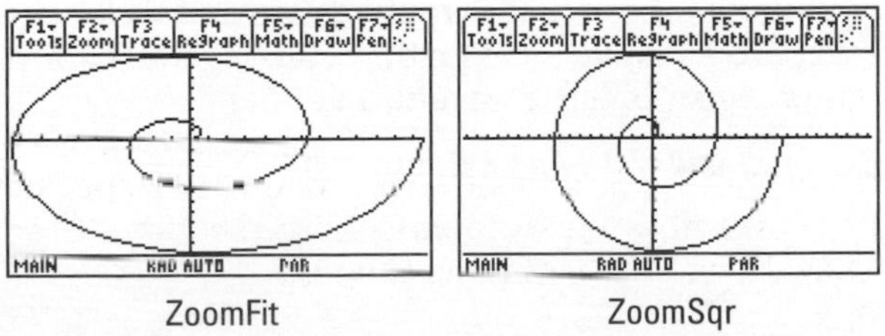

Figure 10-3: A spiral graphed using ZoomFit and then using ZoomSqr.

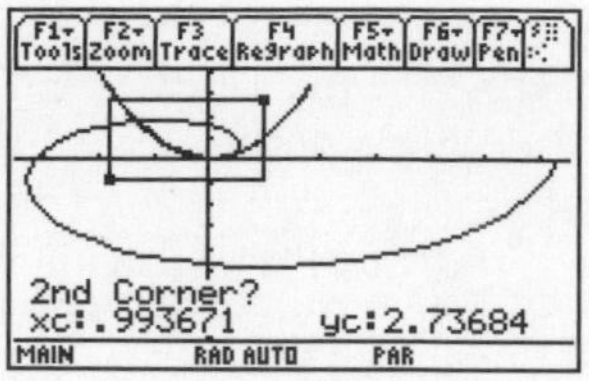

Figure 10-4: Using ZoomBox.

- **Zoom In and Zoom Out:** After the graph is drawn (as I describe earlier in this chapter), these commands allow you to zoom in on a portion of the graph or to zoom out from the graph. They work very much like a zoom lens on a camera. To use these commands:

  1. **Press [F2][2] if you want to zoom in, or press [F2][3] if you want to zoom out.**
  2. **Use the ◀▶▲▼ keys to move the cursor to the spot on-screen from which you want to zoom in or zoom out.**

     This spot becomes the center of the new viewing window.
  3. **Press [ENTER].**

     The graph is redrawn centered at the cursor location.

TIP: If you use a Zoom command to redraw a graph and then want to undo what that command did to the graph, press [F2]▲▲▶[1] to select **ZoomPrev** from the Zoom Memory menu. The graph is then redrawn as it appeared in the previous viewing window.

Tracing a parametric graph

After you graph your parametric equations (as I describe earlier in this chapter), you can have the cursor display the coordinates of the points on the graph by tracing it. To trace a parametric graph, follow these steps:

1. **If necessary, set Coordinates in the Graph Formats menu to RECT or POLAR.**

   If you know that you are already in the rectangular or polar format, you can skip this step. Otherwise press [F1]▲[ENTER]▶ to access the Coordinates option in the Graph Formats menu and then press [1] to have the coordinates displayed in (x, y) rectangular form or press [2] for (r, θ) polar form. Press [ENTER] to save this setting.
2. **Press [F3] and then use ▶ to trace the graph.**

Panning in Parametric mode

When you're tracing parametric equations and the cursor hits an edge of the screen, if **tc** is less than **tmax** and you continue to press ▶, the coordinates of the cursor are displayed at the bottom of the screen. However, because the calculator doesn't automatically adjust the viewing window, you won't see the cursor on the graph. To rectify this situation, make a mental note of the value of **tc** (or jot it down) and then press [ENTER]. The calculator then redraws the graph centered at the location of the cursor at the time you pressed [ENTER]. Then press [F3] to continue tracing the graph.

Unfortunately, after the graph is redrawn, the trace cursor is placed at the beginning of the graph of the first pair of equations appearing in the Y= editor. To get the trace cursor back on the part of the graph displayed in the new viewing window, key in the value of **tc** that you made note of, and then press [ENTER]. If you weren't tracing the first pair of equations appearing in the Y= editor, use ▲ to place the cursor on the graph you *were* tracing. The trace cursor then appears in the middle of the screen, and you can use ◀ or ▶ to continue tracing the graph. When **tc** equals **tmax**, pressing ▶ has no effect because the cursor has reached the end of the graph and can go no farther.

As you investigate a parametric graph by tracing it, here's what you will see:

- **The number of the pair of parametric equations:** The number of the pair of equations in the Y= editor that you're tracing is displayed in the top-right corner of the screen. For example, if you see the number 1, then you're tracing the first pair of equations (xt_1 and yt_1) in the Y= editor. If you've graphed more than one pair of equations and you would like to trace a different pair, repeatedly press ▲ until the number of the desired pair of equations appears.
- **The value of the parameter *t*:** At the bottom of the screen you see a value for **tc**. This is the value of the independent variable (parameter) *t* at the cursor location. Each time you press ▶, the cursor moves to the next plotted point in the graph and updates the value of **tc**. If you press ◀, the cursor moves left to the previously plotted point. And if you press ▲ to trace a different pair of equations, the tracing of that pair starts at the same value of **tc** that was displayed on-screen before you pressed this key.

TIP

When you're using Trace, if you want to start tracing the graph at a specific value of the parameter *t*, just key in that value and press [ENTER]. (The value you assign to *t* must be between **tmin** and **tmax**; if it isn't, you get an error message.) After you press [ENTER], the trace cursor moves to the point on the graph corresponding to that value of *t*. But the calculator doesn't change the viewing window. So if you don't see the cursor, it's on the part of the graph outside the viewing window. The sidebar "Panning in Parametric mode" tells you how to get the cursor and the graph in the same viewing window.

- **The coordinates of the cursor:** On the last line of the screen, you see the coordinates of the cursor location. In rectangular form, these coordinates are denoted by **xc** and **yc**; in polar form they are **rc** and **θc**.

Press [ESC] to terminate tracing the graph. This also removes the number of the pair of parametric equations and the coordinates of the cursor from the screen.

If the number of the pair of equations at the top of the screen and the coordinates at the bottom of the screen are interfering with your view of the graph when you use Trace, increase the height of the window by pressing [◆][2nd], and decreasing the value of **ymin**, and increasing the value of **ymax**.

Displaying Equations in a Table

After you've entered the parametric equations in the Y= editor (as I describe earlier in this chapter), you can have the calculator create one of three different types of tables of values of your parametric equations:

- An automatically generated table that displays the plotted points appearing in the calculator's graphs of the parametric equations
- An automatically generated table that displays the points using the initial *t*-value and *t*-increment that you specify
- A user-defined table in which you enter the *t*-values of your choice and the calculator determines the corresponding values of the parametric equations

The following sections discuss these tables in greater detail.

Creating automatically generated tables

To create a table in which you specify the initial *t*-value and the value by which *t* is incremented, perform the following steps. After the last step, I tell you how to create an automatically generated table displaying the points graphed by the calculator.

1. **Place a checkmark to the left of those equations in the Y= editor that you want to appear in the table; leave the other equations unchecked.**

   Only the equations in the Y= editor that have checkmarks to the left of their names will appear in the table. If only one equation in a pair of equations is checked, only the checked equation appears in the table. For example, the second picture in Figure 10-1 (shown earlier) indicates that only equations xt_2, xt_3, and yt_3 are slated to appear in a table.

To check or uncheck an equation, press [◆][F1], use the ⊖⊖ keys to place the cursor on the equation, and then press [F4] to toggle between checking and unchecking the function.

2. **Press [◆][F4] to access the Table Setup editor (shown in Figure 10-5).**

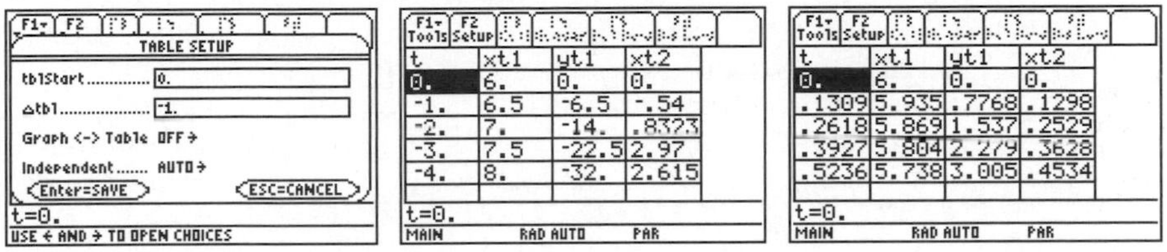

Figure 10-5: Automatically generated tables.

3. **Enter a number in tblStart and press ⊖, and then enter a number in Δtbl and press ⊖.**

   If **Independent** is the only item you see in the Table Setup editor, and if the cursor is on ASK, press ⊙[1] to change this setting to AUTO and then press ⊖⊖⊖ to enter numbers in **tblStart** and **Δtbl**.

   tblStart is the first value of the independent variable *t* to appear in the table, and **Δtbl** is the increment for the independent variable *t*. The calculator constructs the table by adding the value you assign to **Δtbl** to the previous *t*-value and calculating the corresponding values of the parametric equations. In the first picture in Figure 10-5, **tblStart** is assigned the value 0 and **Δtbl** is assigned the value –1 to produce the decreasing values in the *t*-column of the table in the second picture in this figure.

   Use the number keys to enter values in **tblStart** and **Δtbl**. Press ⊖ after making each entry.

4. **Set Graph <–> Table to OFF (as shown in Figure 10-5).**

   You have two choices for the **Graph <–> Table** setting: OFF and ON. When **Graph <–> Table** is set to ON, a table of the points plotted in the calculator's graphs of the equations is created. However, if **Graph <–> Table** is set to OFF, the calculator uses the values in **tblStart** and **Δtbl** to generate the table.

   To change the **Graph <–> Table** setting, use ⊖⊖ to place the cursor on the existing setting, press ⊙ to display the options for that setting, and press the number of the option you want.

5. **Press [ENTER] to save the settings in the Table Setup editor.**

6. **If necessary, press [◆][F5] to display the table.**

   The table appears on the screen, as illustrated in the second picture in Figure 10-5. I explain navigating the table later in this chapter.

To create a table of the plotted points appearing in the calculator's graphs of the parametric equations, follow Steps 1 and 2. If **Independent** is the only item you see in the Table Setup editor, press ▶[1] to change this setting to AUTO and then press ▲▶[2][ENTER] to set **Graph <–> Table** to ON and to save your settings. On the other hand, if you see all four items in this editor, press ▼▼▶[2]▼[1][ENTER] to set **Graph <–> Table** to ON and to save your settings. (With these settings, the calculator ignores the other values in the Table Setup menu.) If necessary, press [◆][F5] to display the table, as shown in the third picture in Figure 10-5.

Creating user-defined tables

To create a user-defined table in which you enter values of the parameter *t* and the calculator figures out the corresponding values of the parametric equations, follow these steps:

1. **Place a checkmark to the left of each function in the Y= editor that you want to appear in the table; leave the other functions unchecked.**

   Only those functions in the Y= editor that have checkmarks to the left of their names will appear in the table. To check or uncheck a function, press [◆][F1], use the ▲▼ keys to place the cursor on the function, and then press [F4] to toggle between checking and unchecking the function.

2. **Press [◆][F4] to access the Table Setup editor.**

3. **Set Independent to ASK and press [ENTER].**

   Independent is the last item in the menu. If it is already set to ASK, press [ENTER] and go to the next step. If it is set to AUTO, press ▼▼▼▶[2] to set **Independent** to ASK and then press [ENTER].

4. **If necessary, press [◆][F5] to display the table. If the table contains entries, press [F4][8][ENTER] to clear the table.**

   You see a table without entries, as in the first picture in Figure 10-6.

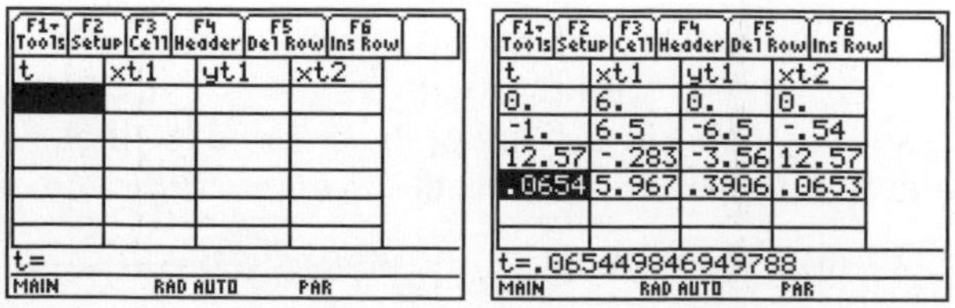

Figure 10-6: Generating a user-defined table.

5. **Enter values for the parameter *t* in the first column.**

   You must enter values consecutively; you cannot skip any cells in the first column of the table. To make the first entry, press [ENTER], use the number keys to enter a value, and then press [ENTER] when you're finished. To make entries after the first entry, press ▼[ENTER], enter a value, and then press [ENTER].

   After you make each entry, the calculator automatically calculates the corresponding values of the checked parametric equations in the Y= editor, as shown in the second picture of Figure 10-6.

Displaying hidden rows and columns

To display more rows in an automatically generated table, repeatedly press ▲ or ▼. If you have constructed a table for more than three equations, repeatedly press ◀ or ▶ to navigate through the columns of the table.

Each time the calculator re-displays a table with a different set of rows, it also automatically resets **tblStart** to the value of *t* appearing in the first row of the newly displayed table. To return the table to its original state, press [F2] to access the Table Setup editor and then change the value that the calculator assigned to **tblStart**. Press [ENTER] to save the value you entered and then press [ENTER] again to return to the table.

TIP

While displaying a table, you can edit the definition of a parametric equation without going back to the Y= editor. To do this, use the ◀▶ keys to place the cursor in the column of the equation you want to redefine and press [F4]. This places the definition of the equation on the command line so that you can edit the definition. (I tell you how to edit expressions in Chapter 1.) When you're finished with your editing, press [ENTER]. The calculator automatically updates the table and the edited definition of the equation in the Y= editor.

Viewing the table and the graph on the same screen

After you graph your parametric equations and generate a table, you can view the graph and the table on the same screen. To do so, follow these steps:

1. **Press [MODE][F2]▶[3] to vertically divide the screen.**
2. **Press ▼▶[5] to place the table in the left screen, and then press ▼▶[4] to place the graph in the right screen.**

   If you prefer to have the graph on the left and the table on the right, just reverse the order of keystrokes in this step by first pressing ▼▶[4] and then pressing ▼▶[5].

3. **Press [ENTER] to save your Mode settings.**

   You now see a screen similar to that in the first picture in Figure 10-7. Notice that the right screen does not contain a graph. It appears in the next step.

4. **Press [2nd][APPS] to toggle between the left and right screens.**

   The right screen is no longer empty, as in the second picture in Figure 10-7.

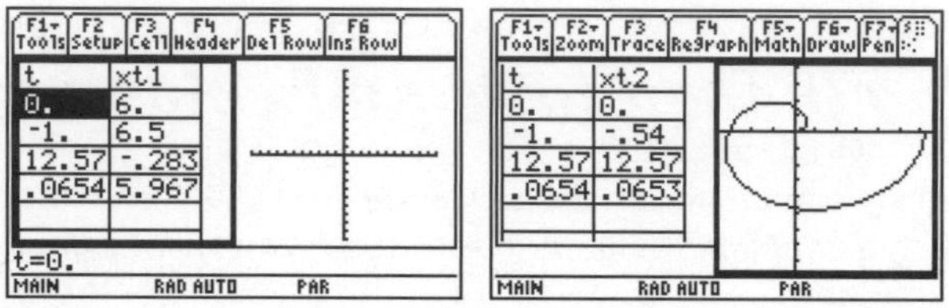

Figure 10-7: Viewing a graph and a table on the same screen.

In a split screen, the half with the bold border is active, and the commands for that application appear at the top of the screen. For example, in the second picture in Figure 10-7, the Graph screen is active. So if you press [F3] while in this application, you can trace the pair of parametric equations graphed in that screen. (I show you how to trace a graph earlier in this chapter.)

TIP

To return to FULL screen mode without making changes in the Mode menu, just press [2nd][ESC]. This automatically sets the **Split Screen** mode on Page 2 of the Mode menu to FULL.

Analyzing Parametric Equations

After graphing parametric equations (as described earlier in this chapter), you can use the options in the [F5] Graph Math menu to find the value of the parametric equations at a specified value of *t* and to evaluate derivatives (*dy/dx*, *dx/dt*, and *dy/dt*) at a specified value of *t*. The calculator will even find the equation of a tangent to the graph at a specified point on the graph. And it will find the length of a portion of the graph or the distance between two points on the same curve or on different curves. This section tells you how to use the Graph Math menu to accomplish these tasks.

Evaluating parametric equations

When you trace the graph, the trace cursor doesn't hit every point on the graph. So tracing isn't a reliable way of finding the values of a pair of parametric equations at a specified value of the parameter *t*. The F5 Graph Math menu, however, contains a command that evaluates a pair of parametric equations at any specified *t*-value. To access and use this command, perform the following steps:

1. **Graph the equations in a viewing window that contains the specified value of *t*.**

   The graph contains a specified value of *t* when that value is between **tmin** and **tmax** in the Window editor. I explain graphing parametric equations and setting **tmin** and **tmax** earlier in this chapter.

2. **If necessary, set the Graph Formats menu to display coordinates in rectangular (*x*, *y*) form or in polar (*r*, θ) form.**

   To do this, press F1▲ENTER▶. Press 1 for rectangular form or press 2 for polar form. Then press ENTER to save these settings.

3. **Press F5 1 to invoke the Value command in the Graph Math menu.**

4. **Enter the specified value of *t*.**

   Use the keypad to enter a value of *t* that is between **tmin** and **tmax**, as illustrated in the first picture in Figure 10-8. As you enter *t*, the value of **tc** at the bottom of the screen is replaced by the number you key in.

5. **Press ENTER to evaluate the parametric equations.**

   After you press ENTER, the number of the first checked pair of parametric equations in the Y= editor appears at the top of the screen, the cursor appears on the graph of those equations at the specified value of *t*, and the coordinates of the cursor appear at the bottom of the screen. This is illustrated in the second picture in Figure 10-8.

6. **Repeatedly press ▲ to see the value of the other graphed parametric equations at your specified value of *t*.**

   Each time you press ▲, the number of the pair of equations being evaluated appears at the top of the screen, and the coordinates of the cursor location appear at the bottom of the screen. This is illustrated in the third picture in Figure 10-8.

Figure 10-8: Evaluating parametric equations at a specified value of *t*.

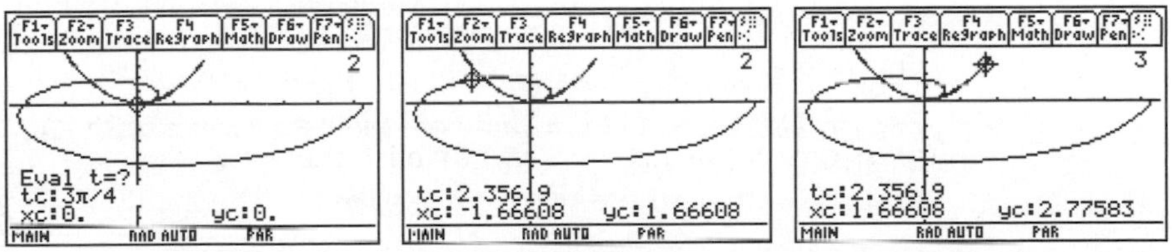

Enter *t* — Press ENTER — Press ▲

After using the **Value** command to evaluate your parametric equations at one value of *t*, you can evaluate them at another value of *t* by keying in the new value and pressing [ENTER]. Pressing any function key (such as [ENTER] or [▶]) *after* evaluating parametric equations deactivates the **Value** command.

TIP

If you plan to evaluate parametric equations at several specified values of *t*, consider constructing a user-defined table of values of parametric equations (as I explain earlier in this chapter).

Finding derivatives

To find the derivative (*dy*/*dx*, *dy*/*dt*, or *dx*/*dt*) of a pair of parametric equations at a specified value of *t*, follow these steps:

1. **Graph the equations in a viewing window that contains the specified value of *t*.**

   The graph contains a specified value of *t* when that value is between **tmin** and **tmax** in the Window editor. I explain graphing parametric equations and setting **tmin** and **tmax** earlier in this chapter.

2. **If necessary, set the Graph Formats menu to display coordinates in rectangular (*x*, *y*) form or in polar (*r*, θ) form.**

   To do this, press [F1][▲][ENTER][▶]. Press [1] for rectangular form or press [2] for polar form. Then press [ENTER] to save these settings.

3. **Press [F5][6] to display the Derivatives submenu of the Graph Math menu and then press [1], [2], or [3] to select the *dy/dx*, *dy/dt*, or *dx/dt* option.**

   Instructions appear at the bottom of the screen, as illustrated in the first picture in Figure 10-9. If these instructions interfere with your view of the graph, create more space at the bottom of the screen by pressing [◆][F2] and decreasing the value of **ymin**. Then press [◆][F3] to redraw the graph and repeat Step 3 to reactivate the Derivatives option.

4. **If necessary, repeatedly press [▲] until the number of the appropriate pair of parametric equations appears at the top of the screen.**

5. **Enter the value (*t*) of the parameter to be used to evaluate the derivative.**

   Use the keypad to enter a value of *t* that is between **tmin** and **tmax**, as illustrated in the first picture in Figure 10-9. As you enter *t*, the value of **tc** at the bottom of the screen is replaced by the number you key in.

6. **Press [ENTER] to evaluate the derivative.**

   After pressing [ENTER], the cursor moves to the point on the graph corresponding to the specified value of *t* and the derivative at that value is displayed at the bottom of the screen, as in the second picture in Figure 10-9.

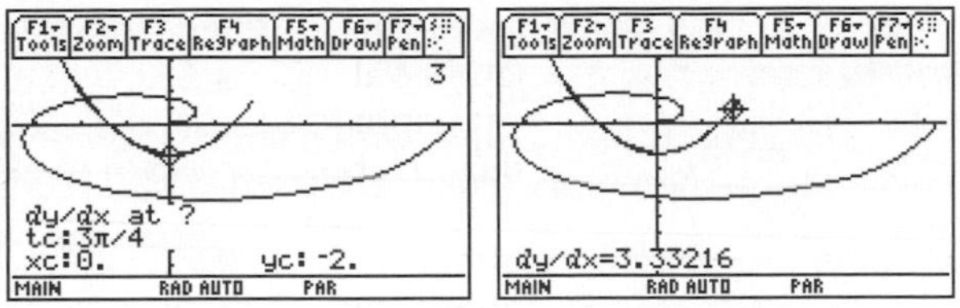

Enter *t*　　Press [ENTER]

Figure 10-9: Evaluating a derivative at a specified value of *t*.

Finding the distance between points

The calculator has a neat feature that draws a line between two points on the same or different parametric curves and calculates the distance between these two points. To use this feature, follow these steps:

1. **Graph the parametric equations in a viewing window that displays the two points between which you want to find the distance.**

   I cover graphing parametric equations and finding an appropriate viewing window earlier in this chapter.

2. **If necessary, set the Graph Formats menu to display coordinates in rectangular (x, y) form or in polar (r, θ) form.**

   To do this, press [F1]⊖[ENTER]⊙. Press [1] for rectangular form or press [2] for polar form. Then press [ENTER] to save these settings.

3. **Press [F5][9] to select the Distance command from the Graph Math menu.**

   If the calculator instructions at the bottom of the screen interfere with your view of the graph, create more space at the bottom of the screen by pressing [♦][F2] and decreasing the value of **ymin**. Then press [♦][F3] to redraw the graph and press [F5][9] to reinstate the **Distance** command.

4. **If necessary, repeatedly press ⊖ until the number of the parametric equations containing the first point appears at the top of the screen.**

5. **Enter the value of the parameter *t* corresponding to the first point and then press [ENTER].**

   Use the keypad to enter a value of *t* that is between **tmin** and **tmax**. As you enter *t*, the value of **tc** at the bottom of the screen is replaced by the number you key in. After you press [ENTER], the cursor moves to that point, as illustrated in the first picture in Figure 10-10.

6. **If the second point is on a different curve, repeatedly press ⊖ until the number of the parametric equations containing the second point appears at the top of the screen.**

   The cursor jumps to the point on the other curve that is defined by the same value *t*, a line is drawn between these points, and you're prompted to enter the second point, as illustrated in the second picture in Figure 10-10.

7. **Enter the value of the parameter *t* corresponding to the second point and then press [ENTER].**

   The calculator connects the two points with a line and displays the distance between these points, as illustrated in the third picture in Figure 10-10.

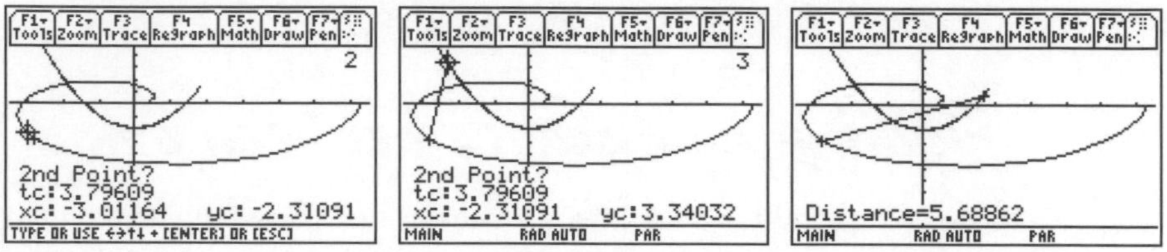

First point Second curve Distance

Figure 10-10: Finding the distance between two points.

Finding the equation of a tangent line

Not only can the calculator draw a tangent to a parametric curve at a specified value of the parameter *t*, but it can also tell you the equation of that tangent. To get the calculator to do this, follow these steps:

1. **Graph the parametric equations in a viewing window that displays the point of tangency.**

   You find out about graphing parametric equations and finding an appropriate viewing window earlier in this chapter.

2. **If necessary, set the Graph Formats menu to display coordinates in rectangular (x, y) form or in polar (r, θ) form.**

   To do this, press [F1][▲][ENTER][▶]. Press [1] for rectangular form or press [2] for polar form. Then press [ENTER] to save these settings.

3. **Press [F5][▲][▲][▲][ENTER] to select the Tangent command from the Graph Math menu.**

   If the calculator instructions at the bottom of the screen interfere with your view of the graph, create more space at the bottom of the screen by pressing [◆][F2] and decreasing the value of **ymin**. Then press [◆][F3] to redraw the graph and press [F5][▲][▲][▲][ENTER] to reinstate the **Tangent** command.

4. **If necessary, repeatedly press [▲] until the number of the parametric equations containing the point of tangency appears.**

5. **Enter the value of the parameter *t* at the point of tangency.**

   Use the keypad to enter a value of *t* that is between **tmin** and **tmax**. As you enter *t*, the value of **tc** at the bottom of the screen is replaced by the number you key in, as illustrated in the first picture in Figure 10-11.

6. **Press [ENTER] to construct the tangent and display its equation.**

   The tangent appears on the graph with the cursor located at the point of tangency, and the equation of the tangent appears at the bottom of the screen, as illustrated in the second picture in Figure 10-11.

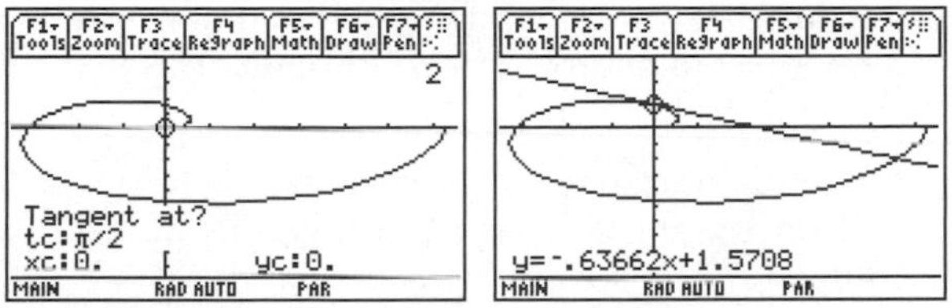

Figure 10-11: Finding the equation of a tangent line to a parametric curve.

Point of tangency — Equation of tangent

A tangent drawn by the **Tangent** command in the Graph Math menu doesn't automatically vanish when you use another Graph Math menu option or when you trace the graph. To erase the tangent, press [F4] to redraw the graph without the tangent line.

Evaluating arc length

The **Arc** command in the Graph Math menu can find the length of a section (arc) of a graphed curve when you specify the beginning and ending point of the section. To get the calculator to find this length, follow these steps:

1. **Graph the parametric equation in a viewing window displaying the part of the curve you want to measure.**

   I explain graphing parametric equations and finding an appropriate viewing window earlier in this chapter.

2. **If necessary, set the Graph Formats menu to display coordinates in rectangular (*x*, *y*) form or in polar (*r*, θ) form.**

   To do this, press [F1]▲[ENTER]▶. Press [1] for rectangular form or press [2] for polar form. Then press [ENTER] to save these settings.

3. **Press [F5]▲▲[ENTER] to select the Arc command from the Graph Math menu.**

   If the calculator instructions at the bottom of the screen interfere with your view of the graph, create more space at the bottom of the screen by pressing [♦][F2] and decreasing the value of **ymin**. Then press [♦][F3] to redraw the graph and press [F5]▲▲[ENTER] to reinstate the **Arc** command.

4. **If necessary, repeatedly press ⊖ until the number of the appropriate pair of parametric equations appears at the top of the screen.**

5. **Enter the value of the parameter *t* corresponding to the first point defining arc length and then press [ENTER].**

   Use the keypad to enter a value of *t* that is between **tmin** and **tmax**. As you enter *t*, the value of **tc** at the bottom of the screen is replaced by the number you key in. After you press [ENTER], the cursor moves to that point and prompts you to enter the second point, as illustrated in the first picture in Figure 10-12.

6. **Enter the value of the parameter *t* corresponding to the second point defining the arc (or use the ▶◀ keys to move the cursor to that point) and press [ENTER].**

   The calculator displays the length of the arc at the bottom of the screen, as illustrated in the second picture in Figure 10-12.

Figure 10-12: Finding arc length of a parametric curve.

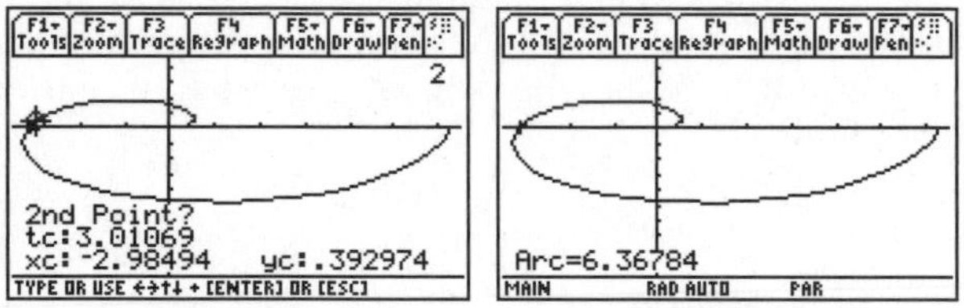

First point Arc length

TIP

When finding arc length, if the value of the parameter *t* that you enter to define the first point of the arc is larger than the value of *t* you enter to define the second point of the arc, then the arc length displayed by the calculator is negative. The true arc length is just the absolute value of the answer displayed by the calculator.

Chapter 11

Polar Equations

In This Chapter

- Converting between polar and rectangular coordinates
- Entering and graphing polar equations
- Constructing a table of values of polar equations
- Evaluating polar equations at a specified value of θ
- Finding derivatives at specified values of θ
- Finding the equation of a tangent to a polar curve

In the polar coordinate system, a point *P* in the plane is described by its distance *r* from the origin and the angle θ made by the positive *x*-axis and the line through the origin and point *P*. The polar coordinates of the point *P* are denoted by (*r*, θ). This is illustrated in Figure 11-1.

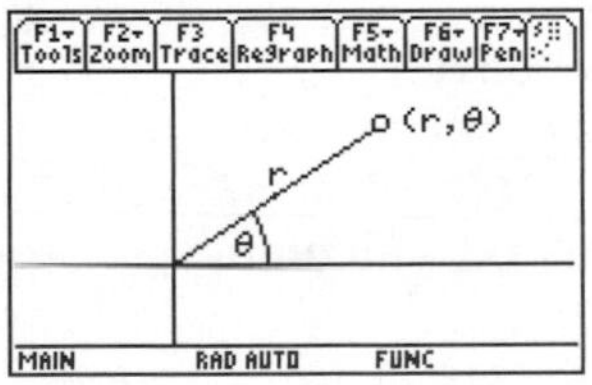

Figure 11-1: Representing a point in polar coordinates.

Converting Coordinates

You're probably used to seeing points in the *xy*-plane displayed in the (*x*, *y*) rectangular form. But when dealing with polar equations, we view these points in their (*r*, θ) polar form. To convert (*x*, *y*) from rectangular to polar form, follow these steps:

1. **Enter the rectangular coordinates surrounded by brackets instead of parentheses, as illustrated in the first entry in Figure 11-2.**

   To insert the brackets, press [2nd][,] and [2nd][÷]. Press [,] to separate the coordinates.

2. **Press 2nd 5 4 to access the MATH Matrix menu.**
3. **Press ▲▶4 to select the Polar option from the Vector Options menu.**
4. **Press ENTER to convert the rectangular coordinates to polar coordinates.**

   If your calculator is in Radian mode, the angle θ is displayed after the angle symbol in radians, as illustrated in the first entry in the first picture in Figure 11-2. If it's in Degree mode, θ appears in degrees, as in the first entry in the second picture of this figure.

   If you don't know the Angle mode of your calculator, look at the bottom of the screen, where you see either RAD (as in the first picture in Figure 11-2) or DEG (as in the second picture), indicating that your calculator is in Radian or Degree mode. I explain setting the mode in Chapter 1.

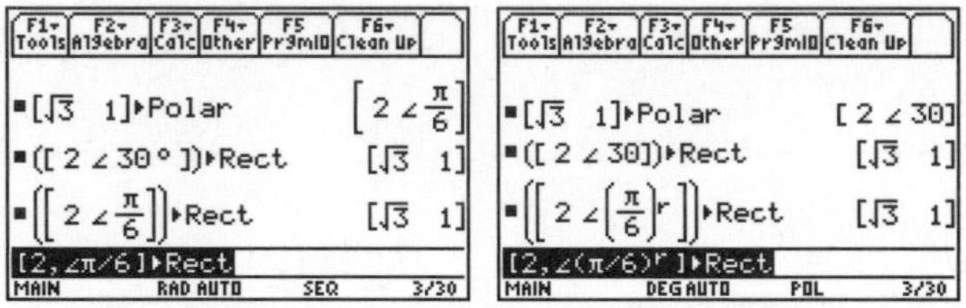

Radian mode   Degree mode

Figure 11-2: Converting between rectangular and polar coordinates.

To convert (*r*, θ) from polar to rectangular coordinates, follow these steps:

1. **Press 2nd , to enter the left bracket and then enter the value of *r* and press ,.**

   Brackets must surround your coordinates.

2. **Press 2nd EE (the angle symbol) to tell the calculator you are about to enter an angle.**

   EE is the third key from the bottom in the left column of calculator keys.

3. **Enter the value of the angle θ.**
4. **If appropriate, press 2nd | to tell the calculator that the angle you entered is measured in degrees or press 2nd 5 2 2 to indicate that the angle is in radians.**

   If the angle you entered in Step 3 is measured in the same Angle mode (Radians or Degrees) as the calculator, you can skip this step. However if the measurement of the angle isn't the same as the Angle mode, you have to tell the calculator about this discrepancy by placing the appropriate symbol after the angle. Entering angles measured in units other than the Angle mode of the calculator is illustrated in the two pictures in Figure 11-2. In the first picture, the calculator is in Radian mode, as indicated by **RAD** on the last line of the screen. So you need the degree symbol when entering an angle measured in degrees (as in the second entry on the screen), but you don't need the radian symbol when entering an angle measured in radians (as in the third entry on the screen). You enter the degree symbol by pressing [2nd][I] after entering the angle.

In the second picture on Figure 11-2, the calculator is in Degree mode, as indicated by **DEG** on the last line of the screen. So when entering an angle that is measured in degrees, no degree symbol is needed, as in the second entry on the screen. But when entering an angle in radians, the radian symbol is necessary, as in the third entry on the screen. The radian symbol is entered by pressing [2nd][5][2][2] after entering the angle.

5. **Press [2nd][÷] to close the brackets and then press [2nd][5][4] to access the MATH Matrix menu.**
6. **Press [▲][▶][5] to select the Rect option from the Vector Options menu.**
7. **Press [ENTER] to convert your polar coordinates to rectangular coordinates.**

   This is illustrated in the second and third entries in both pictures in Figure 11-2.

Entering Polar Equations

The calculator can accommodate up to 99 polar equations. To avoid notational confusion, I denote these polar equations with r_1 through r_{99}; the calculator uses the notation *r*1 through *r*99. For the sake of simplicity, I tell you how to enter the equation r_k. But what is stated in the following for r_k also applies to r_1, r_2, and so on. To enter polar equations into the calculator, follow these steps:

1. **If your calculator isn't in Polar mode, press [MODE][▶][3] to select the POLAR option (as shown in Figure 11-3) and then press [ENTER] to save the Mode settings.**

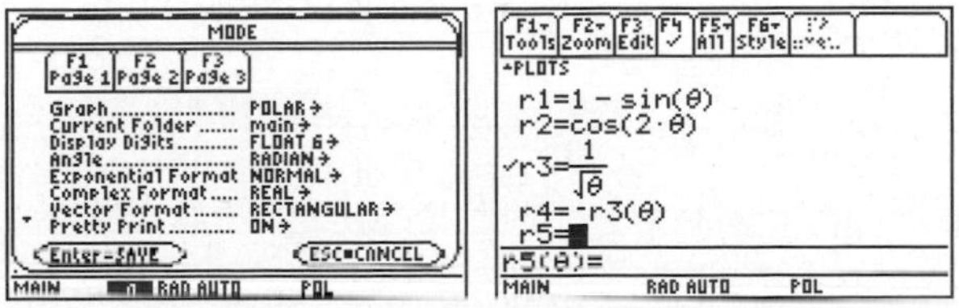

Set mode | Enter equations

Figure 11-3: Entering polar equations.

2. **Press [◆][F1] and enter the definition of r_k. Press [ENTER] to store the definition in the Y= editor.**

   Use the ⊙ and ⊙ keys to place the cursor on r_k and then press [ENTER] to place the cursor on the command line, where you enter the definition. Any previously appearing definition will be overwritten when you key in the new definition. After entering the definition, press [ENTER] to store it in the Y= editor. Examples of definitions of polar equations appear in Figure 11-3.

   When you're defining polar equations, the only symbol the calculator allows for the independent variable is the Greek letter θ. Press [◆][^] to enter this letter in your definitions.

   As a timesaver, when entering polar equations in the Y= editor, you can reference another equation in its definition provided that you enter the equation in the form $r_k(\theta)$ where *k* is the number of the equation. The second picture in Figure 11-3 shows equation r_4 defined as $-r_3$. You enter the definition in r_4 by pressing [(-)][ALPHA][2][3][(][◆][^][)].

3. **If you're graphing more than one polar equation, give each equation a unique graphing style so you can identify it in a graph of multiple polar equations.**

   In Polar mode, you have six style choices: Line, Dot, Square, Thick, Animate, and Path. The Path style uses a circle to indicate the location of the cursor as it draws the graph, and when the graph is complete, the circle disappears and leaves the graph in Line style. The Animate style also uses a circle to indicate a point as it's being graphed, but when the graph is complete, no graph appears on-screen.

   To set the style for a polar equation, use the ⊙⊙ keys to highlight the definition of the equation and press [2nd][F1] to display the style menu. Then press the number of the desired style.

Graphing Polar Equations

After you enter the polar equations into the calculator, as I describe in the preceding section, you can use the following steps to graph the equations:

1. **If necessary, turn off any polar equations and Stat Plots that you don't want to appear in your graph.**

   If you don't know what I mean by this, press [◆][F1] to display the Y= editor. If a checkmark appears to the left of a polar equation in the Y= editor, that equation will be graphed. To prevent this from happening, use the ⊙⊙ keys to highlight the definition of the equation and then press [F4] to remove the checkmark. At a later time, when you do want to graph that equation, repeat this process to reinstate the checkmark. In the second picture in Figure 11-3, only r_3 is checked to be graphed.

The first line in the Y= editor tells you the graphing status of the Stat Plots. (Chapter 20 discusses Stat Plots.) If no number appears after **PLOTS** on the very first line of the Y= editor, as in the second picture in Figure 11-3, then no Stat Plots are selected to be graphed. If one or more numbers do appear, those Stat Plots will be graphed along with the graph of your polar equations.

To prevent an individual Stat Plot from being graphed, use the ⊙⊙ keys to place the cursor on that Stat Plot and press [F4]. ([F4] toggles between deactivating and activating a Stat Plot.) To prevent all Stat Plots from being graphed, press [F5][5].

When you're graphing polar equations, Stat Plots that are active when you don't really want them graphed cause problems that could result in an error message when you attempt to graph your polar equations. So if you aren't planning to graph a Stat Plot along with your equations, make sure all Stat Plots are turned off.

2. **If necessary, set the format for the graph by pressing [F1]⊙[ENTER] and changing the selections in the Graph Formats menu.**

   I explain the Graph Formats menu in Chapter 5. It allows you to have Coordinates displayed in rectangular (x, y) form, polar (r, θ) form, or not displayed at all. Because I see no point in not knowing the coordinates of the cursor location, my recommendation is to display the coordinates in either rectangular or polar form.

   The **Grid** option is your choice, either OFF or ON. And as far as the other options go, my recommendation is to set **Axes** ON, **Leading Cursor** OFF, and **Labels** OFF.

3. **Press [◆][F2] to access the Window editor.**

4. **After each of the first three window variables, enter a numerical value that is appropriate for the equations you're graphing. Press [ENTER] after entering each value.**

   The first picture in Figure 11-4 displays the Window editor when the calculator is in Polar mode. The following list explains the variables you must set in this editor:

   - **θmin:** This setting contains the first value of the independent variable θ that the calculator will use to evaluate all polar equations in the Y= editor. It can be set equal to any real number.

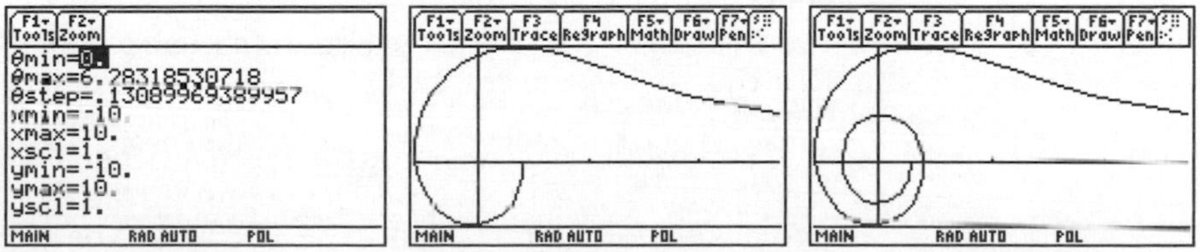

Window θmax = 2π θmax = 4π

Figure 11-4: Graphing polar equations.

- **θmax:** This setting contains the largest value of the independent variable θ that you want the calculator to use to evaluate all polar equations in the Y= editor. It can be set equal to any real number.

  θmax is usually set equal to 2π. However, as illustrated in Figure 11-4, there are occasions when you might want to set it equal to a larger number, such as 4π. You enter π into the calculator by pressing [2nd][^].

- **θstep:** This setting tells the calculator how to increment the independent variable θ as it evaluates the polar equations in the Y= editor and graphs the corresponding points.

  θstep must be a positive real number. You want **θstep** to be a small number, like π/24, but you don't want it to be too small, like 0.001. If **θstep** is too small, it will take a few minutes for the calculator to produce the graph. And if **θstep** is a large number, like 1, the calculator may not graph enough points to show you the true shape of the curve.

  If it's taking a long time for the calculator to graph your polar equations, and you start to regret that very small number you placed in **θstep**, press [ON] to terminate the graphing process. You can then go back to the Window editor and adjust the **θstep** setting.

The remaining items in the Window editor deal with setting a viewing window for your graphs (a procedure I explain in detail in Chapter 5). If you know the dimensions of the viewing window required for your graphs, go ahead and assign numerical values to the remaining variables in the Window editor and advance to Step 7. If you don't know the minimum and maximum *x* and *y* values required for your graphs, the next step tells you how to get the calculator to find them for you.

5. **Press [F2][▲][▲][▲][ENTER] to get ZoomFit to graph your polar equations.**

   After you've assigned values to **θmin**, **θmax**, and **θstep**, **ZoomFit** determines appropriate values for **xmin**, **xmax**, **ymin**, and **ymax,** and then graphs your polar equations. Note, however, that **ZoomFit** graphs polar equations in the smallest possible viewing window, and it doesn't assign new values to **xscl** and **yscl**. This is illustrated in the graphs in Figure 11-4.

   If you like the way the graph looks, you can skip the remaining steps. If you'd like to spruce up the graph, Step 6 gives you some pointers.

6. **Press [♦][F2] and adjust the values assigned to the *x* and *y* settings. Press [ENTER] after entering each new number.**

   Here's how to readjust the viewing window after using **ZoomFit**:

   - **xmin and xmax:** If you don't want the graph to hit the left and right sides of the screen, decrease the value assigned to **xmin** and increase the value assigned to **xmax**. If you want to see the *y*-axis on the graph, assign values to **xmin** and **xmax** so that zero is between these two values.

- **xscl:** Set this equal to a value that doesn't make the *x*-axis look like railroad tracks — that is, an axis with far too many tick marks. Twenty or fewer tick marks makes for a nice-looking axis.
- **ymin and ymax:** If you don't want the graph to hit the top and bottom of the screen, decrease the value assigned to **ymin** and increase the value assigned to **ymax**. If you want to see the *x*-axis on the graph, assign values to **ymin** and **ymax** so that zero is between these two values.

  When you trace a graph, the calculator needs two lines of text to display the coordinates of the cursor. And when you use the options in the Graph Math menu, it needs three lines of text to display its directions. If you don't want this text to be written on top of your graph, decrease the value of **ymin** so that two or three lines of blank space exist between the graph and the bottom of the screen. I cover tracing and using the Graph Math menu later in this chapter.
- **yscl:** Set this equal to a value that doesn't make the *y*-axis look like railroad tracks. Fifteen or fewer tick marks is a nice number for the *y*-axis.

7. **Press [◆][F3] to regraph the polar equations.**

After you've graphed your polar equations, you can save the graph and its entire Window, Y= editor, Mode, and Format settings in a Graph Database. When you recall the Database at a later time, you get more than just a picture of the graph. The calculator also restores the Window, Y= editor, Mode, and Format settings to those stored in the Database. Thus you can, for example, trace the recalled graph. Chapter 5 has the details on saving and recalling a graph database.

Exploring Polar Equations

The calculator lets you explore the graphs of polar equations in three ways: by zooming, by tracing, and by creating tables. Zooming allows you to quickly adjust the viewing window for the graph so that you can get a better idea of the nature of the graph. Tracing shows you the coordinates of the points that make up the graph, and creating a table — well, I'm sure you already know what that shows you. In this section, you find out how to zoom and trace; the next section tells you how to create tables.

Using Zoom commands

After you graph your polar equations (as I explain earlier in this chapter), press [F2] to access the Zoom menu. On this menu, you see all the Zoom commands that are available for graphing functions. Not all of them are useful when graphing polar equations, however. Here are the ones that I use when graphing polar equations:

- **ZoomFit:** This command finds a viewing window for a specified portion of the graph. I tell you how to use **ZoomFit** in the section, "Graphing Polar Equations," earlier in this chapter.
- **ZoomSqr:** Because the calculator screen is rectangular instead of square, circles look like ellipses if the viewing window isn't properly set. **ZoomSqr** adjusts the settings in the Window editor so that circles look like circles. To use **ZoomSqr**, first graph your polar equations and then press [F2][5]. The calculator then redraws the graph in a viewing window that makes circles look like circles. Figure 11-5 illustrates the effect that **ZoomSqr** has on the graph of the cardioid r_1 in Figure 11-3.
- **ZoomBox:** After the graph is drawn (as I describe earlier in this chapter), this command allows you to define a new viewing window for a portion of your graph by enclosing it in a box, as illustrated in Figure 11-6. The box becomes the new viewing window. To construct the box:

  1. **Press [F2][1].**
  2. **Define a corner of the box.**

     To do so, use the ◀▶▲▼ keys to move the cursor to the spot where you want one corner of the box to be located and then press [ENTER]. The calculator marks that corner of the box with a small square.
  3. **Construct the box (as shown in the first picture in Figure 11-6).**

     To do so, press the ◀▶▲▼ keys. As you press these keys, the calculator draws the sides of the box.

     TIP: When you use **ZoomBox**, if you don't like the size of the box, you can use any of the ◀▶▲▼ keys to resize the box. If you don't like the location of the corner you anchored in Step 2, press [ESC] and start over with Step 1. If you're drawing a rather large box, to move the cursor a large distance, press [2nd] and then press an appropriate arrow key.

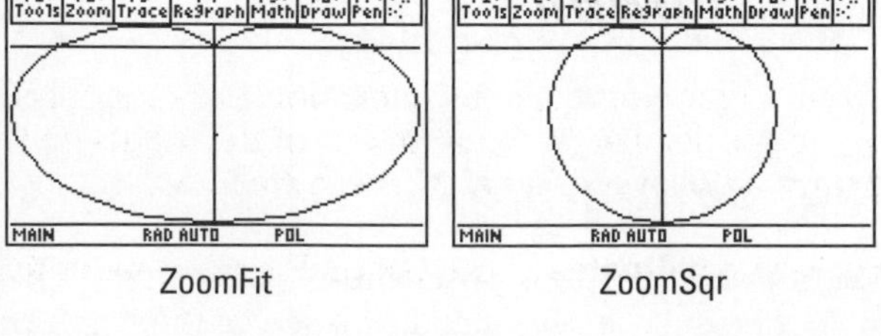

ZoomFit ZoomSqr

Figure 11-5: A cardioid graphed using ZoomFit and then using ZoomSqr.

4. **When you're finished drawing the box, press [ENTER] and the calculator redraws the graph in the viewing window specified by the box (as shown in the second picture in Figure 11-6).**

   When you use **ZoomBox**, [ENTER] is pressed only two times. The first time you press it is to anchor a corner of the zoom box. The next time you press [ENTER] is when you're finished drawing the box and you're ready to have the calculator redraw the graph.

- **Zoom In and Zoom Out:** After the graph is drawn (as I describe earlier in this chapter), these commands allow you to zoom in on a portion of the graph or to zoom out from the graph. They work very much like a zoom lens on a camera. To use these commands:

  1. **Press [F2][2] if you want to zoom in, or press [F2][3] if you want to zoom out.**
  2. **Use the ◀▶▲▼ keys to move the cursor to the spot on-screen from which you want to zoom in or zoom out.**

     This spot becomes the center of the new viewing window.
  3. **Press [ENTER].**

     The graph is redrawn centered at the cursor location.

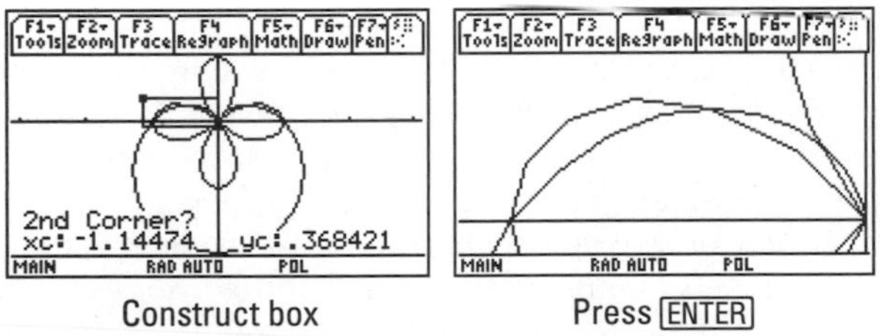

Figure 11-6: Using ZoomBox.

If you used a Zoom command to redraw a graph and then want to undo what that command did to the graph, press [F2]▲▲▶[1] to select **ZoomPrev** from the Zoom Memory menu. The graph is then redrawn as it appeared in the previous viewing window.

Tracing polar graphs

After you graph your polar equations (as I describe earlier in this chapter), you can have the cursor display the coordinates of the points on the graph by tracing it. To trace a polar graph, follow these steps:

1. **If necessary, set Coordinates in the Graph Formats menu to RECT or POLAR.**

If you know that you are already in the rectangular or polar format, you can skip this step. Otherwise press [F1]⊖[ENTER]⊙ to access the **Coordinates** option in the Graph Formats menu and then press [1] to have the coordinates displayed in (x, y) rectangular form or press [2] for (r, θ) polar form. Press [ENTER] to save this setting.

2. **Press [F3] and then use ⊙ to trace the graph.**

As you investigate a polar graph by tracing it, here's what you see:

- **The number of the polar equation:** The number of the polar equation in the Y= editor that you're tracing is displayed in the top-right corner of the screen. For example, if you see the number 1, you're tracing the first polar equation (r_1) in the Y= editor. If you've graphed more than one equation and you would like to trace a different polar equation, repeatedly press ⊖ until the number of the desired equation appears.
- **The value of the parameter θ:** At the bottom of the screen, you see a value for θc. This is the value of the independent variable θ at the cursor location. Each time you press ⊙, the cursor moves to the next plotted point in the graph and updates the value of θc. If you press ⊙, the cursor moves left to the previously plotted point. And if you press ⊖ to trace a different polar equation, the tracing of that equation starts at the same value of θc that was displayed on-screen before you pressed this key.

  TIP: When you're using Trace, if you want to start tracing the graph at a specific value of the parameter θ, just key in that value and press [ENTER]. (The value you assign to θ must be between θ**min** and θ**max**; if it isn't, you get an error message.) After you press [ENTER], the trace cursor moves to the point on the graph corresponding to that value of θ. But the calculator doesn't change the viewing window. So if you don't see the cursor, it's on the part of the graph outside the viewing window. The sidebar "Panning in Polar mode" tells you how to get the cursor and the graph in the same viewing window.
- **The coordinates of the cursor:** On the last line of the screen, you see the coordinates of the cursor location. In rectangular form, these coordinates are denoted by **xc** and **yc**; in polar form, they are **rc** and θ**c**.

Press [ESC] to terminate tracing the graph. This also removes the number of the polar equation and the coordinates of the cursor from the screen.

TIP: If your cursor disappears from the screen while you're tracing a polar graph, the sidebar "Panning in Polar mode" tells you how to rectify this situation.

Panning in Polar mode

When you're tracing a polar equation and the cursor hits an edge of the screen, if θ**c** is less than θ**max** and you continue to press ▶, the coordinates of the cursor are displayed at the bottom of the screen. However, because the calculator doesn't automatically adjust the viewing window, you won't see the cursor on the graph. To rectify this situation, make a mental note of the value of θ**c** (or write it down) and then press [ENTER]. The calculator then redraws the graph centered at the location of the cursor at the time you pressed [ENTER]. Then press [F3] to continue tracing the graph.

The bad news is that after the graph is redrawn the trace cursor appears at the beginning of the graph of the first selected polar equation appearing in the Y= editor. To get the trace cursor back on the part of the graph displayed in the new viewing window, key in that value of θ**c** (hopefully you kept the note handy) and then press [ENTER]. If you weren't tracing the first selected equation appearing in the Y= editor, use ▲ to place the cursor on the graph you *were* tracing. The trace cursor then appears in the middle of the screen, and you can use ◀ or ▶ to continue tracing the graph. When θ**c** equals θ**max**, pressing ▶ has no effect because the cursor has reached the end of the graph.

TIP

If the number of the polar equation at the top of the screen and the coordinates at the bottom of the screen are interfering with your view of the graph when you use Trace, increase the height of the window by pressing [◆][F2] and decreasing the value of **ymin** and increasing the value of **ymax**.

Displaying Equations in a Table

After you've entered polar equations in the Y= editor (as I describe earlier in this chapter), you can have the calculator create one of three different types of tables of values of your polar equations:

- An automatically generated table that displays the plotted points appearing in the calculator's graphs of the polar equations
- An automatically generated table that displays the points using the initial θ-value and θ-increment that you specify
- A user-defined table into which you enter the θ-values of your choice and the calculator determines the corresponding values of the polar equations

The following sections discuss these tables in greater detail.

Creating automatically generated tables

To create a table in which you specify the initial θ-value and the value by which θ is incremented, perform the following steps:

1. **Place checkmarks to the left of the equations in the Y= editor that you want to appear in the table; leave the other equations unchecked.**

   Only the equations in the Y= editor that have checkmarks to the left of their names will appear in the table. To check or uncheck an equation, press ◆F1, use the ▲▼ keys to place the cursor on the equation, and then press F4 to toggle between checking and unchecking the function.

2. **Press ◆F4 to access the Table Setup editor (shown in Figure 11-7).**

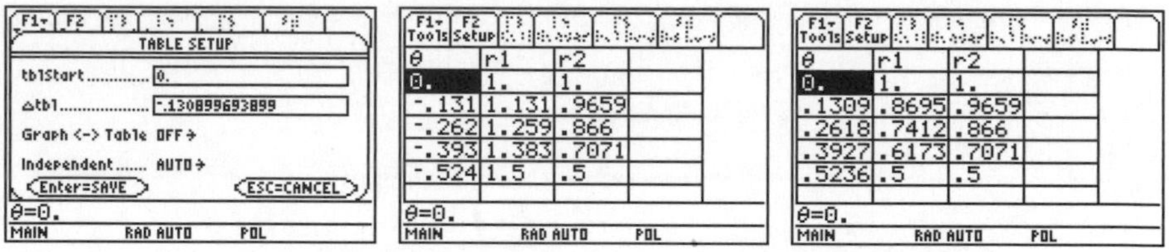

Figure 11-7: Automatically generated tables.

3. **Enter a number in tblStart and press ▼; then enter a number in Δtbl and press ▼.**

   If **Independent** is the only item you see in the Table Setup editor, and if the cursor is on ASK, press ▶1 to change this setting to AUTO and then press ▲▲▲ to enter numbers in **tblStart** and **Δtbl**.

   tblStart is the first value of the independent variable θ to appear in the table, and **Δtbl** is the increment for the independent variable θ. The calculator constructs the table by adding the value you assign to **Δtbl** to the previous θ-value and calculating the corresponding values of the polar equations. In the first picture in Figure 11-7, **tblStart** is assigned the value 0 and **Δtbl** the value $-\pi/24$ to produce the decreasing values in the θ-column of the table in the second picture in this figure.

   Use the number keys to enter values in **tblStart** and **Δtbl**. Press ▼ after making each entry.

4. **Set Graph <–> Table to OFF (as shown in Figure 11-7).**

   You have two choices for the **Graph <–> Table** setting: OFF and ON. When **Graph <–> Table** is set to ON, a table of the points plotted in the calculator's graphs of the equations is created. However, if **Graph <–> Table** is set to OFF, the calculator uses the values in **tblStart** and **Δtbl** to generate the table.

To change the **Graph <–> Table** setting, use ▲▼ to place the cursor on the existing setting, press ▶ to display the options for that setting, and press the number of the option you want.

5. **Press [ENTER] to save the settings in the Table Setup editor.**
6. **If necessary, press [◆][F5] to display the table.**

   The table appears on the screen, as illustrated in the second picture in Figure 11-7. You find out about navigating the table later in this chapter.

To create a table of the plotted points appearing in the calculator's graphs of the polar equations, follow Steps 1 and 2. If **Independent** is the only item you see in the Table Setup editor, press ▶[1] to change this setting to AUTO and then press ▲▶[2][ENTER] to set **Graph <–> Table** to ON and to save your settings. On the other hand, if you see all four items in this editor, press ▼▼▶[2][ENTER] to set **Graph <–> Table** to ON and to save your settings. (With these settings, the calculator will ignore the other values in the Table Setup menu.) If necessary, press [◆][F5] to display the table, as shown in the third picture in Figure 11-7.

Creating a user-defined table

To create a user-defined table in which you enter values of the parameter θ and the calculator figures out the corresponding values of the polar equations, follow these steps:

1. **Place checkmarks to the left of the functions in the Y= editor that you want to appear in the table; leave the other functions unchecked.**

   Only the functions in the Y= editor that have checkmarks to the left of their names will appear in the table. To check or uncheck a function, press [◆][F1], use the ▲▼ keys to place the cursor on the function, and then press [F4] to toggle between checking and unchecking the function.

2. **Press [◆][F4] to access the Table Setup editor.**
3. **Set Independent to ASK and press [ENTER].**

   Independent is the last item in the menu. If it is already set to ASK, press [ENTER] and go to the next step. If it is set to AUTO, press ▲▲▲▶[2] to set **Independent** to ASK and then press [ENTER].

4. **If necessary, press [◆][F5] to display the table. If the table contains entries, press [F1][8][ENTER] to clear the table.**

   You see a table without entries, as in the first picture in Figure 11-8.

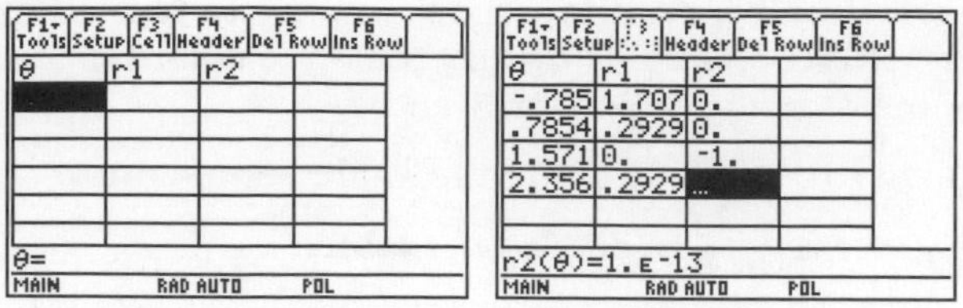

Figure 11-8: Generating a user-defined table.

5. **Enter values for the parameter θ in the first column.**

   You must enter values consecutively; you cannot skip any cells in the first column of the table. To make the first entry, press [ENTER], use the number keys to enter a value, and then press [ENTER] when you're finished. To make entries after the first entry, press ▼[ENTER], enter a value, and then press [ENTER].

   After you make each entry, the calculator automatically calculates the corresponding values of the selected polar equations in the Y= editor, as shown in the second picture of Figure 11-8.

TIP

If you see the ellipsis symbol (. . .) in the table, this means that the calculator doesn't have enough room to display the value in the table. To see the value of that table entry, use the Arrow keys to place the cursor over the entry. The value is then displayed on the command line, as illustrated in the second picture in Figure 11-8.

Displaying hidden rows and columns

To display more rows in an automatically generated table, repeatedly press ▲ or ▼. If you have constructed a table for more than three polar equations, repeatedly press ◀ or ▶ to navigate through the columns of the table.

Each time the calculator redisplays a table with a different set of rows, it also automatically resets **tblStart** to the value of θ appearing in the first row of the newly displayed table. To return the table to its original state, press [F2] to access the Table Setup editor and then change the value that the calculator assigned to **tblStart**. Press [ENTER] to save the value you entered and then press [ENTER] again to return to the table.

TIP

While displaying a table, you can edit the definition of a polar equation without going back to the Y= editor. To do this, use the ◀▶ keys to place the cursor in the column of the equation you want to redefine and press [F4]. This places the definition of the equation on the command line so that you can edit the definition. (See Chapter 1 for more on editing expressions.) When you're finished with your editing, press [ENTER]. The calculator automatically updates the table and the edited definition of the equation in the Y= editor.

Viewing the table and the graph on the same screen

After you graph your polar equations and generate a table, you can view the graphs and the table on the same screen. To do so, follow these steps:

1. **Press [MODE][F2][▶][3] to vertically divide the screen.**
2. **Press [▼][▶][5] to place the table in the left screen and then press [▼][▶][4] to place the graph in the right screen.**

   If you prefer to have the graph on the left and the table on the right, just reverse the order of keystrokes in this step by first pressing [▼][▶][4] and then pressing [▼][▶][5].
3. **Press [ENTER] to save your Mode settings.**

   You now see a screen similar to that in the first picture in Figure 11-9. Notice that the right screen is empty. It appears in the next step.
4. **Press [2nd][APPS] to toggle between the left and right screens.**

   The right screen is no longer empty, as in the second picture in Figure 11-9.

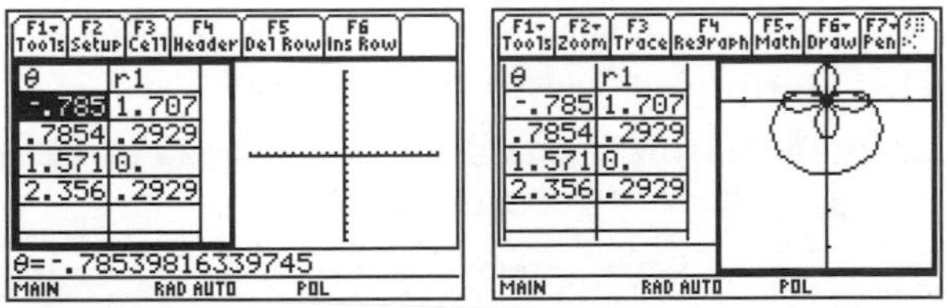

Figure 11-9: Viewing a graph and a table on the same screen.

In a split screen, the screen with the bold border is the active screen, and the commands for the application in that screen appear at the top of the screen. For example, in the second picture in Figure 11-9, the Graph screen is active. So if you press [F3] while in this application, you can trace the polar equations graphed in that screen. (You find out how to trace a graph earlier in this chapter.)

TIP

To return to FULL screen mode without making changes in the Mode menu, just press [2nd][ESC]. This automatically sets the Split Screen mode on Page 2 of the Mode menu to FULL.

Analyzing Polar Equations

After graphing polar equations (as I describe earlier in this chapter), you can use the options in the [F5] Graph Math menu to find the value of the polar equations at a specified value of θ and to evaluate derivatives (*dy/dx* and *dr/d*θ) at a specified value of θ. The calculator will even find the equation of a tangent to the graph at a specified point on the graph. And it will find the length of a portion of the graph or the distance between two points on the same curve or on different curves. This section tells you how to use the Graph Math menu to accomplish these tasks.

Evaluating polar equations

When you trace a graph, the trace cursor doesn't hit every point on the graph. So tracing isn't a reliable way of finding the value of a polar equation at a specified value of θ. The [F5] Graph Math menu, however, contains a command that evaluates polar equations at any specified θ-value. To access and use this command, perform the following steps:

1. **Graph the polar equations in a viewing window that contains the specified value of θ.**

   The graph contains a specified value of θ when that value is between θ**min** and θ**max** in the Window editor. I tell you how to graph polar equations and set θ**min** and θ**max** earlier in this chapter.

2. **If necessary, set the Graph Formats menu to display coordinates in rectangular (*x*, *y*) form or in polar (*r*, θ) form.**

   To do this, press [F1]⏶[ENTER]⏵. Press [1] for rectangular form or press [2] for polar form. Then press [ENTER] to save these settings.

3. **Press [F5][1] to invoke the Value command in the Graph Math menu.**

4. **Enter the specified value of θ.**

   Use the keypad to enter a value of θ that is between θ**min** and θ**max**, as illustrated in the first picture in Figure 11-10. As you enter θ, the value of θ**c** at the bottom of the screen is replaced by the number you key in.

5. **Press [ENTER] to evaluate the first polar equation.**

   After you press [ENTER], the number of the first checked polar equation in the Y= editor appears at the top of the screen, the cursor appears on the graph of that equation at the specified value of θ, and the coordinates of the cursor appear at the bottom of the screen. This is illustrated in the second picture in Figure 11-10.

6. **Repeatedly press (▲) to see the value of the other graphed polar equations at your specified value of θ.**

   Each time you press (▲), the number of the polar equation being evaluated appears at the top of the screen, and the coordinates of the cursor location appear at the bottom of the screen. This is illustrated in the third picture in Figure 11-10.

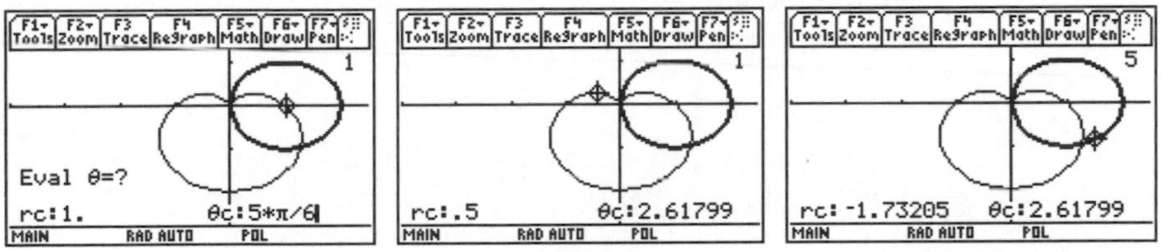

Enter θ — Press [ENTER] — Press (▲)

Figure 11-10: Evaluating polar equations at a specified value of θ.

After using the **Value** command to evaluate your polar equations at one value of θ, you can evaluate them at another value of θ by keying in the new value and pressing [ENTER]. Pressing any function key (such as [ENTER] or (▶)) *after* evaluating polar equations deactivates the **Value** command.

TIP: If you plan to evaluate polar equations at several specified values of θ, consider constructing a user-defined table of values of polar equations (as I explain earlier in this chapter).

Finding derivatives

To find the derivative (*dy/dx* or *dr/d*θ) of a polar equation at a specified value of θ, follow these steps:

1. **Graph the polar equation in a viewing window that contains the specified value of θ.**

   The graph contains a specified value of θ when that value is between **θmin** and **θmax** in the Window editor. Graphing polar equations and setting **θmin** and **θmax** are explained earlier in this chapter.

2. **If necessary, set the Graph Formats menu to display coordinates in rectangular (*x*, *y*) form or in polar (*r*, θ) form.**

   To do this, press [F1](▲)[ENTER](▶). Press [1] for rectangular form or press [2] for polar form. Then press [ENTER] to save these settings.

3. **Press [F5][6] to display the Derivatives submenu of the Graph Math menu and then press [1] or [4] to select the *dy/dx* or the *dr/d*θ option.**

   Instructions appear at the bottom of the screen, as illustrated in the first picture in Figure 11-11. If these instructions interfere with your view of the graph, create more space at the bottom of the screen by pressing [◆][F2] and decreasing the value of **ymin**. Then press [◆][F3] to redraw the graph and repeat Step 3 to reactivate the **Derivatives** option.

4. **If necessary, repeatedly press ⊖ until the number of the appropriate polar equation appears at the top of the screen.**

5. **Enter the value of θ at which the derivative is to be evaluated.**

   Use the keypad to enter a value of θ that is between **θmin** and **θmax**, as illustrated in the first picture in Figure 11-11, in which θ is set to 5π/6. As you enter θ, the value of **θc** at the bottom of the screen is replaced by the number you key in.

6. **Press [ENTER] to evaluate the derivative.**

   After pressing [ENTER], the cursor moves to the point on the graph corresponding to the specified value of θ, and the derivative at that value is displayed at the bottom of the screen, as in the second picture in Figure 11-11.

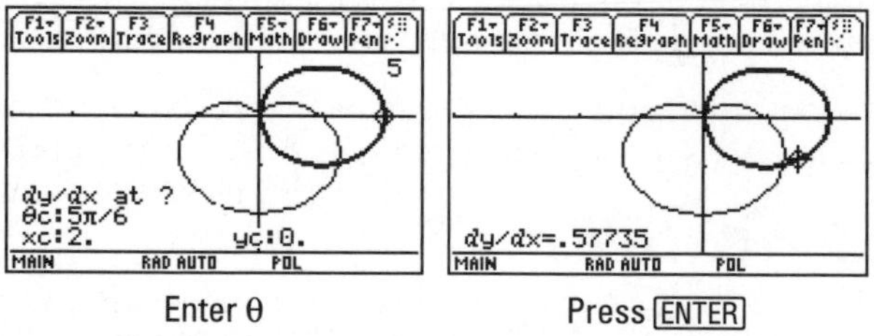

Figure 11-11: Evaluating a derivative at a specified value of θ.

Finding the distance between points

The calculator has a neat feature that draws a line between two points on the same or different polar curves and calculates the distance between these two points. To use this feature, follow these steps:

1. **Graph the polar equations in a viewing window that displays the two points between which you want to find the distance.**

   I tell you about graphing polar equations and finding an appropriate viewing window earlier in this chapter.

2. **If necessary, set the Graph Formats menu to display coordinates in rectangular (*x*, *y*) form or in polar (*r*, θ) form.**

   To do this, press [F1]⏶[ENTER]⏵. Press [1] for rectangular form or press [2] for polar form. Then press [ENTER] to save these settings.

3. **Press [F5][9] to select the Distance command from the Graph Math menu.**

   If the calculator instructions at the bottom of the screen interfere with your view of the graph, create more space at the bottom of the screen by pressing [♦][F2] and decreasing the value of **ymin**. Then press [♦][F3] to redraw the graph and press [F5][9] to reinstate the **Distance** command.

4. **If necessary, repeatedly press ⏶ until the number of the polar equation containing the first point appears at the top of the screen.**

5. **Enter the value of the θ corresponding to the first point and then press [ENTER].**

   Use the keypad to enter a value of θ that is between **θmin** and **θmax**. As you enter θ, the value of **θc** at the bottom of the screen is replaced by the number you key in. After you press [ENTER], the cursor moves to that point and prompts you to enter the value of θ corresponding to the second point, as illustrated in the first picture in Figure 11-12.

6. **If the second point is on a different curve, repeatedly press ⏶ until the number of the polar equation containing the second point appears at the top of the screen.**

   The cursor jumps to the point on the other curve that is defined by the same value θ, a line is drawn between these points, and you are prompted to enter the second point, as illustrated in the second picture in Figure 11-12.

7. **Enter the value of θ corresponding to the second point and then press [ENTER].**

   The calculator connects the two points with a line and displays the distance between these points, as illustrated in the third picture in Figure 11-12.

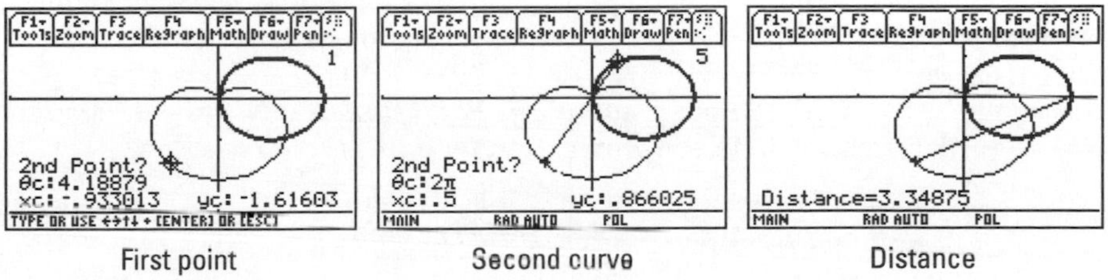

Figure 11-12: Finding the distance between two points.

Finding the equation of a tangent line

Not only can the calculator draw a tangent to a polar curve at a specified value of θ, but it can also tell you the equation of that tangent. To get the calculator to do this, follow these steps:

1. **Graph the polar equation in a viewing window that displays the point of tangency.**

   Graphing polar equations and finding an appropriate viewing window are explained earlier in this chapter.

2. **If necessary, set the Graph Formats menu to display coordinates in rectangular (*x*, *y*) form or in polar (*r*, θ) form.**

   To do this, press [F1][▲][ENTER][▶]. Press [1] for rectangular form or press [2] for polar form. Then press [ENTER] to save these settings.

3. **Press [F5][▲][▲][▲][ENTER] to select the Tangent command from the Graph Math menu.**

   If the calculator instructions at the bottom of the screen interfere with your view of the graph, create more space at the bottom of the screen by pressing [◆][F2] and decreasing the value of **ymin**. Then press [◆][F3] to redraw the graph and press [F5][▲][▲][▲][ENTER] to reinstate the **Tangent** command.

4. **If necessary, repeatedly press [▲] until the number of the polar equation containing the point of tangency appears.**

5. **Enter the value of θ at the point of tangency.**

   Use the keypad to enter a value of θ that is between **θmin** and **θmax**. As you enter θ, the value of **θc** at the bottom of the screen is replaced by the number you key in, as illustrated in the first picture in Figure 11-13.

6. **Press [ENTER] to construct the tangent and display its equation.**

   The tangent appears on the graph with the cursor located at the point of tangency, and the equation of the tangent appears at the bottom of the screen, as illustrated in the second picture in Figure 11-13.

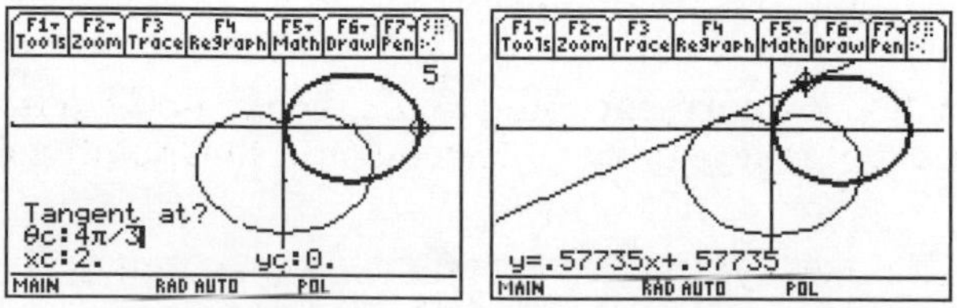

Point of tangency | Equation of tangent

Figure 11-13: Finding the equation of a tangent line to a polar curve.

REMEMBER

A tangent drawn by the **Tangent** command in the Graph Math menu doesn't automatically vanish when you use another Graph Math menu option or when you trace the graph. To erase the tangent, press [F4] to redraw the graph without the tangent line.

Evaluating arc length

The **Arc** command in the Graph Math menu can find the length of a section (arc) of a graphed curve when you specify the beginning and ending point of the section. To get the calculator to find this length, follow these steps:

1. **Graph the polar equation in a viewing window displaying the part of the curve you want to measure.**

   I cover graphing polar equations and finding an appropriate viewing window earlier in this chapter.

2. **If necessary, set the Graph Formats menu to display coordinates in rectangular (*x*, *y*) form or in polar (*r*, θ) form.**

   To do this, press [F1]▲[ENTER]▶. Press [1] for rectangular form or press [2] for polar form. Then press [ENTER] to save these settings.

3. **Press [F5]▲▲[ENTER] to select the Arc command from the Graph Math menu.**

   If the calculator instructions at the bottom of the screen interfere with your view of the graph, create more space at the bottom of the screen by pressing [◆][F2] and decreasing the value of **ymin**. Then press [◆][F3] to redraw the graph and press [F5]▲▲[ENTER] to reinstate the **Arc** command.

4. **If necessary, repeatedly press ▲ until the number of the appropriate polar equation appears at the top of the screen.**

5. **Enter the value of θ corresponding to the first point defining arc length and then press [ENTER].**

   Use the keypad to enter a value of θ that is between **θmin** and **θmax**. As you enter θ, the value of **θc** at the bottom of the screen is replaced by the number you key in. After you press [ENTER], the cursor moves to that point and prompts you to enter the second point, as illustrated in the first picture in Figure 11-14.

6. **Enter the value of θ corresponding to the second point defining the arc (or use the ◀▶ keys to move the cursor to that point) and press [ENTER].**

   The calculator displays the length of the arc at the bottom of the screen, as illustrated in the second picture in Figure 11-14.

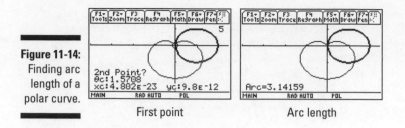

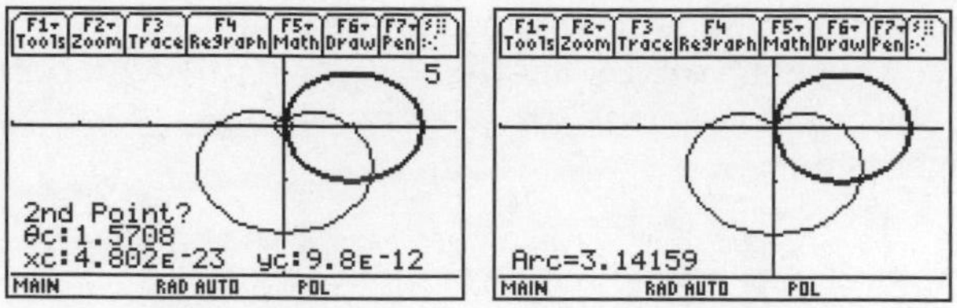

First point

Arc length

Figure 11-14: Finding arc length of a polar curve.

Did the calculator give you a negative value for arc length? No problem — just take the absolute value of the answer given by the calculator to get the true value of the arc length. (The calculator gives a negative value for arc length when the value of θ entered to define the first point of the arc is larger than the value of θ entered to define the second point of the arc.)

Part V
Doing Calculus

"I failed her in Algebra but was impressed with the way she animated her equations to dance across the screen, scream like hyenas, and then dissolve into a clip art image of the La Brea Tar Pits."

In this part . . .

Did you know that a calculator could differentiate, integrate, or evaluate a limit? Well, a TI-89 calculator can! In this part, I tell you how to get your calculator to accomplish these tricks and many others — such as computing the cross product of two vectors and graphing in 3D.

If you're dealing with differential equations, you've come to the right place. In this part, I tell you how to get the calculator to solve them for you. I even tell you how to get the calculator to sketch a slope field.

Chapter 12

Dealing with Differential and Integral Calculus

In This Chapter

- Finding derivatives and partial derivatives
- Evaluating definite and indefinite integrals
- Evaluating one- and two-sided limits
- Finding Taylor and Maclaurin polynomials
- Converting between rectangular, cylindrical, and spherical coordinates
- Dealing with hyperbolic functions

This chapter is just packed with useful information on how to get the calculator to perform a multitude of tasks. In addition to the main topics, this chapter tells you how to evaluate a finite or infinite sum or product, how to find critical points, and how to evaluate arc length. If your calculus needs go beyond the topics that I cover in this chapter, consult Chapters 13, 14, and 15 to find out how to use the calculator to do vector calculus, graph surface curves and contour maps, and solve and graph differential equations.

Using the Calc Menu

Do you need to find the derivative of a function? Evaluate a definite integral? Evaluate a limit? Find the Taylor polynomial of a function? Determine the arc length of a portion of a curve? This section tells you how to accomplish these and other tasks.

The commands that help you do calculus are housed in the Calc menu, as illustrated in the first two pictures in Figure 12-1. You can access this menu in two ways: On the Home screen, press [F3] to display the Calc menu (as shown in the first two pictures in Figure 12-1); or, alternatively, from anywhere in the calculator, you can press [2nd][5]⊖⊖⊖⊖⊕ to display the MATH Calculus menu, which is identical to the Calc menu, as illustrated in the third picture in Figure 12-1.

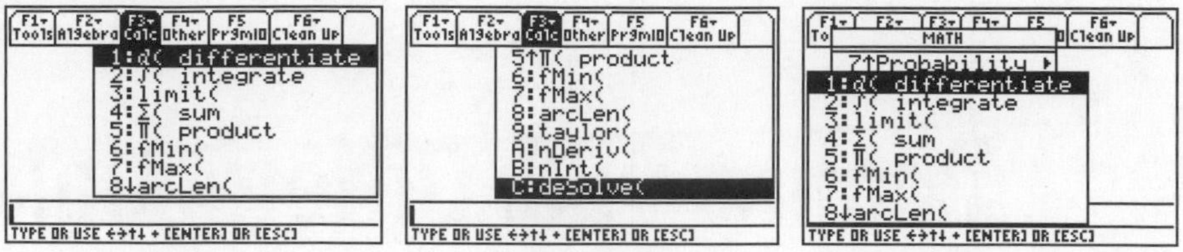

Press [F3] | Press [F3] ⏶ | Press [2nd] [5] ⏶⏶⏶⏶ ⏵

Figure 12-1: The [F3] Calc menu and MATH Calculus menus.

Dealing with derivatives

The topics in this section tell you how to find derivatives and partial derivatives and how to evaluate these derivatives at a specified value of the independent variable.

Finding derivatives

To find a derivative, follow these steps:

1. **If you aren't already on the Home screen, press [HOME].**

   [HOME] is the fourth key from the top in the left column of keys.

2. **Press [F3][1] to select the *d* command from the Calc menu.**

   The command is placed on the command line followed by a left parenthesis.

   TIP: You can also enter the derivative command ***d*** in an expression by pressing [2nd][8].

3. **Enter the function whose derivative you want to find.**

   I cover entering expressions in Chapter 1, and I explain creating user-defined functions in Chapter 3.

   REMEMBER: Use [×] (the times key) when entering the product of two letters that represent different variables or constants. If you don't, the calculator will interpret the two juxtaposed letters as being a two-letter name for a single variable. For example, to the calculator, ax^2 is the square of the single variable with the two-letter name *ax*; whereas $a \cdot x^2$ is viewed by the calculator as the product of the variable *a* and the square of the variable *x*.

4. **Press [,] and enter the variable.**

   Derivatives are always taken with respect to a variable. This is the variable you want to enter. For example, if you are evaluating $\frac{d}{dq}P$, the variable is *q*. I explain entering variable names (words) in Chapter 1.

5. **Press [)] to close the parentheses and then press [ENTER] to evaluate the derivative, as illustrated in the first picture in Figure 12-2.**

Figure 12-2: Finding and evaluating derivatives and partial derivatives.

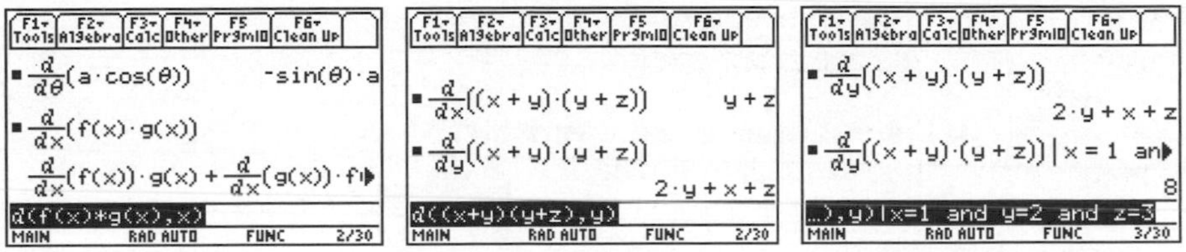

Derivatives | Partial derivatives | Evaluating derivatives

If an answer scrolls off the right side of the screen, as in the last entry in the first picture in Figure 12-2, press ▲ and then repeatedly press ▶ to view the part of the answer not appearing on the screen. To again see the beginning of the answer, press [2nd]◀. When you finish viewing the answer, press ▼ to return to the command line.

Have you forgotten a rule for finding derivatives? No problem — simply use the derivative command to get the calculator to give you that rule, as illustrated in the last entry in the first picture of Figure 12-2, in which the calculator displays the product rule.

Finding partial derivatives

When the calculator finds the derivative with respect to a specified variable, it treats all other variables as being constants. So the ***d*** derivative command can also be used to find partial derivatives. To find the partial derivative of a function, follow the steps in the preceding section for finding a derivative. In Step 4, use the variable for the partial derivative.

The second picture in Figure 12-2 illustrates how you can use the derivative command to find partial derivatives. The first entry in this picture finds $\frac{\partial}{\partial x}(x+y)(y+z)$ and the second finds $\frac{\partial}{\partial y}(x+y)(y+z)$.

Evaluating derivatives and partial derivatives

To evaluate a derivative or partial derivative at specified values of the independent variables, follow these steps:

1. **Place the derivative or partial derivative on the command line.**

   If you haven't yet entered the derivative or partial derivative you want to evaluate, enter it now. The preceding two sections tell you how to do this.

   If the derivative or partial derivative was your last entry and you have already evaluated it, press [2nd]▶ to place the cursor at the end of this entry. The calculator will also deselect the entry by unhighlighting it.

   If you entered the derivative or partial derivative in the calculator before your last entry and before clearing the screen, press [CLEAR] twice to clear the command line and repeatedly press ▲ to highlight the derivative or partial derivative. Then press [ENTER] to place it on the command line.

2. **Press [|] to enter the with command.**

   [|] is in the left column of keys, below the [=] key.

3. **Enter the name of the variable, press [=], and then enter the value of the variable, as illustrated on the command line in the third picture in Figure 12-2.**

4. **If you're evaluating the derivative or partial derivative at more than one independent variable, press [2nd][5][8][8] to select the word *and* from the MATH Test menu and then repeat Step 3 for the next variable.**

   Continue to do this until you've entered all the variables.

5. **Press [ENTER] to evaluate the derivative or partial derivative.**

Evaluating integrals

To evaluate an indefinite integral $\int f(x)\,dx$ or a definite integral $\int f(x)\,dx$, follow these steps:

1. **If you aren't already on the Home screen, press [HOME].**

2. **Press [F3][2] to select the integral command from the Calc menu.**

   The command is placed on the command line, followed by a left parenthesis.

   You can also insert the integral command in an expression by pressing [2nd][7].

3. **Enter the function whose integral you want to find.**

   I explain entering expressions in Chapter 1. See Chapter 3 for more on creating user-defined functions.

   Use [×] (the times key) when entering the product of two letters that represent different variables or constants. If you don't, the calculator will interpret the two juxtaposed letters as being a two-letter name for a single variable. For example, to the calculator, ax^2 is the square of the single variable with the two-letter name *ax*.

4. **Press [,] and enter the variable.**

   Integration is always done with respect to a variable. This is the variable you want to enter. For example, if you're evaluating $\int Pdq$, the variable is *q*. You find out about entering variable names (words) in Chapter 1.

5. **If you're evaluating an indefinite integral (that is, there are no limits of integration), skip this step. If you're evaluating a definite integral, press [,] and enter the lower limit of integration. Then press [,] and enter the upper limit of integration.**

6. **Press [)] to close the parentheses and then press [ENTER] to evaluate the derivative.**

   This process is illustrated in the first picture in Figure 12-3. Because integration is done with respect to *y* in the first entry in this picture, the calculator treats *x* as a constant. Similarly, in the last two entries, *a* is treated as a constant.

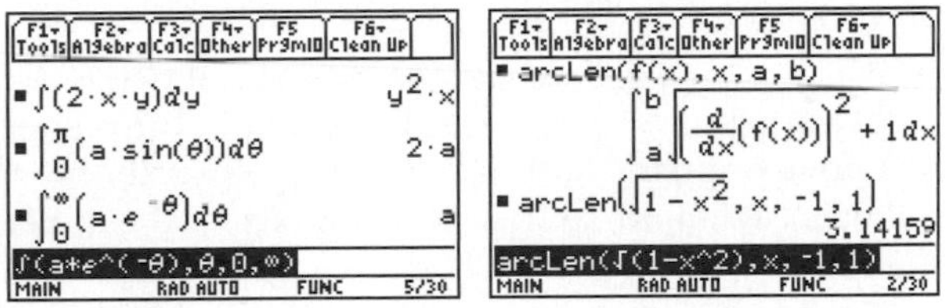

Integration   Arc length

Figure 12-3: Evaluating integrals and finding arc length.

When using the calculator to evaluate an indefinite integral, as in the first entry in the first picture in Figure 12-3, the calculator doesn't tack on the constant of integration. In other words, the answer given by the calculator is the particular solution you get when you set the constant of integration, *C*, equal to zero.

Finding arc length

The **arcLen** command in the Calc menu is used to find the length of the curve *f*(*x*) between $x = a$ and $x = b$. It cannot be used to find the length of a space curve (a three-, or higher, dimensional curve), nor can it be used to find the length of a curve expressed in polar or parametric form.

To use the **arcLen** command to find the length of a rectangular function of one variable, follow all the steps (including Step 5) in the preceding section "Evaluating integrals" with the following exception: In Step 2, press [F3][8] to select the **arcLen** command from the Calc menu. This process is illustrated in the second picture in Figure 12-3, in which the second entry in this picture finds the length of the semicircle of radius 1.

Have you forgotten the formula for finding arc length? No problem — just use the **arcLen** command to get the calculator to give you the formula, as illustrated in the first entry in the second picture in Figure 12-3.

Evaluating limits

Do you need to evaluate a limit such as $\lim_{x \to a} f(x)$ or $\lim_{x \to \infty} f(x)$? Or maybe you want to evaluate a one-sided limit like $\lim_{x \to a^-} f(x)$. If so, the following steps tell you how to evaluate one- and two-sided limits.

1. **If you aren't already on the Home screen, press [HOME].**
2. **Press [F3][3] to select the limit command from the Calc menu.**

   The command is placed on the command line followed by a left parenthesis.
3. **Enter the function whose limit you want to find.**

   I detail entering expressions in Chapter 1. You find out about creating user-defined functions in Chapter 3.

   Use [×] (the times key) when entering the product of two letters that represent different variables or constants. If you don't, the calculator will interpret the two juxtaposed letters as being a two-letter name for a single variable. For example, to the calculator, ax^2 is the square of the single variable with the two-letter name *ax*.
4. **Press [,] and enter the variable.**

   Limits are always taken as a variable goes to a number or to ±∞. This is the variable you want to enter. For example, if you are evaluating $\lim_{q \to a} P$, the variable is *q*. Entering variable names (words) is explained in Chapter 1.
5. **Press [,] and enter the limit of the variable.**

   As an example, when evaluating $\lim_{q \to a} P$, the limit of the variable is *a*. If your limit is a one-sided limit, that will be taken care of in the next step. For example, when evaluating $\lim_{x \to a^-} f(x)$, you would enter *a* in this step.

   To enter ∞, press [♦][CATALOG].
6. **If you're evaluating a two-sided limit, such as those in the first picture in Figure 12-4, skip this step. If you're evaluating a one-sided limit, press [,] and then enter 1 if the limit is from the right or enter –1 if the limit is from the left.**

   This process is illustrated on the command lines of the last two pictures in Figure 12-4.
7. **Press [)] to close the parentheses and then press [ENTER] to evaluate the limit, as illustrated in the pictures in Figure 12-4.**

Sorry folks, but when it comes to calculating limits, you can't always trust the answer the calculator gives you when the function isn't defined as a real number on both sides of the limit. For example, in the third picture in Figure 12-4, the calculator claims that the two-sided limit and both one-sided limits of $\sqrt{x-1}$ are zero as *x* approaches 1. But $\sqrt{x-1}$ isn't defined (as a real number) for *x* to the left of 1. So the two-sided limit and the limit from the left are undefined. But the calculator says they are zero. On the other hand, the calculator gives correct answers for the two-sided and one-sided limits in the second picture in this figure, in which the function is defined as a real number everywhere except at the limit.

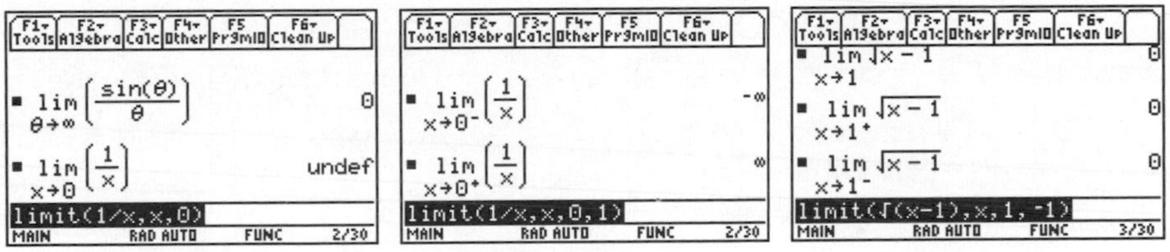

Two-sided limits One-sided limits Erroneous results

Figure 12-4: Evaluating limits.

Finding sums and products

To evaluate a finite or infinite sum $\sum_{n-a}^{b} f(n)$ or product $\prod_{n-a}^{b} f(h)$, follow these steps:

1. **If you aren't already on the Home screen, press [HOME].**
2. **Press [F3][4] to select the sum command from the Calc menu or press [F3][5] to select the product command.**

   The command is placed on the command line, followed by a left parenthesis.
3. **Enter the function whose sum or product you want to find.**

   Entering expressions is explained in Chapter 1. Creating user-defined functions is explained in Chapter 3.

   Use [×] (the times key) when entering the product of two letters that represent different variables or constants. If you don't, the calculator will interpret the two juxtaposed letters as being a two-letter name for a single variable. For example, to the calculator, ax^2 is the square of the single variable with the two-letter name *ax*.
4. **Press [,] and enter the variable.**

   In $\sum_{n-a}^{b} f(n)$ and $\prod_{n-a}^{b} f(h)$, the variable is *n*. You might want to denote this variable by *x*, *y*, *z*, or *t* because fewer keystrokes are needed to enter these letters. I explain entering variable names (words) in Chapter 1.
5. **Press [,] and enter the lower limit of the sum or product. Then press [,] and enter the upper limit.**

   For example, the lower and upper limits of the sum $\sum_{n-a}^{b} f(n)$ are *a* and *b*, respectively. To enter ∞, press [◆][CATALOG].
6. **Press [)] to close the parentheses and then press [ENTER] to evaluate the sum or product, as illustrated in Figure 12-5.**

TIP: When calculating sums or products, if the calculator returns a sum or product, as in the third picture in Figure 12-5, the calculator was unable to come to a conclusion on the convergence or divergence of the sum or product. The sum in this picture is the conditionally convergent alternating harmonic series. But the calculator was unable to compute the sum, so it returned an equivalent expression of the sum.

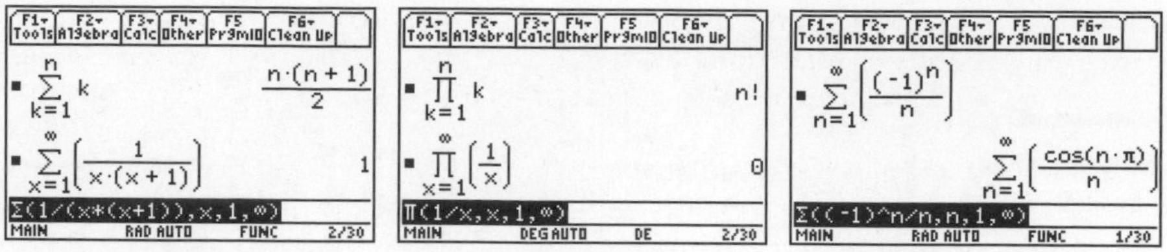

Sums Products Misleading results

Figure 12-5: Evaluating sums and products.

Finding max, min, and inflection points

The best way to find critical points (inflection points, maximum points, and minimum points) is to first graph the function, as I explain in Chapter 5, and then use the Graph Math menu to find the max, min, and inflection points, as I explain in Chapter 7.

But if you're interested only in the *x*-coordinates of the absolute maximum and minimum points, the **fMax** and **fMin** commands in the Calc menu can find these for you. These commands won't always give the *x*-coordinates of relative maximum and minimum points, as illustrated in Figure 12-6. To use the **fMax** and **fMin** commands on the Home screen, follow these steps:

1. **Press [F3][6] to find the *x*-coordinates of the minimum points or press [F3][7] to find the *x*-coordinates of the maximum points.**
2. **Enter the function and then press [,].**
3. **Enter the variable of the function.**
4. **Press [)] to close the parentheses and then press [ENTER].**

   This process is illustrated in the first picture in Figure 12-6.

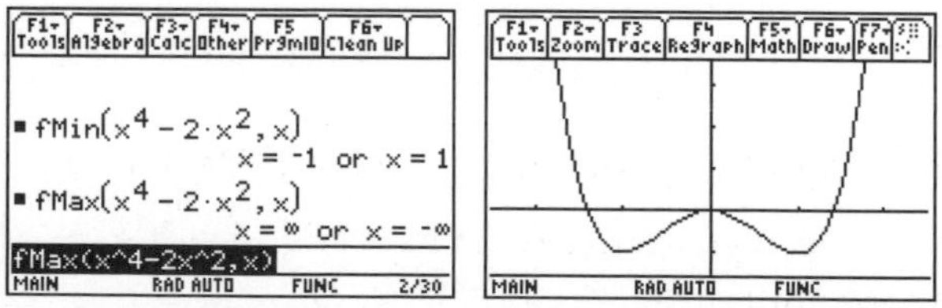

Max and Min Graph

Figure 12-6: Using the fMax and fMin commands in the Calc menu.

Finding Taylor and Maclaurin polynomials

The *n*th Taylor polynomial for *f*(*x*) centered at $x = c$ is the finite sum $P_n(x) = \sum_{k=0}^{n} \frac{f^{(k)}(c)}{k!}(x-c)^k$. When $c = 0$, the Taylor polynomial is called the

Maclaurin polynomial; and when $n = \infty$, the infinite sum is called the Taylor series or Maclaurin series for $f(x)$. The Taylor and Maclaurin polynomials of $f(x)$ are often used to approximate the value of $f(x)$ when x is near the center c.

Although the calculator isn't capable of finding a Taylor or Maclaurin series for a function $f(x)$, it is capable of finding the nth Taylor or Maclaurin polynomial provided the first n derivatives of $f(x)$ exist at $x = c$. To find a Taylor or Maclaurin polynomial, follow these steps:

1. **If you aren't already on the Home screen, press [HOME].**
2. **Press [F3][9] to select the taylor command from the Calc menu.**

   The command is placed on the command line followed by a left parenthesis.
3. **Enter the function.**

   For example, if you're finding the Taylor or Maclaurin polynomial for ln(x), enter the function ln(x), as illustrated in the first picture in Figure 12-7.

   Entering expressions is explained in Chapter 1. Creating user-defined functions is explained in Chapter 3.

   Use [×] (the times key) when entering the product of two letters that represent different variables or constants. If you don't, the calculator will interpret the two juxtaposed letters as being a two-letter name for a single variable. For example, to the calculator, ax^2 is the square of the single variable with the two-letter name ax.
4. **Press [,] and enter the name of variable used in the function you entered in the preceding step.**

   For example, the variable used in the function ln(x) is x, which you enter in the calculator by pressing [X]. You find out about entering variable names (words) in Chapter 1.
5. **Press [,] and enter the order of the Taylor or Maclaurin polynomial.**

   The nth Taylor or Maclaurin polynomial has order n. The value you enter for n must be a finite positive integer.
6. **Press [,] and enter the center of the Taylor or Maclaurin polynomial.**

   The center of the Taylor polynomial $P_n(x) = \sum_{k=\emptyset}^{n} \frac{f^{(k)}(c)}{k!}(x-c)^k$ is c. The value you enter for the center must be a real number. The center of a Maclaurin polynomial is always 0.

   If you're finding a Maclaurin polynomial, you can skip this step and go to the next step. When you make no entry for the center by skipping this step, as in the second picture in Figure 12-7, the calculator assumes that the center is 0.
7. **Press [)] to close the parentheses and then press [ENTER] to find the Taylor or Maclaurin polynomial.**

This process is illustrated in Figure 12-7, in which the first picture shows the third Taylor polynomial for ln(x) centered at x = 1 and the second picture shows the sixth Maclaurin polynomial for e^x.

If the calculator is unable to find a Taylor or Maclaurin polynomial, as in the third picture in Figure 12-7, it returns the command you entered. This usually happens if a derivative doesn't exist, the order specified in Step 5 isn't a finite positive integer, or the center specified in Step 6 isn't a real number.

If an answer scrolls off the right edge of the screen, as in the second picture in Figure 12-7, press ▲ and then repeatedly press ▶ to view the part of the answer not appearing on the screen. To again see the beginning of the answer, press [2nd]◀. When you're finished viewing the answer, press ▼ to return to the command line.

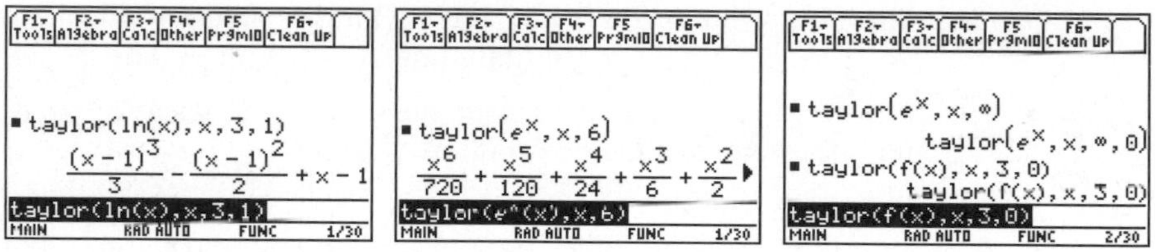

Figure 12-7: Finding Taylor and Maclaurin polynomials.

Converting Coordinates

In two dimensions, points are displayed in rectangular (x, y) or polar (r, θ) coordinates; in three dimensions, they are displayed in rectangular (x, y, z), cylindrical (r, θ, z), or spherical (ρ, θ, ϕ) coordinates, as illustrated in Figure 12-8. Chapter 11 tells you how to convert between rectangular and polar coordinates in two dimensions; this section tells you how to convert between rectangular, cylindrical, and spherical coordinates in three dimensions.

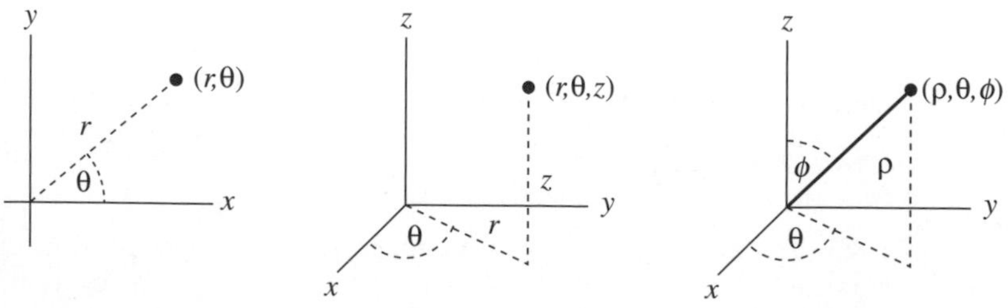

Polar coordinates   Cylindrical coordinates   Spherical coordinates

Figure 12-8: Polar, cylindrical, and spherical coordinates.

To convert between rectangular, cylindrical, and spherical coordinates, follow these steps:

1. **Enter the coordinates to be converted.**

   On the calculator, the coordinates of a point are surrounded by square brackets instead of parentheses, and the coordinates are separated by commas. Moreover, any coordinate that is an angle must be preceded by the angle symbol, ∠, and the angle must agree with the Angle mode of the calculator.

   You enter the brackets by pressing [2nd][,] for the left bracket and [2nd][÷] for the right bracket. You enter the angle symbol by pressing [2nd][EE]. ([EE] is the third key from the bottom of the left column of keys.) For example, on a calculator in Radian mode, you enter the spherical coordinates (4, Π/6, Π/4) by pressing [2nd][,][4][,][2nd][EE][2nd][^][÷][6][,][2nd][EE][2nd][^][÷][4][2nd][÷], as illustrated on the command line of the first picture in Figure 12-9.

   TIP: If you don't know the Angle mode of your calculator, look at the bottom of the screen where you see either RAD (as in the first picture in Figure 12-9) or DEG (as in the second picture) indicating that you are in Radian or Degree mode, respectively. You find out how to set the mode in Chapter 1.

   TIP: You can enter an angle in radians on a calculator that is in Degree mode if you place the radian symbol after the angle, as illustrated on the command line in the second picture in Figure 12-9. You enter the radian symbol by pressing [2nd][5][2][2]. Similarly, you can enter an angle in degrees on a calculator that is in Radian mode by pressing [2nd][|] to place the degree symbol after the angle.

2. **Press [2nd][5][4] to access the MATH Matrix menu and then press [▲][▶] to display the Vector options menu.**

3. **Press [5] if you want to convert to rectangular coordinates, press [6] to convert to cylindrical coordinates, or press [7] to convert to spherical coordinates.**

4. **Press [ENTER] to convert the coordinates.**

   If your calculator is in Radian mode, angles are displayed in radians after the angle symbol, as illustrated in the first picture in Figure 12-9. If it's in Degree mode, the angles are displayed in degrees, as in the second picture of this figure.

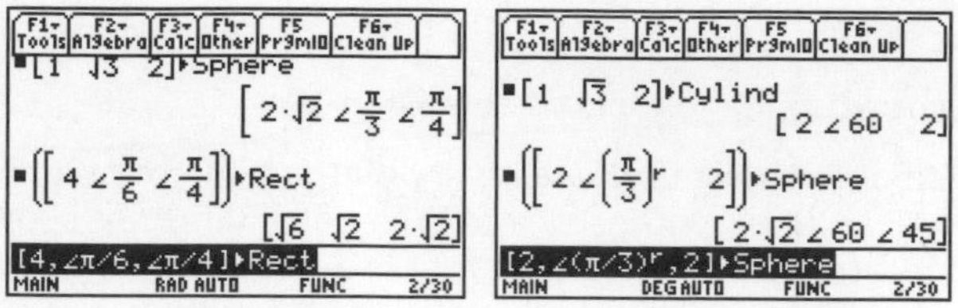

Radian mode   Degree mode

Figure 12-9: Converting between rectangular, cylindrical, and spherical coordinates.

Dealing with Hyperbolic Functions

The commands for the hyperbolic functions are housed in the MATH Hyperbolic menu, which you access by pressing [2nd][5]▲▲▲▶. This menu on a calculator with an up-to-date operating system is displayed in the first picture in Figure 12-10. If on your calculator this menu looks like the second picture in Figure 12-10, your calculator has an out-of-date operating system, and you don't have access to all the hyperbolic functions and their inverses. You can easily correct this situation by upgrading your operating system (see Chapter 21).

Figure 12-10: The MATH Hyperbolic menu and using hyperbolic functions.

You evaluate a hyperbolic function the same way you evaluate any function, as I explain in Chapter 3. For example, to find sinh(0), press [HOME] if you aren't already on the Home screen, press [2nd][5]▲▲▲▶[1] to place the **sinh** command on the command line, press [0] to enter the argument, press [)] to close the parentheses, and then press [ENTER] to evaluate the function. This process is illustrated in the first entry in the third picture in Figure 12-10.

The last two entries in the third picture in Figure 12-10 show that you can select a hyperbolic function from the MATH Hyperbolic menu and find its derivative or integrate it.

Chapter 13

Dealing with Vector Calculus

In This Chapter

- Entering and storing vectors
- Finding the length of a vector and finding unit vectors
- Evaluating dot and cross products
- Creating and recalling user-defined vector operations

The calculator can evaluate the dot or cross product of two vectors, find the length of a vector, and it can find a unit vector in the same direction as a given vector. Other than that, it doesn't have much to offer when it comes to working with vectors. But this chapter does: It tells you how to use these functions and how to work around the calculator's vector-function limitations, such as finding the angle between two vectors.

Entering, Storing, and Recalling Vectors

Textbooks are inconsistent when it comes to denoting vectors: Some use angle brackets, $<v_1, v_2>$, and others use square brackets, $[v_1, v_2]$. Your calculator uses square brackets.

Entering vectors

To enter a vector in the calculator, follow these steps:

1. **If you aren't already on the Home screen, press [HOME].**

   [HOME] is the fourth key from the top in the left column of keys.

2. **Press [2nd][,] to enter the left bracket.**
3. **Enter the first component of the vector.**
4. **Press [,] and enter the next component of the vector.**

5. **Repeat Step 4 until you've entered all components of the vector.**
6. **Press [2nd][÷] to close the brackets, as illustrated in the first two entries in the first picture in Figure 13-1.**

Figure 13-1: Entering, storing, and using vectors.

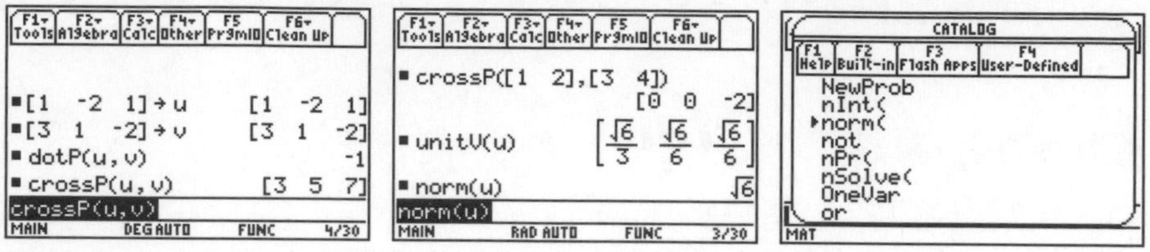

Storing vectors

If you're going to use a vector several times, consider saving (storing) it in a variable name such as *v*. Then, each time you need to use this vector, all you have to do is enter the letter *v* instead of reentering the vector.

To save (store) a vector for future use, follow these steps:

1. **Enter the vector.**

   I explain how to do this in the preceding section.

2. **Press [STO▸], the second key from the bottom in the left column of keys.**
3. **Give your vector a name.**

   To give your vector a one-letter name, press [ALPHA] and then press the key with the letter above the key. For example, if you want to call your vector *v*, as in the second entry in the first picture in Figure 13-1, press [ALPHA][0] because *v* is above the [0] key. (See Chapter 1 for an explanation on how to use the [ALPHA] and [↑] keys to give your vector a multi-character name.)

4. **Press [ENTER] to store the vector, as illustrated in the first two entries in the first picture in Figure 13-1.**

REMEMBER

Values stored under one-letter names can cause problems down the road if you forget they're there. For example, if you stored the number 3 in the letter *x*, the next time you use *x* in an expression, the calculator will replace *x* with 3. To avoid this problem, when you finish with the values stored under one-letter names, press [2nd][F1][1][ENTER] to clear the values stored in all one-letter names. If you'd like to clear all one-letter names and erase the screen, press [2nd][F1][2][ENTER].

Recalling a stored vector

After storing a vector in a one-letter name, you can insert it in an expression simply by pressing [ALPHA] and then pressing the key corresponding to that letter, as illustrated in the first two pictures in Figure 13-1.

Executing Vector Operations

This section tells you how to use the four vector operations found on the calculator: evaluating a dot product, evaluating a cross product, finding the length of a vector, and finding a unit vector in the same direction as a given vector.

Finding unit vectors and evaluating dot and cross products

The Vector operations menu housed in the MATH Matrix menu has commands for finding a unit vector in the same direction as a given vector and commands for finding the dot and cross products of vectors that have the same dimension. To use these commands, follow these steps:

1. **If you're not on the Home screen, press [HOME].**
2. **Press [2nd][5][4] to display the MATH Matrix menu and then press ⊖⊙ to display the Vector operations menu.**
3. **Press [1] to find a unit vector, press [2] to evaluate a cross product, or press [3] to evaluate a dot product.**

   The appropriate command is placed on the command line, followed by a left parenthesis.
4. **If you're finding a unit vector, enter one vector. If you're evaluating a cross or dot product, enter the first vector, press [,], and then enter the second vector.**

   I explain entering vectors earlier in this chapter. If you're evaluating a dot product, the two vectors must have the same dimension. And if you're evaluating a cross product, both vectors must be three-dimensional.
5. **Press [)] to close the parentheses and then press [ENTER] to execute the command, as illustrated in the first two pictures in Figure 13-1 (shown earlier).**

Finding the length of a vector

The **norm** command, which is tucked away in the CATALOG, is used to find the length of a vector. To use this command, follow these steps:

1. **If you're not on the Home screen, press [HOME].**
2. **Select the norm command from the CATALOG.**

   To do this, press [CATALOG] to enter the CATALOG, and then press [6] to view the items in the CATALOG beginning with the letter *N*. Press [2nd]⊝ to view the next page in the CATALOG and repeatedly press ⊝ to place the cursor to the left of the **norm** command, as illustrated in the third picture in Figure 13-1. Then press [ENTER] to place this command on the command line. The calculator supplies the left parenthesis required for this command.
3. **Enter the vector.**

   See "Entering vectors," earlier in this chapter.
4. **Press [)] to close the parentheses and then press [ENTER] to execute the command, as illustrated in the last entry in the second picture in Figure 13-1 (shown earlier).**

Creating User-Defined Vector Operations

Granted, having a calculator that can find only unit vectors, lengths of vectors, and cross and dot products of vectors doesn't give you very much power over vector operations. But if you look at the textbook formulas for the other vector operations, you'll see that the calculator has everything you need — provided you remember the formula for the vector operation you want to execute and are willing to do a little work.

But sometimes doing a little work isn't convenient when you have a big problem to solve. If this is the case, I have two solutions for you:

- **Install the Calculus Tools application on your calculator.**

  This application has many of the vector operations not found on the calculator — like grad, div, and curl. (Chapter 23 tells you how to download and install applications.)

- **Create a custom menu containing all the operations you need, as illustrated in the first two pictures in Figure 13-2.**

  Because the Calculus Tools application doesn't have every vector operation I want, I have resorted to this second solution. Of course I had to write programs for these operations, but they're very short programs, as illustrated in the third picture in Figure 13-2. Appendix A tells you how to write short programs and how to create and display custom menus. If you'd like to use my custom menu, you can download it (and directions on how to use it) from the Wiley Web site for this book.

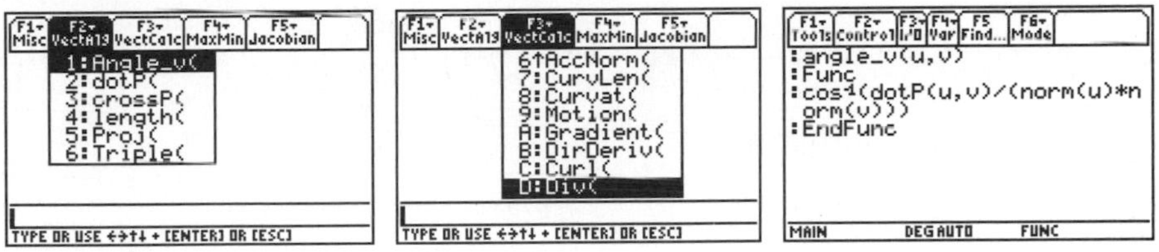

Figure 13-2: A custom menu for vector calculus.

Chapter 14

Graphing Surfaces and Contour Maps

In This Chapter

- Creating three-dimensional graphs
- Graphing level curves
- Drawing contour maps
- Rotating and animating a graph
- Graphing two-dimensional conic sections

Do you know how to get the calculator to draw three-dimensional graphs on that itsy-bitsy calculator screen so they look like the graphs you see in textbooks? And do you know how to animate a three-dimensional graph or construct a contour map? If not, you're looking in the right place because this chapter tells you how.

Graphing Functions of Two Variables

This section tells you how to get your calculator to sketch the various types of graphs of a function of two variables. Later in this chapter, I tell you how to use the calculator to rotate and animate your graph.

To create a 3D graph of a surface, a 3D graph of level curves, or a 2D contour map, follow these steps:

1. **If you haven't already set the Graph mode to 3D, press [2nd][▶][5][ENTER] to do so, as illustrated in the first picture in Figure 14-1.**

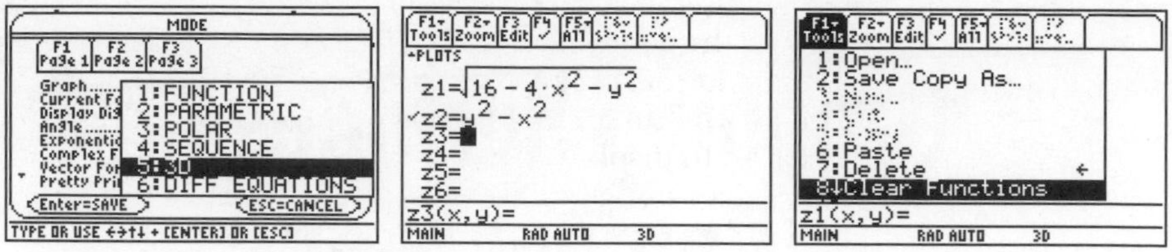

Figure 14-1: Preparing the calculator to graph a function of two variables.

2. **Press [◆][F1] to enter the Y= editor and then uncheck or clear any currently existing functions.**

   The calculator can graph only one 3D function at a time. And only the functions in the Y= editor with checkmarks to the left of their names are selected to be graphed, as illustrated in the second picture in Figure 14-1, in which **z2** will be graphed, but **z1** won't.

   To change the checked status of a function, use the ⊖⊖ keys to highlight the function and then press [F4] to toggle the checkmark between being displayed and not being displayed.

   To clear a single function from the Y= editor, use the ⊖⊖ keys to highlight the function and then press [CLEAR]. To clear all functions currently in the Y= editor, press [F1][8][ENTER], as illustrated in the third picture in Figure 14-1.

3. **Enter the function you want to graph.**

   The function must be in the form $z = f(x, y)$, where the two variables are denoted by the letters x and y, as illustrated in the second picture in Figure 14-1. You enter the letter x by pressing [X], and you enter y by pressing [Y].

   To enter or edit the definition of a function z_n, use the ⊖⊖ keys to place the cursor to the right of the equal sign and press [ENTER] or [F3] to place the cursor on the command line. Then enter or edit the definition of the function and press [ENTER] when you're finished. I explain entering and editing expressions in Chapter 1.

   Because the calculator can graph only one 3D function at a time, after completing this step, *only one* function in the Y= editor should appear with a checkmark next to it. If this isn't the case, the directions in Step 2 tell you how to rectify this situation.

4. **Press [F1]⊖[ENTER] to display the Graph Formats menu and then select the options of your choice.**

   To select options from this menu, use the ⊖⊖ keys to place the cursor on the desired format, press ⊙ to display the options for that format, and then press the number of the option you want. When you're finished setting the various formats for your graph, press [ENTER] to save your settings.

An explanation of the Graph Formats menu options follows:

- **Coordinates:** This format allows you to choose the manner in which the coordinates of the location of the cursor are displayed at the bottom of the graph screen. Select **RECT** to display the location of the cursor as (x, y, z) rectangular coordinates, or select **POLAR** to display 3D points as (r, θ, z) cylindrical coordinates. If, for some strange reason, you don't want to see the coordinates of the cursor when you trace the graph, select **OFF**.

- **Axes:** If you don't want to see the *x*-, *y*-, and *z*-axes on your graph, select **OFF**; if you do want to see them, select **AXES**. These options are illustrated in the first two pictures in Figure 14-2. If you want to see your graph housed in a rectangular box, as in the third picture in Figure 14-2, select **BOX**.

  TIP

  If you change your mind about the display of the axes after graphing the function, simply press [F1]▲[ENTER] to display the Graph Formats menu, change the Axes format, and press [ENTER]. Your graph instantly reappears with the new Axes format.

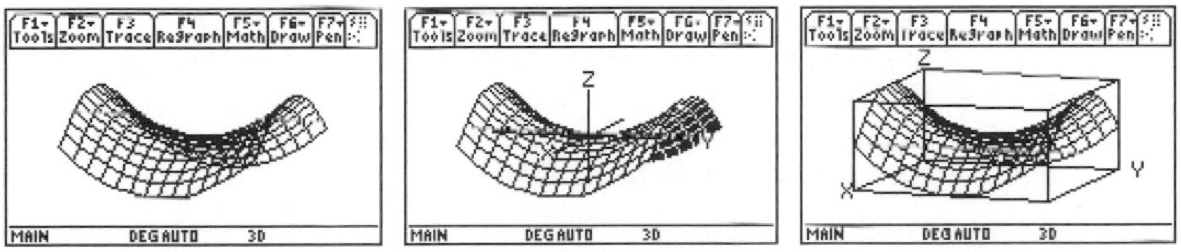

OFF AXES BOX

Figure 14-2: The Axes formats for 3D graphs.

- **Labels:** If you want the *x*-, *y*-, and *z*-axes to be labeled (as in the second two pictures in Figure 14-2), select **ON**; otherwise, select **OFF**. My recommendation is to select **ON** — otherwise you don't know which axis is which when you rotate the graph.

- **Style:** The first picture in Figure 14-2 shows the **WIRE FRAME** style of graphing, and the second picture shows the **HIDDEN SURFACE** style. A wire frame graph is a see-through graph in the sense that it displays surfaces normally hidden from view; a hidden surface graph hides these surfaces and adds a little shading. Most textbook pictures of 3D graphs are displayed in the hidden surface style.

  Select the **CONTOUR LEVELS** option if you want a 3D graph of the level curves, as in the first picture in Figure 14-3, or a 2D contour map, as in the second picture in this figure. Creating custom level curves and contour maps, in which you get to specify which curves appear on the graph, is explained later in this chapter.

  As illustrated in the third picture of Figure 14-3, the **WIRE AND CONTOUR** style of graphing is a composite of the wire frame and contour level styles.

The **IMPLICIT PLOT** style isn't used to produce 3D graphs; it's used to produce a 2D graph of an implicitly defined function of one variable. For example, you can use it to graph the ellipse implicitly defined by the equation $x^2 + 4y^2 + 4x - 8y = 8$. The sidebar in this chapter, "Graphing implicitly defined functions," tells you how to do this.

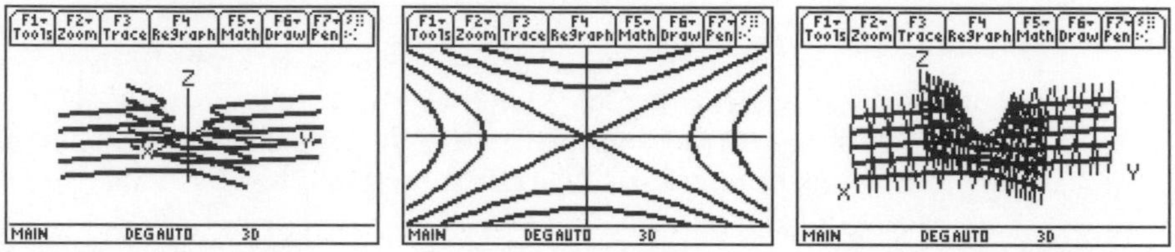

CONTOUR LEVELS CONTOUR LEVELS WIRE AND CONTOUR

Figure 14-3: Style formats for 3D graphs.

5. **If you know the viewing window you want for your graph, go to Step 6. If you haven't the slightest idea what the graph looks like, press [F2][6] (ZoomStd) to graph the function in the Standard viewing window. If you're satisfied with the graph, skip the remaining steps. If you'd like to see more of the graph on the *z*-axis, press [F2][▲][▲][▲][ENTER] (ZoomFit). If you're satisfied with the graph, skip the remaining steps. If you'd like to fine-tune the window to get a better view of your graph, the next step tells you how to do this.**

   ZoomStd graphs the function in the Standard viewing window, in which the *x*-, *y*-, and *z*-coordinates are each between –10 and 10. The first two pictures in Figure 14-5 show the settings of this window, and the first picture in Figure 14-4 displays the graph of $z = y^2 - x^2$ in this window.

   ZoomFit uses the existing limits in the Window editor for the *x*- and *y*-coordinates to determine the limits for the *z*-coordinates needed to display a full graph, as illustrated in the second picture in Figure 14-4.

   TIP: If you were expecting to see a 2D contour map after completing Step 5 but instead see a 3D graph of the level curves, go to the next step and set **eye**θ = –90, **eye**ϕ = 0, and **eye**ψ = 0.

   TIP: You can save some time when you don't know what the 3D graph of a function looks like: Set the first nine items in the Window editor to the values appearing in the first picture of Figure 14-5 and then press [F2][▲][▲][▲][ENTER] to execute the **ZoomFit** command. This procedure saves time by eliminating the need to have the calculator first graph the function in the Standard viewing window, but on the down side, it requires that you remember the first nine Window settings for the Standard viewing window. (You find out about setting the Window in the next step.)

Figure 14-4: The quick and dirty way to display a 3D graph.

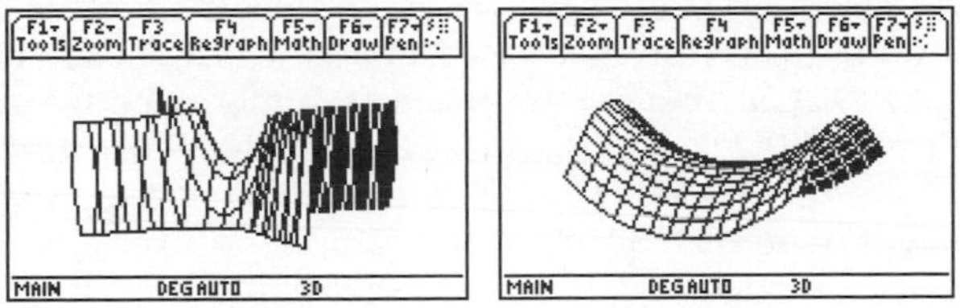

ZoomStd ZoomFit

6. **Press ◆F2 to access the Window editor. After each of the window variables, enter a numerical value that is appropriate for the function you're graphing. Press ENTER or ⊙ to move to the next variable.**

   The first two pictures in Figure 14-5 show the Window editor when the calculator is in 3D Graph mode. The items in this menu determine the viewing window for your graph. The following gives an explanation of the variables you must set in this editor:

   - **eyeθ, eyeϕ, and eyeψ:** These are the angles that orient the coordinate system so that you get the best view of your graph. If you don't know what the best orientation is (which is usually the case), set **eyeθ** = 20, **eyeϕ** = 70, and **eyeψ** = 0 if you are drawing a 3D graph. Set **eyeθ** = –90, **eyeϕ** = 0, and **eyeψ** = 0 if you're creating a 2D contour map. After you graph your function, you can manually rotate it to get a better view of the graph. I explain how to do this later in this chapter.

     The third picture in Figure 14-5 displays the location of the angles θ, ϕ, and ψ. Angle θ is an angle in the *xy*-plane measured from the positive *x*-axis, and angle ϕ is an angle measured from the positive *z*-axis. These two angles determine the line of sight from your eye to the origin of the coordinate system. Angle ψ rotates the graph around this line of sight. If you want to, you can fine-tune the angle settings. To change the value of one of these angles, enter the degree value of the angle even if your calculator is in Radian mode. *Do not* place the degree symbol after this value.

   - **xmin, xmax, ymin, and ymax:** These are, respectively, the smallest and largest values of *x* in view on the *x*-axis and the smallest and largest values of *y* in view on the *y*-axis.

Figure 14-5: Setting the viewing window.

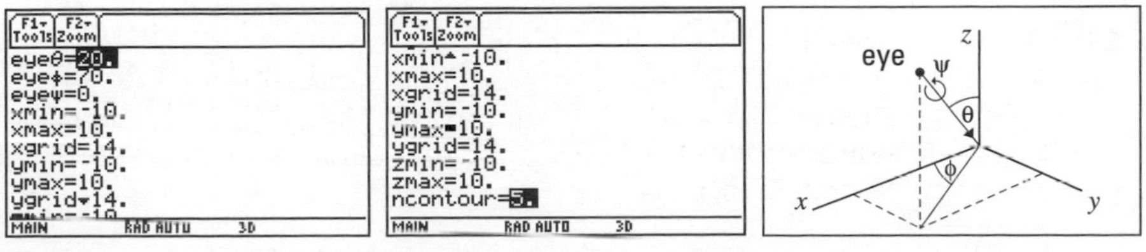

- **xgrid and ygrid:** These settings determine the number of wires in the *x*- and *y*- directions used to display the 3D graph, as illustrated in Figure 14-6. Setting each of these to numbers between 14 and 20 is recommended. Fewer than 14 wires in either direction might not give you a very good idea of the shape of the graph. And if you have a lot more than 20 wires, the calculator takes a long time to graph the function, and the resulting graph is almost solid black. The default setting is 14 for each of these, as illustrated in the second picture in Figure 14-4.

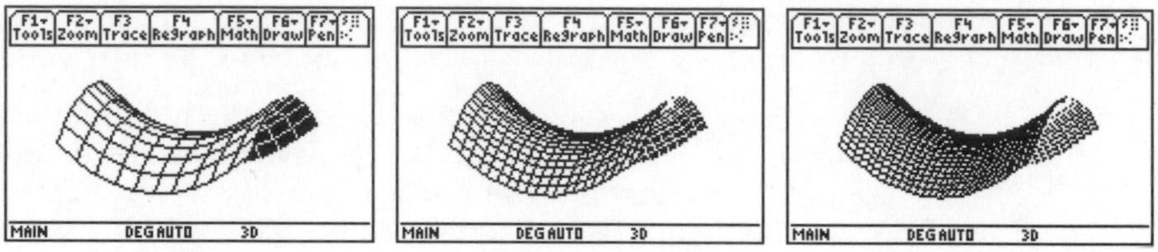

xgrid = ygrid = 10 xgrid = ygrid = 20 xgrid = ygrid = 30

Figure 14-6: Grid settings for 3D graphs.

- **zmin and zmax:** These are, respectively, the smallest and largest values of *z* in view on the *z*-axis.

  If you have assigned values to **xmin**, **xmax**, **ymin**, and **ymax**, but don't know what values to assign to **zmin** and **zmax**, press [F2][▲][▲][▲][ENTER] to invoke the **ZoomFit** command. This command uses the **xmin**, **xmax**, **ymin**, and **ymax** settings to determine the settings for **zmin** and **zmax** needed to display a full graph, and then it automatically draws the graph.

- **ncontour:** This setting determines the number of level curves that will be graphed when the graph style is set to CONTOUR LEVELS or WIRE AND CONTOUR, as I explain in Step 4. The default setting of 5 produces five level curves, as illustrated earlier in Figure 14-3. A setting between 5 and 10 is ideal. When there are fewer than 5 level curves, discerning the shape of the graph might be difficult, as in the first picture in Figure 14-7. And when there are more than 10 level curves, distinguishing between these curves might be tricky, as in the third picture in this figure.

  If you want to create custom level curves or contour maps in which you specify which curves appear on the graph, set **ncontour** to 0 so no curves appear on the graph. I explain how to then graph your custom level curves or contour map later in this chapter, in the section "Creating Custom Level Curve Graphs and Contour Maps."

7. **Press [♦][F3] to graph the function.**

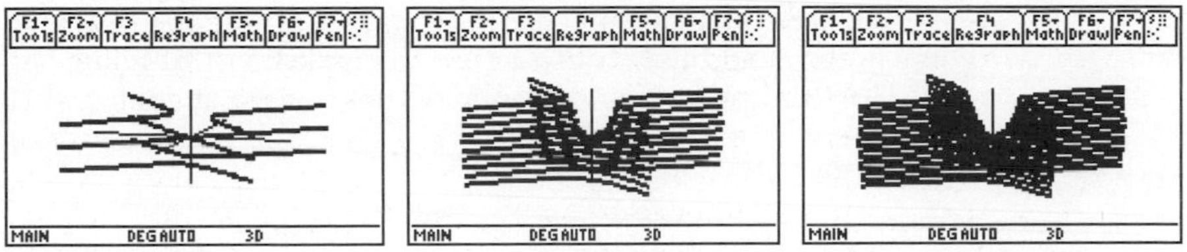

ncontour = 3 ncontour = 10 ncontour = 15

Figure 14-7: Graphing level curves.

TIP

If the calculator takes too long to draw a 3D graph and you'd like to abort the process, press [ON].

TIP

After graphing your function, you can always return to the Window editor and change any of your settings. To do this, press [♦][F2] to display the Window editor, make the changes you want, and then press [♦][F3] to redraw the graph in the new window. It might take the calculator awhile to perform some calculations before it displays the graph.

Changing the Display of a Graph

You can change the way a graph is displayed by zooming in on the graph to get a better view of a small portion of it, or by zooming out to get a global picture. You can also change the display by rotating the graph or by changing the Graph style. And you can change the way you see the graph by looking at it along a coordinate axis. This section tells you how to do these things.

Zooming in and out

After graphing a function of two variables, which I explain earlier in this chapter, you can zoom in on a portion of the graph or zoom out from the graph. This works very much like a zoom lens on a camera.

To zoom in or out while on the Graph screen, press [F2][2] to zoom in or press [F2][3] to zoom out. Then use the ◀▶▲▼ keys to move the cursor to the spot on the screen from which you want to zoom in or zoom out. Then press [ENTER]. The graph is redrawn centered at the cursor location. After zooming in or out, you might want to adjust the settings in the viewing window to get a better view of the graph. To do this, press [♦][F2] to display the Window editor, make the changes you want, and then press [♦][F3] to redraw the graph in the new window.

This process is illustrated in Figure 14-8. The first picture in this figure is the original graph, and the second picture is the result of zooming out from the origin. The third picture is the zoomed-out graph in an adjusted viewing window.

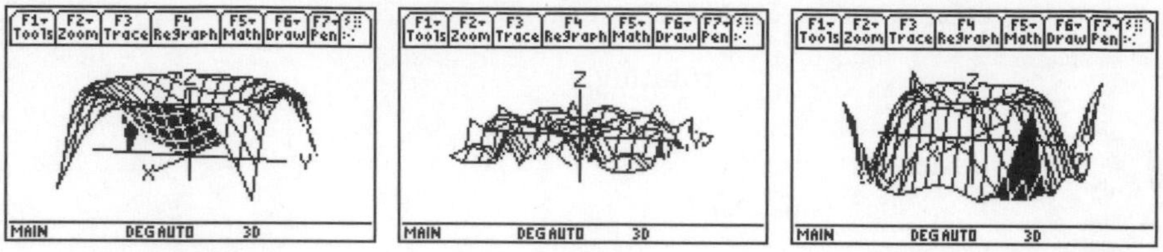

Original graph Zoom out Adjusted window

Figure 14-8: Zooming out from a 3D graph.

TIP: If you used a Zoom command to redraw a graph and then want to undo what that command did to the graph, press [F2][▲][▲][▶][1] to select **ZoomPrev** from the Zoom Memory menu. This command redraws the graph as it appeared in the previous viewing window.

TIP: A quick way to zoom in on a graph from the center of the graph is to press [×] (the times sign) to see the graph in what is called *expanded view,* as illustrated in Figure 14-9. To return to the original graph, press [×] again to see the graph in the normal view.

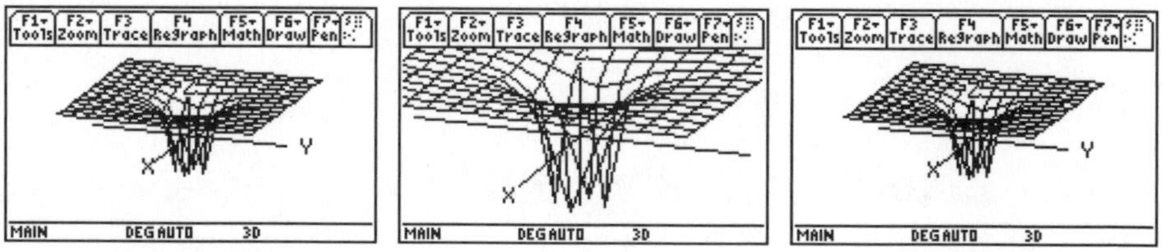

Original graph Press [×] Press [×] again

Figure 14-9: Toggling between normal and expanded view.

Rotating a 3D graph

You can use the Arrow keys to rotate a graph: [◀] and [▶] rotate the graph horizontally, to the left and to the right, respectively; [▲] and [▼] rotate the graph vertically.

REMEMBER: If you want to put a graph back in its original position after using the Arrow keys to rotate it, press [0].

Animating a 3D graph

To make your graph move continuously in one direction, press and hold the appropriate Arrow key until the graph starts moving. For example, if you press and hold ◀, the graph will continuously rotate horizontally in the clockwise direction.

When the graph is rotating, you can press and hold a different Arrow key to change the direction of rotation. To stop the animation, press [ENTER]. To return the graph to its original position, press [0].

Viewing a graph along a coordinate axis

The first picture in Figure 14-10 displays the graph of a saddle curve. The second picture shows what this saddle looks like when you view it by looking directly down the *x*-axis, and the third picture shows what it looks like when viewed along the *y*-axis.

Figure 14-10: Viewing a graph along the *x*- and *y*-axes.

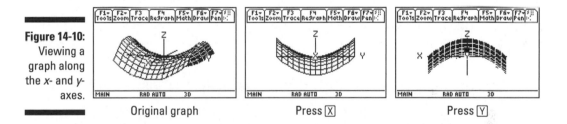

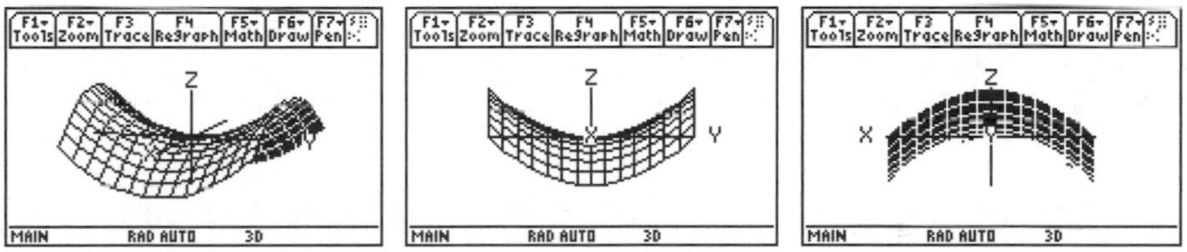

Original graph — Press [X] — Press [Y]

You can view any 3D graph along any of the three coordinate axes by pressing the key corresponding to the letter of that axis. For example, press [X] to view the graph along the *x*-axis, and then press [Y] to see the view along the *y*-axis. To return the graph to its original view, press [0].

TIP

Pressing [Z] to view a 3D graph of level curves along the *z*-axis is a quick way of converting the graph to a contour map, as illustrated in the first two pictures in Figure 14-11.

REMEMBER

After viewing a 3D graph along a coordinate axis, you can press [×] (the times sign) to enlarge the view, as illustrated in the last picture in Figure 14-11. To return the graph to the previous view along the coordinate axis, press [×] again.

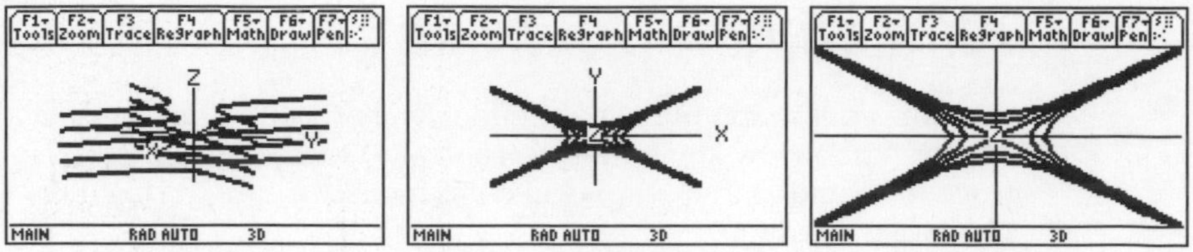

Original graph Press [Z] Press [×]

Figure 14-11: Converting a graph of level curves to a contour map.

Changing the style of a 3D graph

After creating a 3D graph (see "Graphing Functions of Two Variables" earlier in this chapter), you can easily change its graphing style by repeatedly pressing [I] to cycle the graph style through the four possible graphing styles: WIRE FRAME, HIDDEN SURFACE, CONTOUR LEVELS, and WIRE AND CONTOUR.

Figure 14-12 illustrates what the 3D graph in the first picture in Figure 14-8 looks like in different graph styles. Because the function in Figure 14-8 was initially graphed in the HIDDEN SURFACE style, pressing [I] displays the graph in the next style, which is the CONTOUR LEVELS style (the first picture in Figure 14-12). Pressing [I] again displays it in the WIRE AND CONTOUR style (the second picture in Figure 14-12), and pressing [I] a third time displays the graph in the WIRE FRAME style (the third picture). Pressing [I] a fourth time takes you back to the HIDDEN SURFACE style of the initial graph.

TIP

If you originally graphed your function in the WIRE FRAME or HIDDEN SURFACE style of graphing, the calculator takes a few moments to evaluate the contours needed to change the graph style to the CONTOUR LEVELS or WIRE AND CONTOUR style of graphing.

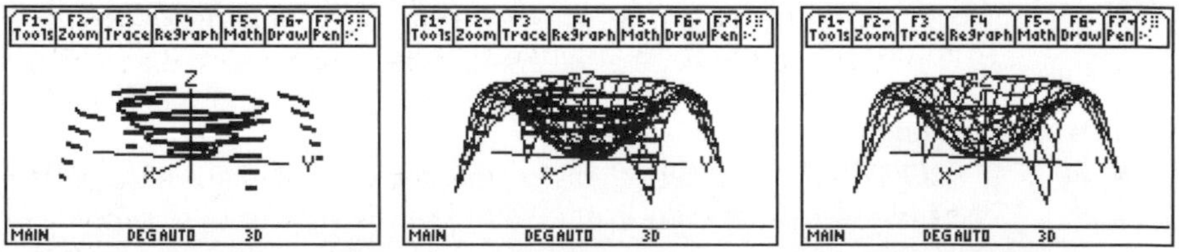

Press [I] Press [I] Press [I]

Figure 14-12: Changing the graphing style of a HIDDEN SURFACE graph.

Tracing a 3D Graph

If your 3D graph is graphed with the WIRE FRAME, HIDDEN SURFACE, or WIRE AND CONTOUR style of graphing, as I describe in the preceding section, you can press [F3] (Trace) and then use the Arrow keys to get the calculator to show the coordinates of the points on the wire frame used to display the graph. Although the calculator cannot trace the level curves in the CONTOUR LEVELS style of graphing, you can press [I] to convert such a graph to the WIRE AND CONTOUR style so the calculator can trace the wire frame.

To trace a WIRE FRAME, HIDDEN SURFACE, or WIRE AND CONTOUR style graph, press [F3] and then use the Arrow keys to move the cursor along the wire frame that constitutes the graph. Here's how the Arrow keys move:

- **◀ and ▶:** These keys move the cursor from lattice point to lattice point in the direction of the x-axis, with ▶ moving in the positive x direction and ◀ moving in the negative x direction. When the x-coordinate of the cursor location is **xmin**, pressing ◀ produces no movement in the cursor because the calculator hasn't computed the points on the graph when the x-coordinate is less than **xmin**. Similarly, when the x-coordinate of the cursor location is **xmax**, pressing ▶ results in no movement.
- **▲ or ▼:** These keys move the cursor from lattice point to lattice point in the direction of the y-axis, with ▲ moving in the positive y direction and ▼ moving in the negative y direction. When the y-coordinate of the cursor location is **ymin**, pressing ▼ produces no movement in the cursor because the calculator hasn't computed the points on the graph when the y-coordinate is less than **ymin**. In the same way, when the y-coordinate of the cursor location is **ymax**, pressing ▲ does not move the cursor.

The following list describes what you see as you trace a graph:

- **The number of the function:** The number of the function in the Y= editor that you're tracing is displayed in the top-right corner of the screen. For example, if you see the number 3, you're tracing the third graph (z_3) in the Y= editor.
- **The values of x, y, and z:** At the bottom of the screen, you see the coordinates that define the cursor location displayed in either rectangular (x, y, z) or polar (r, θ, z) form with, for example, the rectangular x-coordinate denoted by **xc** or the polar r-coordinate denoted by **rc**. If no number appears after **zc**, the function isn't defined as a real number at the values displayed for **xc** and **yc**.

Press [CLEAR] to terminate tracing the graph. This also removes the number of the function and the coordinates of the cursor from the screen.

TIP

When you're using Trace, if you want to start tracing your function at a specific value of the independent variables x and y, just key in that value for x, press [ENTER], key in that value for y, and press [ENTER]. (The value you assign to x must be between **xmin** and **xmax**, and the value you assign to y must be between **ymin** and **ymax**; if they aren't, you get an error message.) After you press [ENTER], the trace cursor moves to the point on the graph that has the x- and y-coordinates you just entered. If that point isn't on the portion of the graph appearing on the screen, press [ENTER] to center the graph at the cursor location. Then press [F3] to continue tracing the graph.

Creating Custom Level Curve Graphs and Contour Maps

A customized graph of level curves lets you specify which level curves appear in your 3D graph. And a customized contour map is simply a projection onto the *xy*-plane of a customized graph of level curves. To create either type of graph, follow these steps:

1. **Graph the function containing the level curves.**

   Graphing the function first ensures that you have an appropriate viewing window for the level curves or contour map you will graph in the following steps. For example, to produce the graphs of custom level curves and contour maps in Figures 14-13 and 14-14, I first graphed the function $z = y^2 - x^2$ as it appears in the second picture in Figure 14-4.

   The first section in this chapter explains how to graph a function of two variables.

2. **If you haven't already done so, set the Graph style to CONTOUR LEVELS.**

   To do this, press [F1][▲][ENTER] to enter the GRAPH FORMATS menu, press [▼][▼][▼][▶] to display the Style menu, press [3] to select CONTOUR LEVELS, and then press [ENTER] to save your setting.

3. **Press [◆][F2] to enter the Window editor and set eyeθ = 20 and eyeϕ = 70 if you are graphing custom 3D level curves, or set eyeθ = –90 and eyeϕ = 0 if you are graphing a custom 2D contour map. Then set eyeψ = 0 and ncontour = 0. Leave the other entries as they are.**

   eyeθ, **eyeϕ**, and **eyeψ** are the first three items in the Window editor, and **ncontour** is the last item. I explain setting the Window in Step 6 in the section "Graphing Functions of Two Variables," earlier in this chapter.

   TIP: If you want to see both the level curves graph and the contour map, set the "eye" settings for a graph of 3D level curves. In the following section, I tell you how to quickly convert between these two types of graphs.

4. **Press [◆][F3] to display the graph of the coordinate axes.**

   If you set the Window for a 3D graph, you see a screen similar to the first picture in Figure 14-13. If you set the Window for a 2D contour map, you see the *x*- and *y*-axes with *z* at the origin if labels are displayed.

   TIP: Pressing [×] toggles the size of the picture on your calculator's screen between large and small. For example, if what you see on the screen looks much larger than what you see in the first picture in Figure 14-13, press [×]. Press [×] again to return the screen to the original size.

TIP

If axis labels aren't displayed, press [F1][▲][ENTER][▼][▼][▶][2][ENTER] if you want to display them.

5. **Press [2nd][F1][8] to select the DrwCtour command from the Draw menu.**

   The calculator places the **DrwCtour** command on the command line of the Home screen, as illustrated in the second picture in Figure 14-13.

REMEMBER

   The **DrwCtour** and **Draw Contour** commands in the Graph Draw menu both draw a contour or level curve on your graph, but they're used in different situations. To draw the level curve passing through the point (x, y, z), you use the **DrwCtour** command when you know the value of the z-coordinate, and you use the **Draw Contour** command when you know the values of the x- and y-coordinates.

6. **Enter the z-value of the level curve or contour you want to graph. To graph more than one level curve or contour, enter the z-values surrounded by braces and separated by commas, as illustrated in the second picture in Figure 14-13. Then press [ENTER] to graph the level curves or contours.**

   Graphing level curves is illustrated in the third picture in Figure 14-13.

   A level curve or contour of the 3D function $z = f(x, y)$ is a 2D graph of $f(x, y) = c$ (or equivalently $z = c$) where c is a constant. When the 2D graph $f(x, y) = c$ is graphed as a 3D level curve, it is graphed in the $z = c$ plane; when it is graphed as a contour in a 2D contour map, it is graphed in the xy-plane. You enter these constants c (the z-values) in this step.

   For example, the third picture in Figure 14-13 displays the 3D level curves $z = 3$, $z = 5$, and $z = 8$ for the 3D function $z = y^2 - x^2$. The level curve $z = 3$, for example, is the 2D curve $y^2 - x^2 = 3$ graphed in the $z = 3$ plane. The third picture in Figure 14-14 displays these same curves, plus another curve that was added later, graphed in a custom contour map.

   Press [2nd][(] to enter a left brace and [2nd][)] to enter a right brace.

TIP

The section later in this chapter titled "Graphing a level curve or contour that passes through a specified point" tells you how to graph a level curve when you don't know its z-value.

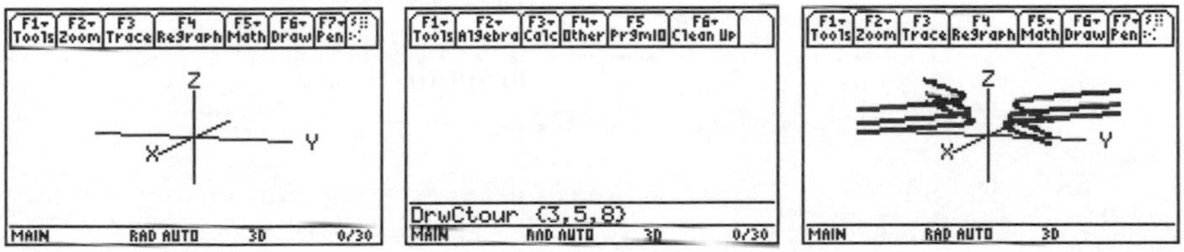

Figure 14-13: Graphing custom level curves.

Graphing implicitly defined functions

A function in one variable is implicitly defined when it is *not* written explicitly in the form $y = f(x)$. Most equations of conic sections, such as the ellipse $x^2 + 4y^2 + 4x - 8y = 8$, are implicitly defined functions. Such functions cannot be graphed in Function Graph mode, but they can be graphed in 3D Graph mode. Here's how you graph implicitly defined functions:

1. **If necessary, press [MODE](▶)[5][ENTER] to put the calculator in 3D Graph mode.**
2. **Press [♦][F1] to enter the Y= editor and then uncheck any existing functions by using the (▲)(▼) keys to highlight the function and pressing [F4].**
3. **Use the (▲)(▼) keys to highlight an undefined function and press [ENTER]. Set the implicitly defined equation equal to zero and key in the nonzero side of the equation. Then press [ENTER] to save this definition in the Y= editor.**
4. **Press [F1](▲)[ENTER] to display the Graph Formats menu and press (▼)(▼)(▼)(▶)[5] to select the Implicit Plot style. Then press [ENTER] to save your settings.**
5. **Press [♦][F2] to enter the Window editor and set xmin, xmax, ymin, and ymax to the values of your choice. (If you don't know what settings to use, press [F2][6] to use the Standard viewing window and skip the next step.) Leave the other settings as they are.**
6. **Press [♦][F3] to graph the function.**

After graphing the implicitly defined function, you can, if you so desire, press [♦][F2] and adjust the settings for **xmin**, **xmax**, **ymin**, and **ymax**. Then press [♦][F3] to re-graph the function.

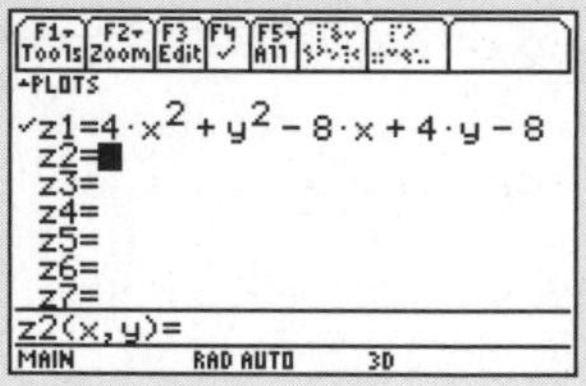

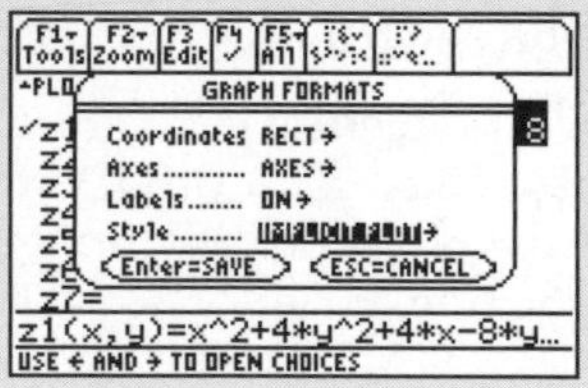

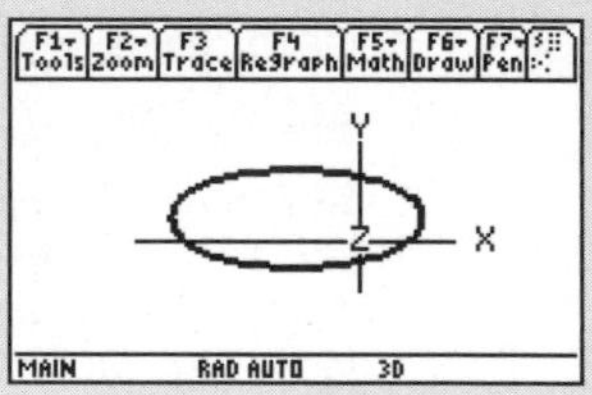

TIP: After graphing custom level curves or contours, you can add more level curves or contours to the graph by repeating Steps 5 and 6.

Converting between level curves and contour maps

To convert a 3D graph of level curves to a 2D contour map, simply press [Z] to view the graph along the *z*-axis, as illustrated in the second picture in Figure 14-11. You can then toggle the size of the screen display between small and large by pressing [×], as illustrated in the third picture in this figure. And pressing [0] returns you to the 3D graph of level curves.

To convert a 2D contour map to a 3D graph of level curves, press [◆][F2], set **eyeθ** = 20, **eyeϕ** = 70, and **eyeψ** = 0 and then press [◆][F3] to display the 3D graph of level curves.

Graphing a level curve or contour that passes through a specified point

If you don't know the *z*-value of the level curve or contour you want to graph, but you do know the *x*- and *y*-coordinates of a point on the level curve or contour, here's how you graph that level curve or contour:

1. **Follow Steps 1 through 4 in the previous section, "Creating Custom Level Curve Graphs and Contour Maps."**
2. **Press [2nd][F1][7] to select the Draw Contour command from the Draw menu.**

   The **Contour At?** instruction appears at the bottom of the screen, as illustrated in the first picture in Figure 14-14. The number in the upper-right corner of the screen is the number of the function in the Y= editor that the calculator will use to graph the level curve or contour you specify in the next step.
3. **Enter the *x*-coordinate of the point and then press [ENTER], as illustrated in the second picture in Figure 14-14.**
4. **Enter the *y*-coordinate of the point and press [ENTER].**

   The calculator, after taking a few moments to calculate the contour, displays the level curve or contour on the graph, as illustrated in the third picture in Figure 14-14, where the level curve through the point (3, 2, *z*) is added to the graph in the third picture from Figure 14-13.

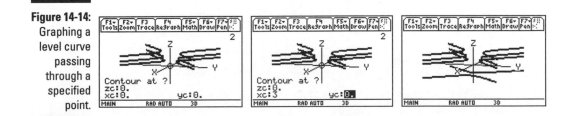

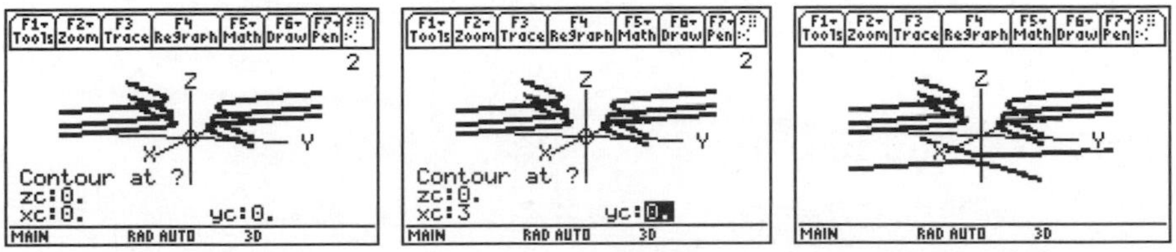

Figure 14-14: Graphing a level curve passing through a specified point.

Chapter 15

Dealing with Differential Equations

In This Chapter

- Solving ODEs and linear systems of ODEs
- Solving initial value and boundary value problems
- Graphing ODEs and systems of ODEs
- Graphing slope fields and phase portraits

If you've ever solved a linear system of ordinary differential equations (ODEs) by hand, you're going to love how quickly you can solve such a system with the calculator. And what about that painstaking task of graphing a slope field by hand — wouldn't you rather have the calculator do that for you? If so, this chapter is for you.

Solving Ordinary Differential Equations

The **deSolve** command in the Calc menu is used to solve first and second order ordinary differential equations (ODEs). It is capable of finding general solutions, particular solutions, solutions to initial value problems, and solutions to boundary value problems. In the following two sections, I tell you how to get the **deSolve** command to find these different types of solutions.

TIP

Although the **deSolve** command is capable of solving only first and second order ODEs, you aren't left high and dry when it comes to solving higher order ODEs. The sidebar in this chapter, "Dealing with ODEs of large order," tells you how to convert an *n*th order ODE to a system of *n* first order ODEs. Later in this chapter, in the section "Solving Linear Systems of ODEs," I tell you how to solve the resulting system of first order ODEs.

Finding the general solution to an ODE

The general solution to the first order ODE $\frac{dy}{dx} = 2x$ is $y = x^2 + C$, where C is an arbitrary real number. When you substitute a number for C, you get a particular solution to the ODE — for example, $y = x^2 + 2$. But in real-world applications, you usually can't find particular solutions by simply replacing C with a number; instead, you find them by specifying an initial condition. For example, one initial condition that gives the particular solution $y = x^2 + 2$ is $y(0) = 2$; another is $y(1) = 3$. When you specify an initial condition, the problem is called an *initial value problem.* An example of a first order initial value problem is $\frac{dy}{dx} = 2x,\ y(1) = 3$.

In this section, I tell you how to find the general solution to a first or second order ODE, and in the next section, I tell you how to find a particular solution by specifying the initial conditions. Here are the steps you use to find the general solution:

1. **If you aren't already on the Home screen, press [HOME].**
2. **Press [2nd][F1][1][ENTER] to clear any numbers stored in the letters *a* through *z*, or press [2nd][F1][2][ENTER] to clear these numbers and also clear the Home screen.**

   If, for example, the number 5 is stored in the letter *t* and you then use the letter *t* when defining your differential equation, the calculator will replace *t* with the number 5. Because we don't always remember what is stored in which letters, clearing the contents of these letters is always a good idea.
3. **Press [F3][▲][ENTER] to place the deSolve command on the command line.**
4. **Enter the differential equation, as illustrated in the first picture in Figure 15-1.**

   The calculator requires that derivatives be denoted by using the prime symbol ('). For example, on the calculator, $\frac{dy}{dx}$ is denoted by y'. Press [2nd][=] to enter the prime symbol used to denote the first derivative; press [2nd][=][2nd][=] to enter the double prime needed to denote the second derivative.

   REMEMBER

   Don't use juxtaposition to denote multiplication — use the [×] key. The expression xy', in which x and y' are juxtaposed, is interpreted by the calculator as the derivative of the single variable having the two-letter name xy. It also interprets $y(x + 2)$ as the function y evaluated at $x + 2$.
5. **Press [,] and enter the independent variable. Then press [,] and enter the dependent variable.**

   When, for example, the derivative is denoted by $\frac{dy}{dx}$, the independent variable is x and the dependent variable is y.

6. **Press) to close the parentheses and then press ENTER to find the general solution to the differential equation, as illustrated in the first two pictures in Figure 15-1.**

   Textbooks use C_1, C_2, C_3, and so on, to denote constants of integration; your calculator uses @1, @2, @3, and so on. Each time the calculator needs a new constant of integration, it changes the value of the integer after the @ symbol. As an example, the general solution to the differential equation in the second entry in the first picture in Figure 15-1 is displayed in a textbook as $y = C_1e^{3x} + C_2e^{x/3}$.

   If the calculator displays the general solution in implicit form, as in the first entry in the second picture in Figure 15-1, you can use the **solve** command in the Algebra menu to find, if possible, the explicit form of the general solution. When the calculator can't write the general solution in explicit form, it at least displays the solution in a more meaningful form, as in the last entry in the second picture in Figure 15-1.

   To use the **solve** command after the calculator has given the general solution in implicit form, press F2 1 to enter the **solve** command on the command line and then press 2nd (-) to tell the calculator that the equation you want to solve is the previous answer (the implicit form of the general solution). Then press , and enter the dependent variable. Finally, press) to close the parentheses, and press ENTER to solve for the dependent variable, as illustrated in the second entry in the second picture in Figure 15-1.

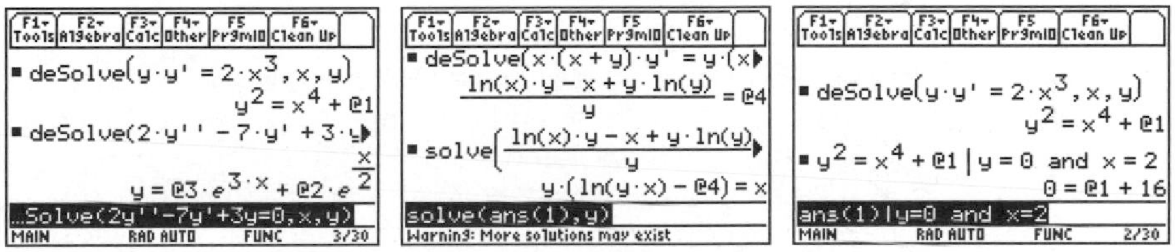

Figure 15-1: Solving first and second order differential equations.

The next section tells you how to find particular solutions to a first or second order differential equation. But with a first order equation, it is sometimes faster to use the **with** command to solve for the constant of integration and then plug that solution into the general solution. To solve for the constant of integration after finding the general solution to a first order differential equation, follow these steps:

1. **Press 2nd (-) to tell the calculator that you want to evaluate the last answer (the general solution) and then press | to enter the with command.**
2. **Key in the name of one of the variables, press =, and key in the value you want to assign to that variable.**

3. **Press [2nd][5][8][8] to enter the word *and*.**

4. **Finally, key in the name of the other variable, press [=], key in the value you want to assign to that variable, and press [ENTER] to evaluate the constant of integration, as illustrated in the third picture in Figure 15-1.**

Solving initial value and boundary value problems

The general solution of a second order ODE $y'' = f(x, y, y')$ has the form $y = C_1y_1(x) + C_2y_2(x)$, where C_1 and C_2 are real numbers and y_1 and y_2 are functions of x. Because there are two constants of integration, namely C_1 and C_2, you need to supply the calculator with two pieces of information so it can find the particular solution you want. When those two pieces of information are in the form $y(a) = b$ and $y'(a) = c$, the problem is called an *initial value* problem. But when the two pieces of information are in the form $y(a) = b$ and $y(c) = d$, it's called a *boundary value* problem. To solve an initial value or boundary value problem, follow these steps:

1. **Follow steps 1 through 4 in the preceding section to place the deSolve command on the command line and to enter the differential equation.**

   TIP

   If you have already found the general solution to the differential equation, to find a particular solution, you can simply edit the command that generated the general solution. To do this, press [CLEAR] twice to erase the command line, use the [▲] key to highlight the **deSolve** statement used to find the general solution, and then press [ENTER] to place that statement on the command line. Use the [▶] key to place the cursor after the definition of the differential equation and before the comma, follow the directions in Steps 2 and 3 in this step list, and then press [ENTER] to find the particular solution. You can safely skip Step 4.

2. **Press [2nd][5][8][8] to select the word *and* from the MATH Test menu and enter the first initial condition or boundary value.**

   When the dependent variable is denoted by y, for example, the initial condition or boundary value is entered in the form $y(a) = b$, as illustrated on the command line in the first picture in Figure 15-2.

3. **If you're finding a particular solution to a second order differential equation, press [2nd][5][8][8] to select the word *and* from the MATH Test menu and enter the second initial condition or boundary value. If you're finding a particular solution to a first order differential equation, skip this step.**

   If you're solving an initial value problem and the dependent variable is denoted by y, for example, you must enter the second initial condition in the form $y'(a) = c$, where the value a of the independent variable is the same value of the independent variable you used in Step 2. If it isn't, you get an error message. Entering this second initial condition is illustrated on the command line of the second picture of Figure 15-2.

As illustrated in the third picture of Figure 15-2, when you're solving a boundary value problem, you enter the second boundary value in the form *y*(*c*) = *d*, where the value *c* of the independent *is not* the same as the value of the independent variable you used in Step 2.

4. **Follow Steps 5 and 6 in the preceding section to first tell the calculator the names of the independent and dependent variables and to then find the particular solution to your differential equation, as illustrated in Figure 15-2.**

Figure 15-2: Finding particular solutions to first and second order differential equations.

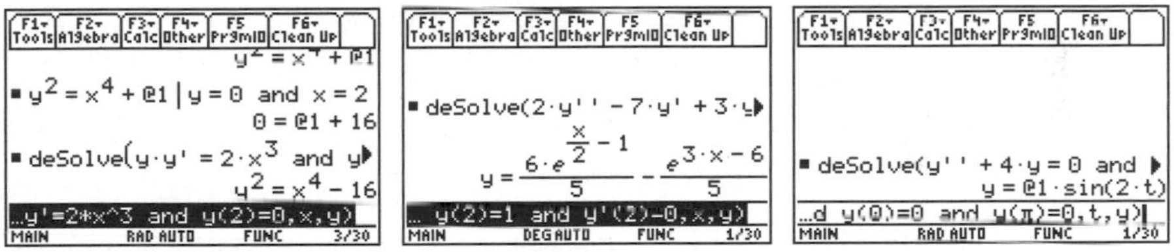

First order ODE — Initial value problem — Boundary value problem

Solving Linear Systems of ODEs

In this section, I tell you how to use matrix methods to solve a linear system of first order differential equations. But before doing this, you need to understand the terminology and notation I use — it's standard textbook terminology and notation, so if I don't go into enough detail, feel free to consult any textbook on differential equations.

The first two systems of equations displayed in Figure 15-3 are first order systems because they use only the first derivatives of the functions *x* and *y*. And they're linear because the coefficients of these functions are real numbers. The first system in this figure is called *homogeneous* because there are no nonzero constant terms in the equations in the system. The second system is *nonhomogeneous* because the equations in the system contain nontrivial constant terms. The third system of equations in this figure displays the general form of a first order, nonhomogeneous, linear system of two differential equations. This general form is expressed in matrix form by $\mathbf{x}'(t) - \mathbf{A}\mathbf{x}(t) + \mathbf{g}(t)$, where

$$x'(t) = \begin{bmatrix} x'(t) \\ y'(t) \end{bmatrix} \quad \mathbf{A} = \begin{bmatrix} a_{11} & a_{12} \\ a_{21} & a_{22} \end{bmatrix} \quad \mathbf{x}(t) = \begin{bmatrix} x(t) \\ y(t) \end{bmatrix} \quad \mathbf{g}(t) = \begin{bmatrix} g_1(t) \\ g_2(t) \end{bmatrix}$$

Figure 15-3: Examples of first order linear systems of differential equations.

Homogeneous	Nonhomogeneous	General form
$x' = 4x + 2y$ $y' = 3x - y$	$x' = 4x + 2y - 15e^{-2t}$ $y' = 3x - y - 4e^{-2t}$	$x' = a_{11}x + a_{12}y + g_1(t)$ $y' = a_{21}x + a_{22}y + g_2(t)$

Solving first order homogeneous systems

In matrix form, a first order homogeneous linear system of n differential equations is expressed as $\mathbf{x}'(t) = \mathbf{A}\mathbf{x}(t)$, where $\mathbf{A}$ is an $n \times n$ matrix having real number entries. The general solution to this system is $\mathbf{x}(t) = c_1\mathbf{x}_1(t) + c_2\mathbf{x}_2(t) + \ldots + c_n\mathbf{x}_n(t)$, where $c_1, c_2, \ldots, c_n$ are constants and $\mathbf{x}_1(t), \mathbf{x}_2(t), \ldots, \mathbf{x}_n(t)$ are n linearly independent solutions to the system.

As an example, when matrix $\mathbf{A}$ has n distinct real eigenvalues $\omega_1, \omega_2, \ldots, \omega_n$, the general solution to $\mathbf{x}'(t) = \mathbf{A}\mathbf{x}(t)$ is

$$\int \frac{x^2 + x + 1}{(x-1)^3}\,dx$$

where $\mathbf{v}_i$ is an eigenvector associated with the eigenvalue ω_i. The next section tells you how to find the eigenvalues and eigenvectors of matrix $\mathbf{A}$, and the four sections after that tell you how to use these eigenvalues and eigenvectors to find n linearly independent solutions $\mathbf{x}_i(t)$ to the homogeneous system.

Finding eigenvalues and eigenvectors

You use the eigenvalues and eigenvectors of matrix $\mathbf{A}$ to find the linearly independent solutions to the homogeneous system $\mathbf{x}'(t) = \mathbf{A}\mathbf{x}(t)$. To find the eigenvalues and eigenvectors of $\mathbf{A}$, follow these steps:

1. **Press [MODE] and set the Angle mode to Radian, the Complex Format mode to Rectangular, and the Vector Format mode to Rectangular. Then press [ENTER] to save these settings and exit the Mode menu.**

   See Chapter 1 for details on setting the mode.

2. **Enter and store the coefficient matrix A in the calculator.**

   If **A** is bigger than a 3×3 matrix, using the Data Matrix editor to store **A** in the calculator is easier. Chapter 16 tells you how to do this.

Matrices that are entered on the Home screen must be enclosed in square brackets, as illustrated on the command line of the first picture in Figure 15-4. The elements in the rows of the matrix are separated by commas, and the rows are delineated by semicolons. You enter the square brackets by pressing [2nd][,] and [2nd][÷], and you enter the semicolon by pressing [2nd][9]. After entering your matrix, press [STO▶][ALPHA], press the key that corresponds to the letter you want to use to name your matrix, and then press [ENTER] to store the matrix.

The matrix in the first picture in Figure 15-4 is the coefficient matrix of the homogeneous system of equations in Figure 15-3.

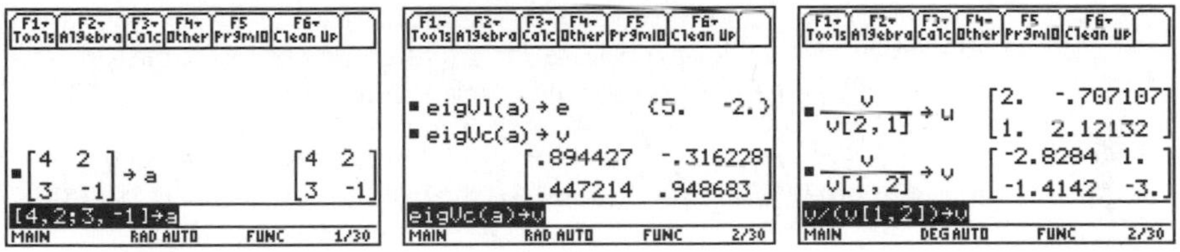

Figure 15-4: Finding the eigenvalues and eigenvectors of a matrix.

3. **If you entered A in the Data Matrix editor, press [HOME] to return to the Home screen.**
4. **Press [2nd][5][4][9] to select the eigVl (eigenvalue) command from the MATH Matrix menu.**

   The calculator places the command on the command line, followed by a left parenthesis.

   TIP: After pressing [2nd][5][4] to enter the MATH Matrix menu, you can repeatedly press ⊙ to highlight the **eigVl** command and then press [ENTER] to place it on the command line.
5. **Press [ALPHA] and then press the key that corresponds to the letter of the name you gave your matrix in Step 2. Then press [)] to close the parentheses.**
6. **Press [STO▶][ALPHA] and then press the key that corresponds to the letter you want to use to name the list of eigenvectors the calculator is about to find. (Don't use the same letter you used in Step 2.)**

   You might need to reference this list of eigenvectors later, so it's a good idea to save it in a one-letter name now.
7. **Press [ENTER] to find the eigenvalues of your matrix, as illustrated in the first entry in the second picture in Figure 15-4.**

   The eigenvalues appear enclosed in braces; unfortunately, they're separated by spaces instead of commas. The first entry in the second picture in Figure 15-4 shows that the eigenvalues of the homogeneous system in Figure 15-3 are 5 and –2.

8. **Press [2nd][5][4][ALPHA][=] to select the eigVc (eigenvector) command from the MATH Matrix menu, repeat Step 5 to enter the name of the coefficient matrix, and then press [)] to close the parentheses.**

9. **Press [STO▸][ALPHA] and then press the key that corresponds to the letter you want to use to name the eigenvector matrix. (Don't use the same letters you used in Steps 2 and 6.)**

   You might need to reference this matrix later, so it's a good idea to save it in a one-letter name now.

10. **Press [ENTER] to find the eigenvectors, as illustrated in the last entry in the second picture in Figure 15-4.**

    The eigenvectors are displayed in a matrix. The first column of this matrix is the eigenvector associated with the first eigenvalue found in Step 7. The second column is the eigenvector associated with the second eigenvalue found in Step 7, and so on.

    Eigenvectors aren't unique. In fact, any multiple of an eigenvector is also an eigenvector. The next step shows you how to use this fact to express your eigenvectors in a more meaningful form.

11. **Express the components of an eigenvector as integers (where possible) by dividing the eigenvector matrix you found in Step 10 by the component in that eigenvector with the smallest absolute value, as illustrated in the third picture in Figure 15-4. Save each result in a different one-letter name.**

    As an example, the first eigenvector in the eigenvector matrix in the second picture in Figure 15-4 is in the first column of this matrix. The entry in this column with the smallest absolute value is 0.447214. This is the number you want to divide the eigenvector matrix by in order to express the components of the first eigenvector as integers, as illustrated in the first entry in the third picture in this figure.

    You don't actually have to key in the value of the component in the eigenvector that has the smallest absolute value — you can use the format *MatrixName*[*row*, *column*] to reference it by its row and column placement in the eigenvector matrix you found in Step 10. As an example, because 0.447214 is in the 2nd row, 1st column of eigenvector matrix **v**, the command that divides this matrix by 0.447214 is **v**/**v**[2,1]. The keystrokes that enter this command and store it in the one-letter name *u* are [ALPHA][0][÷][ALPHA][0][2nd][,][2][,][1][2nd][÷][STO▸][ALPHA][+][ENTER], as illustrated in the first entry in the third picture in Figure 15-4.

    You might need to repeat this process for each eigenvector (column) in the matrix you found in Step 10. For example, the component with the smallest absolute value in the second eigenvector in matrix **v** (see the second picture of Figure 15-4) is in the 1st row, 2nd column. Executing the command **v**/**v**[1, 2] divides the matrix by this entry, resulting in a more meaningful expression of this eigenvector, as illustrated in the second entry of the third picture in this figure.

Dealing with distinct real eigenvalues

When matrix **A** has *n* distinct real eigenvalues $\omega_1, \omega_2, \ldots, \omega_n$, the general solution to $\mathbf{x}'(t) = \mathbf{A}\mathbf{x}(t)$ is

$$\int \frac{x^2+x+1}{(x-1)^3}\,dx$$

where $\mathbf{v}_i$ is an eigenvector associated with the eigenvalue ω_i. As an example, Figure 15-4 shows that the eigenvalues of the homogeneous system in Figure 15-3 are 5 and –2, and the associated eigenvectors are respectively

$$\begin{bmatrix}2\\1\end{bmatrix} \text{ and } \begin{bmatrix}1\\-3\end{bmatrix}$$

The solution to the system in matrix form is

$$x(t) = c_1\begin{bmatrix}2\\1\end{bmatrix}e^{5t} + c_2\begin{bmatrix}1\\-3\end{bmatrix}e^{-2t}$$

If you want to use the solution to solve an initial value problem, store the solution matrix for the homogeneous system in the calculator. And if you want to use the solution to the homogeneous system to solve a nonhomogeneous system, store the fundamental matrix for the homogeneous system in the calculator. (The fundamental matrix is the matrix that houses the linearly independent solutions to the homogeneous system in its columns.) In the sections "Solving a first order nonhomogeneous linear system" and "Solving initial value problems" later in this chapter, I explain how to use the solution matrix and the fundamental matrix to solve initial value problems and nonhomogeneous systems. To store the solution matrix and the fundamental matrix in the calculator, follow these steps:

1. **If you haven't already done so, find the eigenvalues and eigenvectors of the coefficient matrix of the homogeneous system, as I explain in the preceding section.**
2. **Store the linearly independent solutions to the system in different one-letter names, as illustrated in the first picture in Figure 15-5.**

   When the eigenvalues are real and distinct, the linearly independent solutions to the system have the form $x_i(t) = v_i e^{\omega_i t}$, where $\mathbf{v}_i$ is the eigenvector associated with the eigenvalue ω_i. The associated eigenvector is in the same column of the eigenvector matrix as the real eigenvalue in the list of eigenvalues. For example, if the eigenvalue is the first eigenvalue in the list of eigenvalues, then the associated eigenvector is in the first column of the eigenvector matrix.

To enter and store one linearly independent solution in a one-letter name, follow these steps:

a. Enter the eigenvector.

You have two ways to enter the eigenvector: You can enter it as a column vector or you can reference its column in the eigenvector matrix. If the eigenvector matrix is larger than a 2 × 2 matrix and if you stored this matrix in a one-letter name, it takes fewer keystrokes to reference its column in the eigenvector matrix than it does to enter it as a column vector.

If you enter the eigenvector as a column vector, it must be enclosed in square brackets, as illustrated at the beginning of the command line in the first picture in Figure 15-5. The rows of the column vector are separated by semicolons. You enter the square brackets by pressing [2nd][,] and [2nd][÷], and you enter the semicolon by pressing [2nd][9].

Entering the eigenvector by referencing its column in the eigenvector matrix is a bit more complicated because the calculator allows you to reference a row of a matrix, but not a column. This obstacle is easily overcome by first transposing the matrix, referencing the row, and then transposing the result so you have a column matrix. To do this, press [(][ALPHA], press the key that corresponds to the letter of the name you gave this matrix, and press [2nd][5][4][1] to transpose the matrix. Then press [2nd][,] to enter a left bracket, enter the number of the column in which the eigenvector appears in the eigenvector matrix, and press [2nd][÷] to close the brackets. Finally, press [)] to close the parentheses and then press [2nd][5][4][1] to transpose the row matrix into a column matrix, as illustrated at the beginning of the command line in the first picture in Figure 15-7 (shown later).

TIP

If you frequently use the calculator to solve systems of more than two differential equations, consider putting the **MatCol** function in the Main folder of your calculator. This function allows you to reference a column of a matrix by simply specifying the name of the matrix and the number of the column. I explain how to create and use this function in Appendix A.

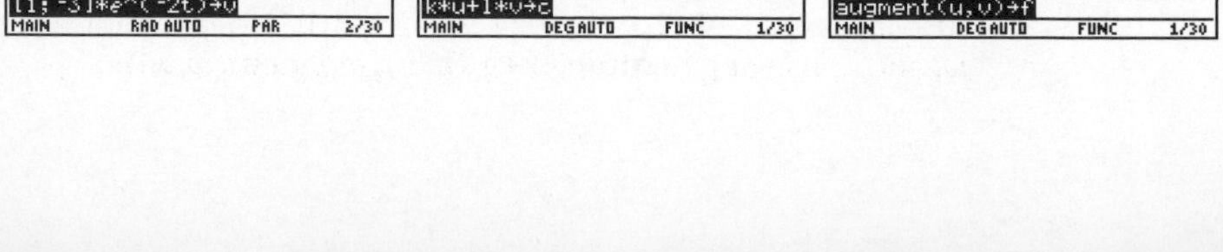

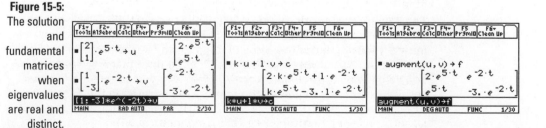

Figure 15-5: The solution and fundamental matrices when eigenvalues are real and distinct.

b. **Press [×][◆][X] to multiply the eigenvector by *e* raised to a power and then enter the eigenvalue associated with the eigenvector you entered in the preceding step.**

c. **Press [×] and enter the independent variable. Then press [)] to close the parentheses.**

Although the independent variable in all examples in this chapter is *t*, you may have denoted it by some other letter.

d. **Press [STO▸][ALPHA] and press the key that corresponds to the letter of the name you want to give this solution and then press [ENTER], as illustrated in the first picture in Figure 15-5.**

3. **Construct the solution matrix and store it in a one-letter name, as illustrated in the second picture in Figure 15-5 and in the first picture in Figure 15-8.**

The solution matrix for a homogeneous linear system of *n* differential equations has the form $\mathbf{x}(t) = c_1\mathbf{x}_1 + c_2\mathbf{x}_2 + \ldots + c_n\mathbf{x}_n$ where $\mathbf{x}_1, \mathbf{x}_2, \ldots, \mathbf{x}_n$ are the linearly independent solutions you entered in the last step, and $c_1, c_2, \ldots, c_n$ are real variables. Enter the right side of this equation by using different one-letter names for the real variables. Don't use the same names you used in Step 2 to name the linearly independent solutions. After entering a real variable, press [×] and then enter the name of the linearly independent solution.

After entering the solution matrix, press [STO▸][ALPHA] and then press the key that corresponds to the name you want to give this matrix. Don't use the same name you used to define any part of the solution matrix. Then press [ENTER] to store the solution matrix, as illustrated in the second picture in Figure 15-5.

4. **Construct the fundamental matrix and save it in a one-letter name, as illustrated in the third picture in Figure 15-5 and in the second picture in Figure 15-8.**

The fundamental matrix for a homogeneous system of differential equations is the matrix that houses the linearly independent solutions to the system in its columns. This matrix is constructed by augmenting the linearly independent solutions you constructed in Step 2 into one matrix. But they have to be augmented two at a time, as illustrated in the third picture in Figure 15-5 and in the second picture in Figure 15-8.

To augment two matrices, press [2nd][5][4][7] to select the **augment** command from the MATH Matrix menu. Then enter the name of the first matrix, press [,], enter the name of the second matrix, and then press [)] to close the parentheses, as illustrated in the third picture in Figure 15-5. You can augment three matrices the same way you augment two matrices: by nesting the **augment** command, as illustrated in the second picture in Figure 15-8. To augment four linearly independent solutions, *u*, *v*, *w*, and *x*, use **augment(augment(***u*, *v*), **augment(***w*, *x*)).

After constructing the fundamental matrix, press STO▶ ALPHA and then press the key that corresponds to the letter of the name you want to give this matrix — use a letter that you have *not* used in the preceding steps. Then press ENTER to store the fundamental matrix.

Dealing with repeated eigenvalues and distinct eigenvectors

The first picture in Figure 15-6 shows a coefficient matrix **A** with a repeated eigenvalue, namely 2. But as the second picture in this figure shows, **A** has a complete set of distinct eigenvectors. Because these eigenvectors are linearly independent, the solution to the homogeneous system $\mathbf{x}'(t) = \mathbf{A}\mathbf{x}(t)$ in this figure is $\mathbf{x}(t) = c_1\mathbf{v}_1e^{9t} + c_2\mathbf{v}_2e^{2t} + c_3\mathbf{v}_3e^{2t}$ where $\mathbf{v}_1$ is the eigenvector in the first column of the matrix in the second picture in this figure, and $\mathbf{v}_2$ and $\mathbf{v}_3$ are respectively the eigenvalues in the second and third columns in the matrix in the third picture in this figure. (You discover how to find eigenvalues and eigenvectors in the earlier section, "Finding eigenvalues and eigenvectors.")

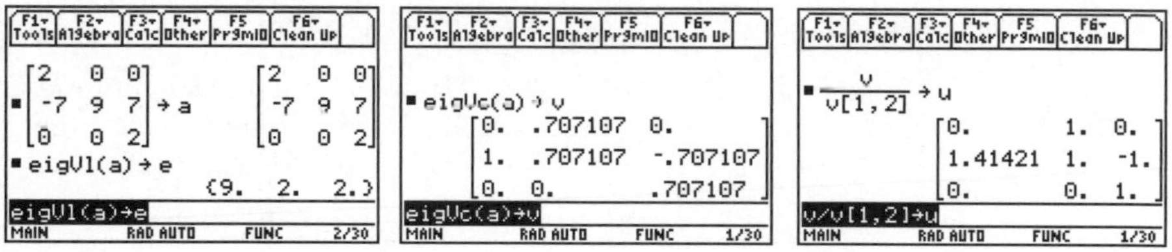

Figure 15-6: A repeated eigenvalue with a complete set of eigenvectors.

If you want to use the solution to solve an initial value problem, store the solution matrix for the homogeneous system in the calculator. And if you want to use the solution to the homogeneous system to solve a nonhomogeneous system, store the fundamental matrix for the homogeneous system in the calculator. (The fundamental matrix is the matrix that houses the linearly independent solutions to the homogeneous system in its columns.) Later in this chapter, I explain how to use the solution matrix and the fundamental matrix to solve initial value problems and nonhomogeneous systems. To store the solution matrix and the fundamental matrix in the calculator, follow the steps in the preceding section. The process of following these steps using the homogeneous system in Figure 15-6 is illustrated in Figures 15-7 and 15-8.

Dealing with repeated eigenvalues and repeated eigenvectors

The first picture in Figure 15-9 shows a coefficient matrix **A** with a repeated eigenvalue, namely 4. But as the second picture in this figure shows, **A** also has a repeated eigenvector. (For more information on finding eigenvalues and eigenvectors, see the earlier section "Finding eigenvalues and eigenvectors.")

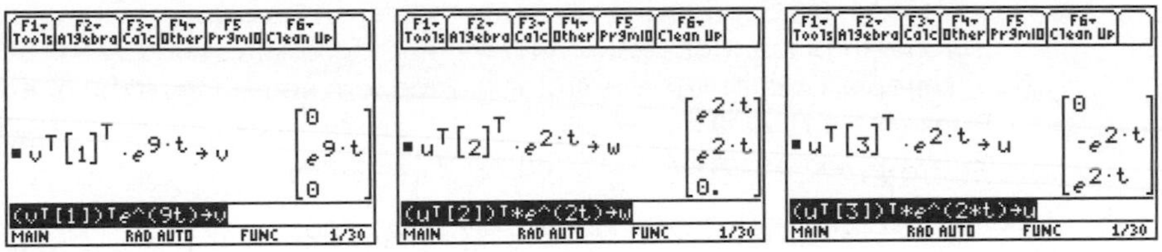

Figure 15-7: Solutions for repeated eigenvalues and distinct eigenvectors.

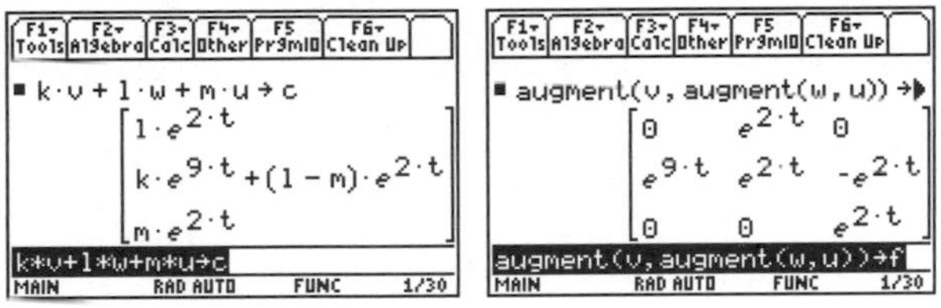

Figure 15-8: The solution matrix and fundamental matrix for repeated eigenvalues and distinct eigenvectors.

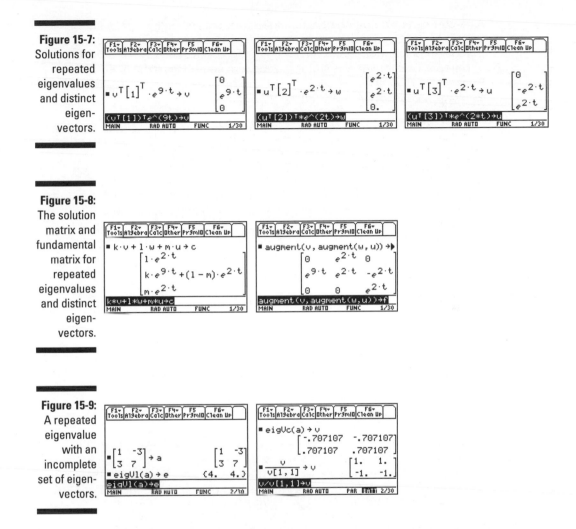

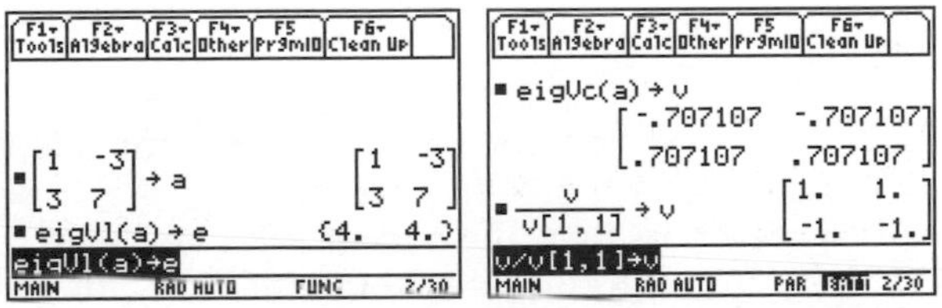

Figure 15-9: A repeated eigenvalue with an incomplete set of eigenvectors.

Finding the n linearly independent solutions needed to solve a homogeneous system of n differential equations gets quite complicated when you have both repeated eigenvalues and repeated eigenvectors. In this section, I tell you how to use the calculator to find two linearly independent solutions when the eigenvalue and associated eigenvector are both repeated twice. For the more general case (when eigenvectors are repeated more than twice), consult a textbook on differential equations. Although I don't tell you the mathematics needed to solve the general case, the way you use the calculator to solve the general case is quite similar to the way you use it to solve the case I do tell you about.

As any textbook on differential equation tells you, when an eigenvalue ω and its associated eigenvector are each repeated twice, the two linearly independent solutions generated by that eigenvalue are $x_1(t) = ue^{\omega t}$ and $x_2(t) = (ut + v)e^{\omega t}$ where **u** and **v** are nonzero solutions to the matrix equations $(\mathbf{A} - \omega\mathbf{I})^2\mathbf{v} = \mathbf{0}$ and $(\mathbf{A} - \omega\mathbf{I})\mathbf{v} = \mathbf{u}$. To use the calculator to find the solutions for the vectors **u** and **v**, follow these steps:

1. **Store the matrix A – ωI in a one-letter name, as illustrated in the first entry in the first picture in Figure 15-10.**

   A is the coefficient matrix of the homogeneous linear system of differential equations, and ω is the repeated eigenvalue. You enter the identity matrix **I** by pressing [2nd][5][4][6], then entering the number of rows in the identity matrix, and then pressing [)] to close the parentheses.

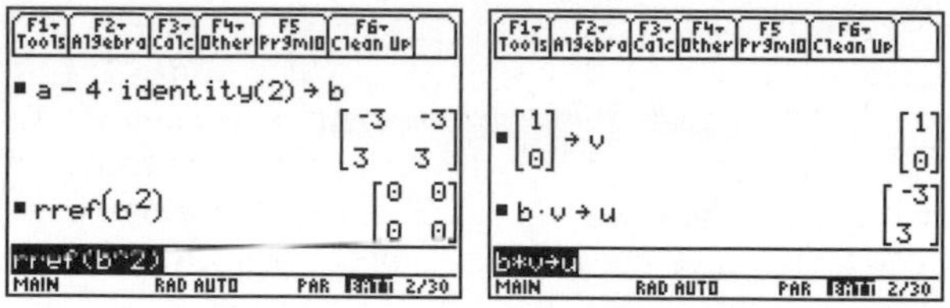

Figure 15-10: Solving a system with an incomplete set of eigenvectors.

2. **Press [2nd][5][4][4] to select the rref (reduced row-echelon form) command from the MATH Matrix menu.**

3. **Press [ALPHA] and press the key that corresponds to the letter of the name you gave matrix A – ωI in Step 1. Then press [^][2] to square that matrix. Finally, press [)] to close the parentheses and then press [ENTER] to express $(\mathbf{A} - \omega\mathbf{I})^2$ in reduced row-echelon form, as illustrated in the second entry in the first picture in Figure 15-10.**

4. **Find, by hand, a nonzero solution to $(\mathbf{A} - \omega\mathbf{I})^2\mathbf{v} = \mathbf{0}$. Enter this solution as a column vector and store it in a one-letter name, as illustrated in the first entry in the second picture in Figure 15-10.**

   $(\mathbf{A} - \omega\mathbf{I})^2\mathbf{v} = \mathbf{0}$ never has a unique solution, so the calculator cannot find a nonzero solution for you. But finding that solution by hand is so easy you rarely even have to reach for a pencil. Here's why: $(\mathbf{A} - \omega\mathbf{I})^2\mathbf{v} = \mathbf{0}$ is a homogeneous system of equations stated in matrix form, and $\mathbf{rref}(\mathbf{A} - \omega\mathbf{I})^2\mathbf{v} = \mathbf{0}$ is the equivalent reduced row-echelon form of this system. So finding a nonzero solution to $(\mathbf{A} - \omega\mathbf{I})^2\mathbf{v} = \mathbf{0}$ is equivalent to finding a nonzero solution to $\mathbf{Mv} = \mathbf{0}$, where **M** is the matrix you found in Step 3. You can usually just look at the matrix to find a nonzero solution.

For example, in the second entry in the first picture in Figure 15-10, **rref** $(\mathbf{A} - \omega\mathbf{I})^2$ is the zero matrix. So any value of **v** is a solution to $(\mathbf{A} - \omega\mathbf{I})^2\mathbf{v} = \mathbf{0}$. In particular, $\mathbf{v} = [1\ 0]^T$ is a nonzero solution. Other nonzero solutions are $\mathbf{v} = [0\ 1]^T$, $\mathbf{v} = [1\ 1]^T$, and so on. But all you need is one nonzero solution. ($[a\ b]^T$ denotes the transpose of the vector $[a\ b]$. For example, $[1\ 0]^T = \begin{bmatrix}1\\0\end{bmatrix}$.)

As a nontrivial example, if **rref**$(\mathbf{A} - \omega\mathbf{I})^2$, the matrix you found in Step 3, is $\begin{bmatrix}1 & 2\\0 & 0\end{bmatrix}$ and $\mathbf{v} = \begin{bmatrix}a\\b\end{bmatrix}$, then finding a nonzero solution to $(\mathbf{A} - \omega\mathbf{I})^2\mathbf{v} = \mathbf{0}$ is equivalent to finding a nontrivial solution to $\begin{bmatrix}1 & 2\\0 & 0\end{bmatrix}\begin{bmatrix}a\\b\end{bmatrix} = \begin{bmatrix}a+2b\\0\end{bmatrix} = \begin{bmatrix}0\\0\end{bmatrix}$. You can pick any nonzero values for a and b that satisfy the equation $a + 2b = 0$. For example, if you pick $a = 2$ and $b = -1$, then $\mathbf{v} = [2\ {-1}]^T$ is a nonzero solution to $(\mathbf{A} - \omega\mathbf{I})^2\mathbf{v} = \mathbf{0}$.

To enter and store your solution for **v**, first enter **v** as a column vector. (A column vector must be enclosed in square brackets. The rows of the column vector are separated by semicolons. You enter the square brackets by pressing [2nd][,] and [2nd][÷], and you enter the semicolon by pressing [2nd][9].) Then press [STO▶][ALPHA] and press the key that corresponds to the letter of the name you want to give this vector. Then press [ENTER] to store the vector, as illustrated in the first entry in the second picture in Figure 15-10.

5. **To find vector u, first press [ALPHA] and then press the key that corresponds to the letter of the name you gave matrix A – ωI in Step 1. Then press [×][ALPHA] and press the key that corresponds to the one-letter name you gave the vector you found in Step 4. Finally, press [STO▶][ALPHA], press the key that corresponds to the letter of the name you want to give this vector, and then press [ENTER] to display the vector, as illustrated in the second entry in the second picture in Figure 15-10 (shown earlier).**

The vectors found in the second picture in Figure 15-10 show that the general solution to the homogeneous system of differential equations that have the coefficient matrix in the first picture of Figure 15-9 is

$$\mathbf{x}(t) = c_1\begin{bmatrix}-3\\3\end{bmatrix}e^{4t} + c_2\left(\begin{bmatrix}-3\\3\end{bmatrix}t + \begin{bmatrix}1\\0\end{bmatrix}\right)e^{4t}.$$

If you want to use the solution to solve an initial value problem, store the solution matrix for the homogeneous system in the calculator. And if you want to use the solution to the homogeneous system to solve a nonhomogeneous system, store the fundamental matrix for the homogeneous system in the calculator. (The fundamental matrix is the matrix that houses the linearly independent solutions to the homogeneous system in its columns.) Later in this chapter, I explain how to use the solution matrix and the fundamental matrix to solve initial value problems and nonhomogeneous systems.

To store the solution matrix and the fundamental matrix in the calculator, follow these steps:

1. **If you haven't already done so, follow the steps at the beginning of this section to find vectors v and u.**

   When an eigenvalue ω and its associated eigenvector are each repeated two times, the two linearly independent solutions generated by that eigenvalue are $\mathbf{x}_1(t) = \mathbf{u}e^{\omega t}$ and $x_2(t) = (\mathbf{u}t + v)e^{\omega t}$, where **u** and **v** are nonzero solutions to the matrix equations $(\mathbf{A} - \omega\mathbf{I})^2\mathbf{v} = \mathbf{0}$ and $(\mathbf{A} - \omega\mathbf{I})\mathbf{v} = \mathbf{u}$. These are the vectors you should have found already. You might have given them different names, but in the following steps I call them **v** and **u**.

2. **Store the solution $x_2(t) = (\mathbf{u}t + v)e^{\omega t}$ in a one-letter name, as illustrated in the first picture in Figure 15-11.**

   To do this, press [(], enter the name you gave to vector **u**, press [×], and enter the independent variable. (I denote the independent variable by *t*, you may have used a different letter.) Then press [+], enter the name you gave to vector **v**, and press [)] to close the parentheses. Then press [×][◆][X], enter the eigenvalue, press [×], and enter the independent variable. Finally, press [STO▸][ALPHA], enter the name you want to give this solution, and then press [ENTER] to store the solution. Because you don't need vector **v** anymore, you can give this solution the same name you gave vector **v**.

3. **Store the solution $\mathbf{x}_1(t) = \mathbf{u}e^{\omega t}$ in a one-letter name, as illustrated in the first entry in the second picture in Figure 15-11.**

   To do this, press [ALPHA] and press the key that corresponds to the name you gave vector **u**. Then press [×][◆][X], enter the eigenvalue, press [×], and enter the independent variable. Finally, press [STO▸][ALPHA], enter the name you want to give this solution, and then press [ENTER] to store the solution in the calculator. Because you don't need vector **u** anymore, you can give this solution the same name you gave vector **u**.

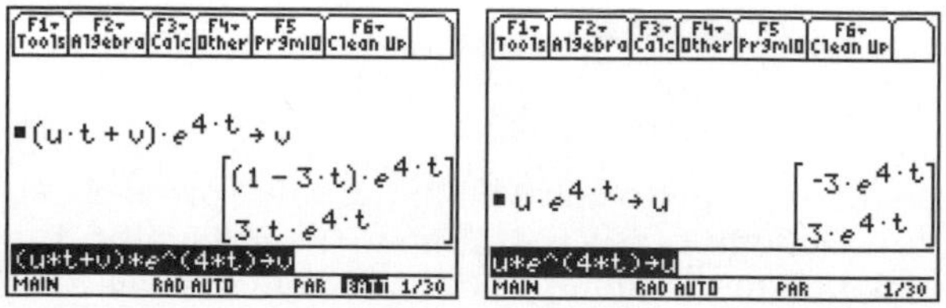

Figure 15-11: Linearly independent solutions when eigenvectors are repeated.

4. **To construct the solution matrix and the fundamental matrix, follow Steps 3 and 4 in the section titled "Dealing with distinct real eigenvalues," earlier in this chapter.**

   The first picture in Figure 15-12 shows the solution matrix for the system of differential equations with the coefficient matrix in the first picture in Figure 15-9 (shown earlier), and the second picture in Figure 15-12 displays the fundamental matrix for this system.

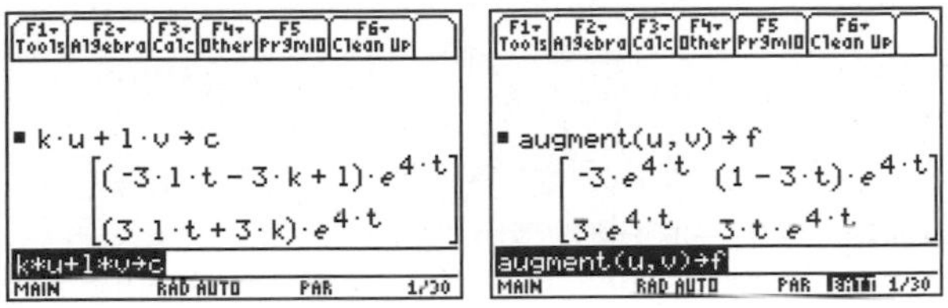

Figure 15-12: The solution and fundamental matrices when eigenvectors are repeated.

Dealing with complex eigenvalues and eigenvectors

Complex eigenvalues of the coefficient matrix **A** of a linear system of differential equations always come in complex conjugate pairs, as illustrated in the first picture in Figure 15-13. Their associated eigenvectors are also conjugates, as illustrated in the second picture in this figure. (I explain finding eigenvalues and eigenvectors earlier in this chapter.)

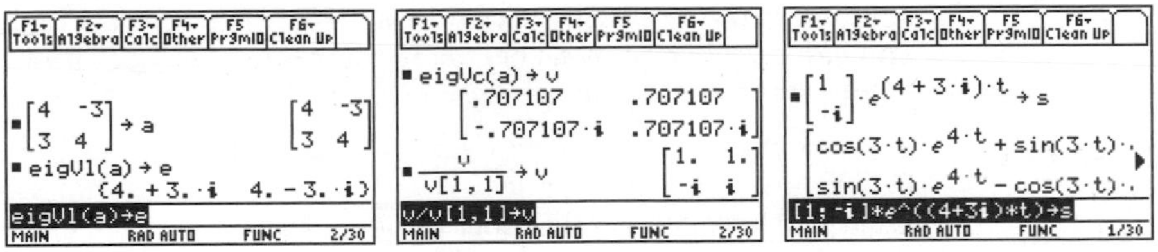

Figure 15-13: A matrix with complex eigenvalues and eigenvectors.

As any textbook on differential equations tells you, when the coefficient matrix **A** of a homogeneous linear system of differential equations has a complex eigenvalue ω with associated eigenvector **v**, the real and imaginary parts of the complex solution to $\mathbf{x}(t) = ve^{\omega t}$ are linearly independent, real solutions to the homogeneous linear system of differential equations. Because the complex solution to $\mathbf{x}(t) = \overline{\mathbf{v}}e^{\varpi t}$ is the conjugate of the solution to $\mathbf{x}(t) = ve^{\varpi t}$, they both have the same real and imaginary parts. To get the calculator to find the real and imaginary parts of $\mathbf{x}(t) = ve^{\varpi t}$, follow these steps:

1. **If you found the complex eigenvalues and eigenvectors by following the steps in the earlier section, "Finding eigenvalues and eigenvectors," go to Step 2. If not, follow Steps 1 and 11 in "Finding eigenvalues and eigenvectors."**

   Step 1 in that section tells you how to set the Mode to get the calculator to properly display the solutions you're about to find, and Step 11 tells you how to display and store the eigenvectors in a simplified form, as illustrated in the last entry in the second picture in Figure 15-13.

2. **Enter the eigenvector associated with the complex eigenvalue.**

   The associated eigenvector is in the same column of the eigenvector matrix as the complex eigenvalue in the list of eigenvalues. For example, if the complex eigenvalue is the first eigenvalue in the list of eigenvalues, then the associated eigenvector is in the first column of the eigenvector matrix.

   There are two ways to enter this eigenvector: You can enter it as a column vector, or you can reference its column in the eigenvector matrix. If the eigenvector matrix is larger than a 2×2 matrix and if you stored this matrix in a one-letter name, it takes fewer keystrokes to reference its column in the eigenvector matrix than it does to enter it as a column vector.

   If you enter the eigenvector as a column vector, you must enclose it in square brackets, as illustrated at the beginning of the command line in the third picture in Figure 15-13. The rows of the column vector are separated by semicolons. You enter the square brackets by pressing [2nd][,] and [2nd][÷], you enter the semicolon by pressing [2nd][9], and you enter the complex number *i* by pressing [2nd][CATALOG].

   Entering the eigenvector by referencing its column in the eigenvector matrix is a bit more complicated because the calculator allows you to reference a row of a matrix, but not a column. You can easily overcome this obstacle by first transposing the matrix, referencing the row, and then transposing the result so you have a column matrix. To do this, press [(][ALPHA], press the key that corresponds to the letter of the name you gave this matrix, and press [2nd][5][4][1] to transpose the matrix. Then press [2nd][,] to enter a left bracket, enter the number of the column in which the eigenvector appears in the eigenvector matrix, and press [2nd][÷] to close the brackets. Finally, press [)] to close the parentheses and then press [2nd][5][4][1] to transpose the row matrix into a column matrix, as illustrated at the beginning of the command line in the first picture in Figure 15-14.

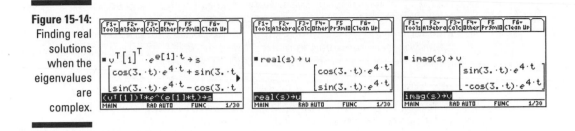

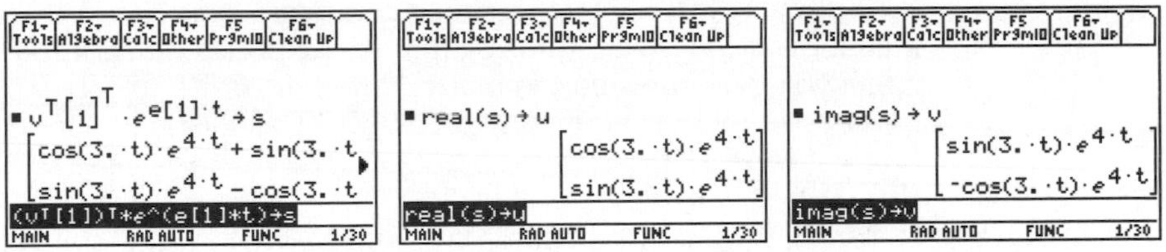

Figure 15-14: Finding real solutions when the eigenvalues are complex.

3. **Press [×][◆][X] to multiply the eigenvector by *e* raised to the eigenvalue that you enter in the next step.**

4. **Enter the complex eigenvalue associated with the eigenvector you entered in Step 2.**

   You have two ways to enter this eigenvalue: You can enter it as a complex number or, if you stored the list of eigenvalues in a one-letter name, you can reference its location in this list. If you enter the eigenvalue as a complex number, you must surround it with parentheses, as illustrated on the command line of the third picture in Figure 15-13. You enter the complex number *i* by pressing [2nd][CATALOG].

   To enter the complex eigenvalue by referencing its location in the named list of eigenvalues, press [ALPHA] and then press the key that corresponds to the letter of the name you gave to this list. Then press [2nd][,], enter the location of the eigenvalue in the list, and press [2nd][÷] to close the brackets, as illustrated on the command line in the first picture in Figure 15-14.

5. **Press [×] and enter the independent variable. Then press [)] to close the parentheses.**

   The independent variable in all examples in this chapter is *t*. You might have denoted it by some other letter.

6. **Press [STO▸][ALPHA] and then press the key that corresponds to the letter of the name you want to give the complex solution matrix. Then press [ENTER] to find the matrix form of the complex solution to the system of differential equations, as illustrated in the third picture in Figure 15-13 and the first picture in Figure 15-14.**

7. **Find the real and imaginary parts of the solution matrix. Store each in a one-letter name.**

   To find the real part, press [2nd][5][5][2] to select the **real** command from the MATH Complex menu, and then press [ALPHA] and the key that corresponds to the letter of the name you gave the solution matrix in Step 6. Press [)] to close the parentheses, press [STO▸][ALPHA], and then press the key that corresponds to the one-letter name you want to give this vector. Finally, press [ENTER] to display and store the vector, as illustrated in the second picture in Figure 15-14. Similarly, to find the imaginary part, press [2nd][5][5][3] to select the **imag** command from the MATH Complex menu, press [ALPHA] and the key that corresponds to the letter of the name you gave the solution matrix. Then press [)][STO▸][ALPHA], enter a letter to name the vector, and press [ENTER], as illustrated in the third picture in Figure 15-14.

The vectors in the last two pictures in Figure 15-14 show that the general solution to the homogeneous system of differential equations with the coefficient matrix in the first picture of Figure 15-13 is $\mathbf{x}(t) = c_1\begin{bmatrix} e^{4t}\ \cos 3t \\ e^{4t}\ \sin 3t \end{bmatrix} + c2\begin{bmatrix} e^{4t}\ \sin 3t \\ e^{4t}\ \cos 3t \end{bmatrix}$.

If you want to use this solution to solve an initial value problem, store the real solution matrix for the homogeneous system in the calculator. And if you want to use the solution to the homogeneous system to solve a nonhomogeneous system, store the fundamental matrix for the homogeneous system. (The fundamental matrix is the matrix that houses the linearly independent solutions to the homogeneous system in its columns.) Later in this chapter, I explain how to use the solution matrix and the fundamental matrix to solve initial value problems and nonhomogeneous systems.

To construct the real solution matrix and the fundamental matrix, follow Steps 3 and 4 in the section titled "Dealing with distinct real eigenvalues," which appears earlier in this chapter. The first picture in Figure 15-15 shows the solution matrix for the system of differential equations with the coefficient matrix in the first picture in Figure 15-13, and the second picture in Figure 15-15 displays the fundamental matrix for this system.

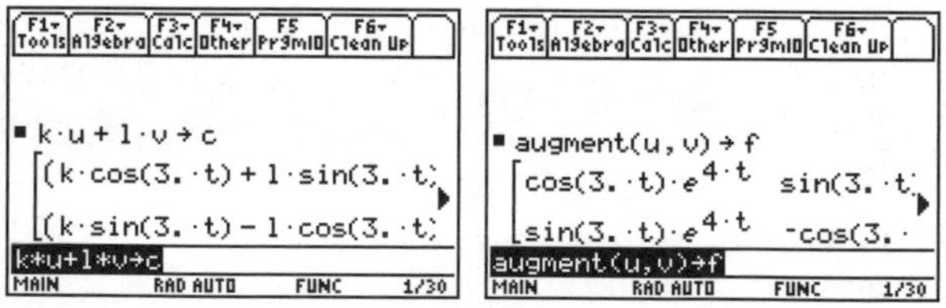

Figure 15-15: The solution and fundamental matrices when eigenvalues are complex.

Solving a first order nonhomogeneous linear system

Without going into great mathematical detail, the general solution to a first order nonhomogeneous linear system $\mathbf{x}'(t) = \mathbf{A}\mathbf{x}(t) + \mathbf{g}(t)$ is $\mathbf{x}(t) = \mathbf{x}_c(t) + \mathbf{x}_p(t)$, where $\mathbf{x}_c(t)$ is the general solution to the associated homogeneous system $\mathbf{x}'(t) = \mathbf{A}\mathbf{x}(t)$, and $\mathbf{x}_p(t)$ is a particular solution to the original nonhomogeneous system. And the method of variations of parameters states that $\mathbf{x}_p(t) = \Phi(t)\int \Phi(t)^{-1}\mathbf{g}(t)\,dt$ is a particular solution to the nonhomogeneous system. In this formula, $\Phi(t)$ is the fundamental matrix of the associated homogeneous system; that is, it is the matrix whose columns contain the linearly independent solution vectors of the associated homogeneous system.

I explain how to find the general solution to the associated homogeneous system earlier in this chapter. To get the calculator to find a particular solution, follow these steps:

1. **If you haven't already done so, construct the fundamental matrix for the associated homogeneous system and store it in a one-letter name.**

   You find out how to do this in the earlier section, "Solving first order homogeneous systems." In Figure 15-16, the fundamental matrix is the matrix **f** in the third picture in Figure 15-5. This is the fundamental matrix for the homogeneous system having the coefficient matrix in the first picture in Figure 15-4.

2. **Press [ALPHA] and then press the key that corresponds to the name you gave the fundamental matrix in Step 1.**

3. **Press [×][2nd][7][ALPHA], press the key that corresponds to the name you gave the fundamental matrix, and then press [^][(-)][1].**

4. **Press [×] and enter g(*t*), as illustrated on the command line in Figure 15-16.**

   In Figure 15-16, the fundamental matrix is named *f*, and $\mathbf{g}(t)$ is $\begin{bmatrix} -15 \\ -4 \end{bmatrix} e^{-2t}$.

   A column vector must be enclosed in square brackets, and the rows of the column vector are separated by semicolons. You enter the square brackets by pressing [2nd][,] and [2nd][÷], and you enter the semicolon by pressing [2nd][9].

5. **Press [,], enter the independent variable, and then press [)] to close the parentheses.**

   The independent variable in all examples in this chapter is *t*. However, you might have denoted it by some other letter.

6. **If you plan to use the solution to the nonhomogeneous system to solve an initial value problem, press [STO▸][ALPHA] and then press the key that corresponds to the letter of the name you want to give the particular solution, as illustrated at the end of the command line in Figure 15-16.**

7. **Press [ENTER] to display the particular solution, as illustrated in Figure 15-16.**

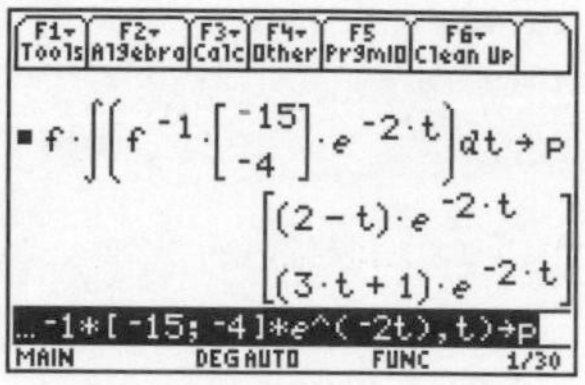

Figure 15-16: Finding a particular solution to a nonhomogeneous system.

Solving initial value problems

An initial value problem for a linear system of differential equations has the form $\mathbf{x}'(t) = \mathbf{A}\mathbf{x}(t) + \mathbf{g}(t)$, $\mathbf{x}(t_0) = \mathbf{x}_0$, where t_0 is the initial value of the independent variable *t*. To solve such a problem, follow these steps:

1. **If you haven't already done so, construct the solution matrix for the associated homogeneous system and store it in a one-letter name.**

   I explain how to do this earlier, in the section "Solving first order homogeneous systems." In Figure 15-17, the solution matrix is the matrix **c** in the second picture in Figure 15-5. This is the solution matrix for the homogeneous system with the coefficient matrix in the first picture in Figure 15-4.

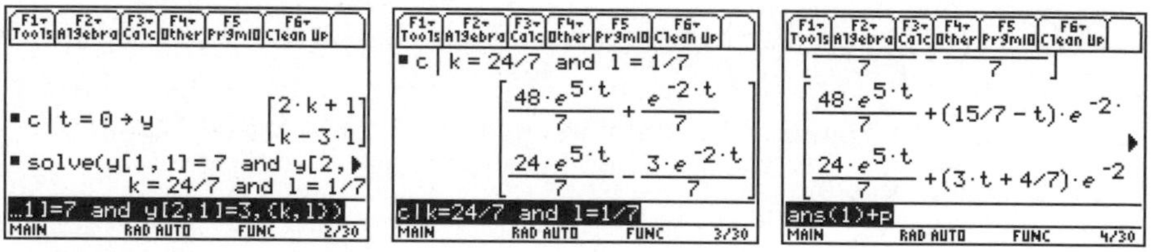

Figure 15-17: Solving an initial value problem.

2. **Evaluate the solution matrix at the initial value of the independent variable and store the result in a one-letter name, as illustrated in the first entry in the first picture in Figure 15-17.**

   To do this, follow these steps:

   a. **Press [ALPHA], press the key that corresponds to the name you gave the solution matrix, and then press [|] (the with command).**

   b. **Enter the independent variable, press [=], and then enter the initial value of the independent variable.**

   The initial condition used in Figure 15-17 is $\mathbf{x}(0)\begin{bmatrix}7\\3\end{bmatrix}$. So the initial value of the independent variable in this figure is 0.

c. **Press [STO▸][ALPHA] and press the key that corresponds to the letter you want to use to name this matrix.**

Use a letter you haven't used before while solving the homogeneous or nonhomogeneous system.

d. **Press [ENTER] to evaluate the solution matrix at the initial value of the independent variable.**

3. **Solve the system of equations obtained by setting the matrix in Step 2 equal to the initial condition, as illustrated in the second entry in the first picture in Figure 15-17.**

To do this, follow these steps:

a. **Press [F2][1] to place the solve command on the command line.**

b. **Enter the left side of the first equation, press [=], and then enter the right side of the same equation.**

The left side of the first equation is the first row in the matrix you stored in the calculator in Step 2. To enter it in the calculator, press [ALPHA], press the key that corresponds to the name you gave the matrix in Step 2, and then press [2nd][,][1][,][1][2nd][÷] to tell the calculator that you want the entry in the first row, first column of this matrix.

The right side of this equation is in the first row of the initial condition matrix $\mathbf{x}_0$. For example, in Figure 15-17, the initial condition is $\mathbf{x}(0)\begin{bmatrix}7\\3\end{bmatrix}$, so 7 is the right side of the first equation.

c. **Press [2nd][5][8][8] to select the word *and* from the MATH Test menu. Then enter the left side of the next equation, press [=], and enter the right side of the same equation, as illustrated on the command line in the first picture in Figure 15-17.**

To enter the left side of the next equation, press [ALPHA], press the key that corresponds to the name you gave the matrix in Step 2, press [2nd][,], and then press the number key that corresponds to the number of the equation. (For example, press [2] for the second equation, [3] for the third equation, and so on.) Finally, press [,][1][2nd][÷] to tell the calculator that you want the entry in the first — and only — column of this matrix.

If you're entering the *n*th equation, the right side of this equation is in the *n*th row of the initial condition matrix $\mathbf{x}_0$. For example, in Figure 15-17 the initial condition is $\mathbf{x}(0)\begin{bmatrix}7\\3\end{bmatrix}$, so 3 is the right side of the second equation.

d. **Repeat the preceding step until you've entered all the equations.**

e. Press [,] and then enter the list of constants enclosed in braces and separated by commas, as illustrated on the command line in the first picture in Figure 15-17.

The constants you enter in this step are the constants you used to define the solution matrix. For example, the solution matrix **c** in the second picture in Figure 15-5 denotes these constants by *k* and *l*. You enter the braces by pressing [2nd][(] and [2nd][)].

f. Press [)] to close the parentheses and then press [ENTER] to solve the system of equations, as illustrated in the second entry in the first picture in Figure 15-17.

4. Solve the initial value problem for the homogeneous system by evaluating the solution matrix at the values obtained in the preceding step, as illustrated in the second picture in Figure 15-17.

To do this, enter the name you gave the solution matrix in Step 1 and press [|] to insert the **with** command. Then press ⊝ to highlight the solutions you found in the preceding step and press [ENTER] to place these solutions on the command line. Finally, press [ENTER] to evaluate the solution matrix.

The second picture in Figure 15-17 shows that the solution to the homogeneous system in Figure 15-3 that has initial conditions $x(0) = 7$ and $y(0) = 3$ is $x(t) = \frac{1}{7}(48e^{5t} + e^{-2t})$ and $x(t) = \frac{1}{7}(24e^{5t} + 3e^{-2t})$.

5. To solve the initial value problem for the nonhomogeneous system, press [+][ALPHA] and then press the key that corresponds to the name you gave the particular solution to the nonhomogeneous system, as illustrated in the third picture in Figure 15-17.

I explain how to find the particular solution to a nonhomogeneous system in the preceding section. The third picture in Figure 15-17 shows that the solution to the nonhomogeneous system in Figure 15-3 with initial conditions $x(0) = 7$ and $y(0) = 3$ is $x(t) = \frac{1}{7}\left[48e^{5t} + (15 - 7t)e^{-2t}\right]$ and $x(t) = \frac{1}{7}\left[24e^{5t} + (21t - 4)e^{-2t}\right]$.

TIP

If the information on the calculator's screen scrolls off the right edge of the screen, as in the third picture in Figure 15-17, press ⊝ and then repeatedly press ⊙ to view the rest of the information. To see the left side of the screen again, press [2nd]⊙. When you're finished viewing the information on the screen, press ⊝ to return the cursor to the command line.

Solving systems of higher order ODEs

The method of using eigenvalues and eigenvectors to solve a linear system of first order ODEs (ordinary differential equations), which I explain earlier in this chapter, can also be used to solve a linear system of higher order ODEs if you first transform each equation in the system to a system of first order ODEs. I explain how to do this in the sidebar in this chapter.

As an example, the last example in the sidebar transforms the system $x'' = -3x + y$ and $y'' = 2x - 2y + 40\sin(3t)$ to the system $\mathbf{x}' = \mathbf{A}\mathbf{x} + \mathbf{g}(t)$, where

$$\mathbf{x} = \begin{bmatrix} x(t) \\ x'(t) \\ y(t) \\ y(t) \end{bmatrix} \quad \mathbf{A} = \begin{bmatrix} 0 & 1 & 0 & 0 \\ -3 & 0 & 1 & 0 \\ 0 & 0 & 0 & 1 \\ 2 & 0 & -2 & 0 \end{bmatrix} \quad \mathbf{g}(t) = \begin{bmatrix} 0 \\ 0 \\ 0 \\ 40\sin 3t \end{bmatrix}$$

I explain how to solve this nonhomogeneous system of first order equations earlier in this chapter. The solutions to the system $x'' = -3x + y$ and $y'' = 2x - 2y + 40\sin(3t)$ are $x(t)$ and $y(t)$, which are respectively housed in the first and third rows of the column vector **x**.

Graphing ODEs and Systems of ODEs

The calculator can graph slope fields for only first order ODEs, but it can graph the solution to an initial value problem for any order ODE. It can also graph direction fields and phase portraits in the phase plane for a system of two ODEs. The following three sections explain how to get the calculator to do these things.

Graphing slope fields and solutions to first order ODEs

The first order differential equation $\frac{dy}{dx} = F(x, y)$ is a formula for the slope of the tangent to the solution curve $y = f(x)$ at an arbitrary point (x, y). The slope field of this first order differential equation is a graph consisting of short line segments representing the tangent, and thus the slope, at the points (x, y). To get the calculator to graph a slope field, follow these steps:

1. **Press [MODE][▸][6][ENTER] to set the Graph mode to DIFF EQUATIONS.**
2. **Press [◆][F1] to enter the Y= editor and uncheck or clear any currently existing functions.**

   REMEMBER

   The calculator can graph *only one* slope field at a time. And only those differential equations in the Y= editor with checkmarks to the left of their names can have their slope fields graphed, as illustrated in the first picture in Figure 15-18, in which the slope field of **y1** can be graphed, but those of **y2** and **y3** can't.

   To change the checked status of a differential equation, use the [▲][▼] keys to highlight the differential equation and then press [F4] to toggle the check mark between being displayed and not being displayed.

To clear a single differential equation from the Y= editor, use the ▲▼ keys to highlight the equation and then press [CLEAR]. To clear all differential equations currently in the Y= editor, press [F1][8][ENTER].

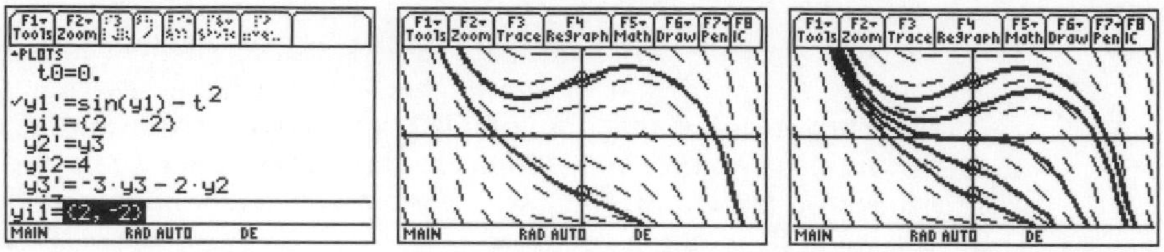

Figure 15-18: Graphing slope fields and solutions curves for first order ODEs.

3. **If you're graphing only the slope field, set t0 equal to 0. If you're graphing the slope field and a solution curve, set t0 equal to the value of the independent variable in the initial condition for the solution curve.**

   For example, if you want to graph the initial value problem $\frac{dy}{dx} = F(x, y)$ with $y(a) = b$, set **t0** equal to a.

   t0 is the first entry in the Y= editor. To edit the value assigned to this variable, repeatedly press ▲ to highlight the current value, press [ENTER] to place the cursor on the command line, use the number keys to enter the value you want to assign to **t0**, and then press [ENTER].

4. **Enter your first order differential equation.**

   When entering first order differential equations in the Y= editor, the independent variable must be denoted by *t*, and the dependent variable must be entered in the form *yn* where *n* is the number of the differential equation. For example, if the differential equation $y' + x^2 = \sin(y)$ is entered in **y1'**, you must enter it in the form $\sin(y1) - t^2$, as illustrated in the first picture in Figure 15-18.

   To enter or edit the definition of a differential equation in ***yn*'**, use the ▲▼ keys to place the cursor to the right of the equal sign and press [ENTER] or [F3] to place the cursor on the command line. Then enter or edit the definition of the differential equation and press [ENTER] when you're finished. (See Chapter 1 for the details on entering and editing expressions.) To enter **y1**, for example, press [Y][1]. To enter the independent variable, press [T].

   Because the calculator can graph only one slope field at a time, after completing this step, *only one* differential equation in the Y= editor should be checked. If this isn't the case, you get an error message when you try to graph the slope field. The directions in Step 2 tell you how to deselect differential equations in the Y= editor.

5. **If you want to graph only the slope field of the differential equation you entered in y*n*', leave yi*n* blank. If you want to graph the slope field and a solution curve, set yi*n* equal to the value of the dependent variable in the initial condition for the solution curve.**

   For example, if you want to graph the initial value problem $\frac{dy}{dx} = F(x, y)$ with $y(a) = b$, set **yi*n*** equal to *b*.

   To leave **yi*n*** blank or to edit its current value, use the ▲▼ keys to highlight the current value in **yi*n***. To leave **yi*n*** blank, press [CLEAR]. To edit its current value, press [ENTER] to place the cursor on the command line, use the number keys to enter the value you want to assign to **yi*n***, and then press [ENTER].

   TIP: You can graph more than one solution curve by entering the initial conditions for the curves in a list, as illustrated on the command line of the first picture in Figure 15-18. The graph of these curves is shown in the second picture of this figure. A list must be enclosed in braces and its elements must be separated by commas. You enter the braces by pressing [2nd][(] for the left brace and [2nd][)] for the right.

6. **Press [F1]▲[ENTER] to display the Graph Formats menu and set the last item (Fields) to SLPFLD. Set the other items to the options of your choice.**

   To select options from this menu, use the ▲▼ keys to place the cursor on the desired format, press ▶ to display the options for that format, and then press the number of the option you want. When you're finished setting the various formats, press [ENTER] to save your settings.

   An explanation of the Graph Formats menu options follows:

   - **Coordinates:** This format gives you a choice between having the coordinates of the location of the cursor displayed at the bottom of the graph screen in three ways. Select RECT to display the location of the cursor as (x, y) rectangular coordinates, or select POLAR to display points as (r, θ) polar coordinates. If for some strange reason you don't want to see the coordinates of the cursor when you trace the graph, select OFF.
   - **Grid:** Set this to ON if you want the calculator to graph the intersection points of the tick marks on the horizontal and vertical axes. Because the slope field is typically graphed over these points, my recommendation is to select OFF.
   - **Axes:** If you don't want to see the axes on your graph, select OFF; if you do want to see them, select ON. I recommend selecting ON.
   - **Leading Cursor:** Set this to ON if you want the calculator to animate the cursor as it graphs a solution curve. My recommendation is to select OFF.
   - **Labels:** If you want the *x*- and *y*-axes to be labeled, select ON; otherwise, select OFF. My recommendation is to select OFF — you know what the axes are and the labels are rarely placed near the axes.

- **Solution Method:** The calculator uses the RK (Runge-Kutta) or EULER numerical method to graph slope fields and solution curves. RK is more accurate, but it takes more time than EULER. I recommend selecting RK.

  If the calculator takes a long time to produce a graph by using the RK solution method, press [ON] to stop the calculation and then press [F1]▲[ENTER]▼▼▼▼▼▼▶[2][ENTER] to change to the EULER solution method. The calculator then draws the graph by using the faster Euler method.

- **Fields:** Select SLPFLD to graph first order differential equations. If you try to graph first order differential equations by using any other Fields option, you get an error message.

7. **Press [♦][F2] to enter the Window editor, and then set xmin, xmax, xscl, ymin, ymax, and yscl to an appropriate window size. Set ncurves to an integer between 0 and 10, inclusive. Then press [♦][F3] to draw the graph.**

   To edit the value assigned to a Window variable, use the ▲▼ keys to highlight the current value, key in the new value, and then press [ENTER]. After the graph is drawn, the calculator displays a circle around the initial conditions, as illustrated in the last two pictures in Figure 15-18.

   xmin, **xmax**, **ymin**, and **ymax** are, respectively, the smallest and largest values of *x* in view on the *x*-axis and the smallest and largest values of *y* in view on the *y*-axis. **xscl** and **yscl** give the spacing between tick marks on the respective axes.

   ncurves should be set to 0 if you're graphing only the slope field or if you assigned one or more values to **yi*n*** in Step 5. If **yi*n*** is blank and **ncurves** is set to 5, for example, the calculator automatically draws 5 solution curves, as illustrated in the third picture in Figure 15-18.

   If you don't know an appropriate size for the viewing window, press [F2][6] to draw the graph in the standard viewing window ($-1 \le x \le 10$, $-10 \le y \le 10$) and then, if necessary, repeat this step to find a better viewing window.

   If you know how you want to set the *x*-axis in your viewing window but you don't know how to set the *y*-axis, enter your values for **xmin**, **xmax**, and **xscl** and then press [F2]▲▲▲[ENTER] to have the calculator figure out the appropriate *y*-values and draw the graph in the resulting window. Then, if necessary, repeat this step to adjust **ymin**, **ymax**, and **yscl**.

TIP

To trace a solution curve, press [F3] and use the ◀▶ keys to display the coordinates of the points on the solution curve. If you have more than one solution curve, use the ▲▼ keys to move the cursor to another solution curve and then use the ◀▶ keys to trace that curve.

TIP

After graphing a slope field, with or without solution curves, you can add solution curves to the graph. To do this, first press [2nd][F3], as illustrated in the first picture in Figure 15-19. If you know the coordinates of a point through which the curve passes, enter the *t*-coordinate of that point, press [ENTER], enter the *y*-coordinate, and then press [ENTER]. If you don't know the exact coordinates of a point on the solution curve, use the Arrow keys to move the cursor to a point on the slope field, as illustrated in the second picture in Figure 15-19, and then press [ENTER] to draw the solution curve that passes through that point, as illustrated in the third picture in this figure. To erase any solution curves you added to the slope field, press [F4]. Solution curves added in this fashion cannot be traced.

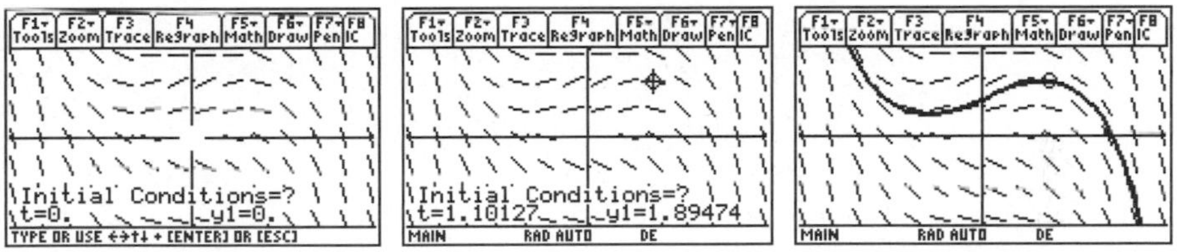

Press [2nd] [F3] — Move cursor — Press [ENTER]

Figure 15-19: Adding solution curves to a slope field.

Graphing solutions to higher order ODEs

For differential equations of order greater than one, the calculator is capable of graphing only one solution to an initial value problem. To graph such a solution, follow these steps:

1. **Convert your higher order initial value problem to a system of first order differential equations.**

   Because the calculator can graph only first order differential equations, you must convert higher order differential equations to a system of first order differential equations. The sidebar in this chapter tells you how to do this.

   As an example, the second order initial value problem

   $$y'' + 3y' + 2y = 0,\ y(0) = 4,\ y'(0) = 1$$

   is equivalent to the system

   $$y'_1 = y'_2, \qquad y_1(0) = 4$$
   $$y'_2 = 3y_2 - 2y_1, \quad y_2(0) = 1$$

2. **If necessary, press [MODE][▶][6][ENTER] to set the Graph mode to DIFF EQUATIONS.**

3. **Press [◆][F1] to enter the Y= editor and then set t0 equal to the value of the independent variable in the initial condition for the solution curve.**

   For example, if you want to graph the initial value problem $y''+f\left(t,\ y,\ y'\right),\ y(a)=b,\ y'(a)=c$, set **t0** equal to a.

   t0 is the first entry in the Y= editor. To edit the value assigned to this variable, repeatedly press ⊖ to highlight the current value, press [ENTER] to place the cursor on the command line, use the number keys to enter the value you want to assign to **t0**, and then press [ENTER].

4. **Enter your system of first order differential equations and initial conditions, as illustrated in the first picture in Figure 15-20.**

   Steps 4 and 5 in the preceding section give details on entering first order differential equations and initial conditions. Equations **y2'** and **y3'** in the first picture of Figure 15-18 show you how to enter the equations in the first picture of Figure 15-20 when there are other differential equations in the Y= editor.

Figure 15-20: Graphing solutions to higher order ODEs.

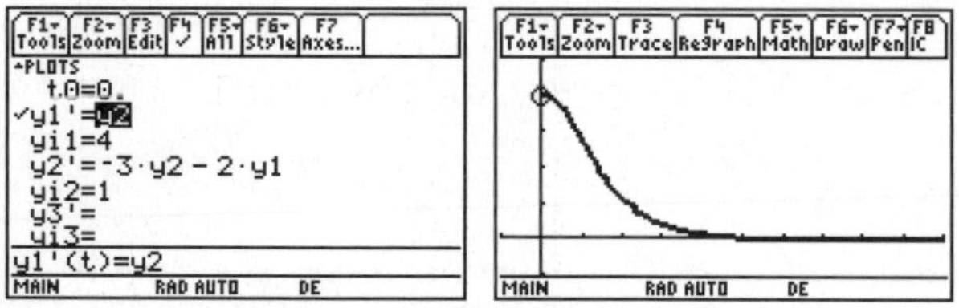

5. **Press [F5][1] to uncheck all differential equations in the Y= editor. Then use the ⊖⊖ cursor keys to highlight the first equation in your system of equations and press [F4] to place a checkmark next to it, as illustrated in the first picture in Figure 15-20.**

   The solution to the original higher order differential equation is y. When you converted this higher order equation to a system of first order ODEs, you set $y_1 = y$. So the solution you want to find is the solution to only the first equation in the system.

6. **Press [F1]⊖[ENTER] to display the Graph Formats menu and set the last item (Fields) to FLDOFF. Set the other items to the options of your choice.**

   Step 6 in the preceding section gives details on how to set the items in the Graph Formats menu. If the Fields item in this menu isn't set to FLDOFF, you get an error message or unexpected results.

7. **Press ◆F2 to enter the Window editor and set xmin, xmax, xscl, ymin, ymax, and yscl to an appropriate window size. Then press ◆F3 to draw the graph, as illustrated in the second picture in Figure 15-20.**

   Should you need it, Step 7 in the preceding section gives details on how to set these items in the Window editor. It also gives tips on what to do if you don't know what values to give these items.

To trace a solution curve, press F3 and use the ◀▶ keys to display the coordinates of the points on the solution curve.

After graphing a higher order initial value problem, you can graph another solution that has different initial conditions. To do this, first press 2nd F3, as illustrated in the first picture in Figure 15-21. Because you're graphing the first equation in your system as a function of *t*, press ENTER to accept the default setting. If you know the coordinates of a point through which the curve passes, enter the *t*-coordinate of that point, press ENTER, enter the *y*-coordinate, and then press ENTER. If you don't know the exact coordinates of a point on the solution curve, use the Arrow keys to move the cursor to a point on the slope field, as illustrated in the second picture in Figure 15-21, and then press ENTER to draw the solution curve that passes through that point, as illustrated in the third picture. To erase any solution curves you added to the slope field, press F4. Solution curves added in this fashion can't be traced.

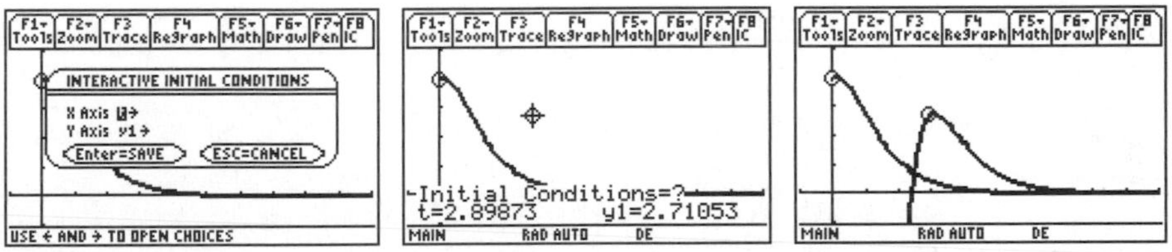

Figure 15-21: Graphing a higher order ODE with a different initial condition.

Graphing phase planes and portraits

The *phase plane* of a system of two differential equations $y_1' = f(t, y_1, y_2)$ and $y_2' = g(t, y_1, y_2)$ is the plane in which the *x*-axis is y_1 and the *y*-axis is y_2. A *phase portrait* is a graph, in parametric form $(y_1(t), y_2(t))$, of this system subject to given initial conditions. The calculator can graph the slope field (direction field) of the phase plane and it can graph a phase portrait. To do this, follow these steps:

1. **If necessary, press MODE ▶ 6 ENTER to set the Graph mode to DIFF EQUATIONS.**

2. **Press [◆][F1] to enter the Y= editor and set t0 equal to the value of the independent variable in the initial condition for the solution curve.**

   For example, if you want to graph the initial value problem

   $$y'+f\left(t,\, y_1,\, y_2\right),\ y_1(a)=b$$
   $$y'_2+f\left(t,\, y_1,\, y_2\right),\ y_2(a)=c,'$$

   set **t0** equal to *a*.

   t0 is the first entry in the Y= editor. To edit the value assigned to this variable, repeatedly press ⊖ to highlight the current value, press [ENTER] to place the cursor on the command line, use the number keys to enter the value you want to assign to **t0**, and then press [ENTER].

3. **Enter your system of two first order differential equations and initial conditions, as illustrated in the first picture in Figure 15-22.**

   Steps 4 and 5 in the earlier section, "Graphing slope fields and solutions to first order ODEs," give details on entering first order differential equations and initial conditions.

   REMEMBER: The calculator usually doesn't understand juxtaposition (implied multiplication), so when in doubt, press [×] to tell the calculator you are multiplying. For example, the calculator views **y1y2**, the juxtaposition of the two variables **y1** and **y2**, as a single variable with a four-character name. To tell the calculator that you are multiplying these two variables, enter **y1[×]y2**.

   TIP: If you want to see only the direction field, leave the initial conditions blank.

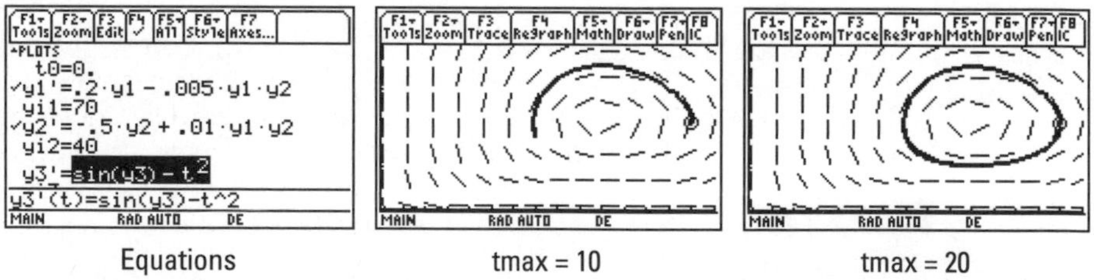

Figure 15-22: Graphing phase planes and phase portraits.

4. **Make sure that the only differential equations in the Y= editor that are checked are the two equations in the system you're graphing.**

   This is illustrated in the first picture of Figure 15-22, in which the first two equations in the Y= editor are the equations in the system that will be graphed.

   To change the checked status of a differential equation, use the ⊖⊖ keys to highlight the differential equation and then press [F4] to toggle the checkmark between being displayed and not being displayed.

5. **Press F1↑ENTER to display the Graph Formats menu and set the last item (Fields) to DIRFLD. Set the other items to the options of your choice.**

   Step 6 in the earlier section titled "Graphing slope fields and solutions to first order ODEs" gives details on how to set the items in the Graph Formats menu. If the Fields item in this menu is not set to DIRFLD, you will get an error message or unexpected results.

6. **Press ◆F2 to enter the Window editor and set xmin, xmax, xscl, ymin, ymax, and yscl to an appropriate window size. Then press ◆F3 to draw the graph, as illustrated in the second picture in Figure 15-22.**

   Should you need it, Step 7 in the earlier section, "Graphing slope fields and solutions to first order ODEs," gives details on how to set these items in the Window editor. It also gives tips on what to do if you don't know what values to give these items.

7. **If you don't get a complete graph, as in the second picture in Figure 15-22, press ◆F2 and increase the value assigned to tmax in the Window editor. Then press ◆F3 to redraw the graph, as illustrated in the third picture in this figure.**

TIP

To trace a solution curve, press F3 and use the ◀▶ keys to display the coordinates of the points on the solution curve. At the bottom of the screen, **tc** gives the value of the independent variable *t*, and **xc** and **yc** are respectively the values of the solutions to the first and second equations in your system at this value of *t*.

TIP

After graphing a phase plane, with or without a phase portrait, you can add phase portraits to the graph. To do this, first press 2nd F3, as illustrated in the first picture in Figure 15-23. If you know the initial values you want to assign the two equations in your system, enter the value for the first equation in the system, press ENTER, enter the value for the second equation, and then press ENTER. If you don't know the initial values, use the Arrow keys to move the cursor to a point on the phase plane, as illustrated in the second picture in Figure 15-23, and then press ENTER to draw the phase portrait, as illustrated in the third picture. To erase any phase portraits you added to the direction field, press F4. Phase portraits added in this fashion cannot be traced.

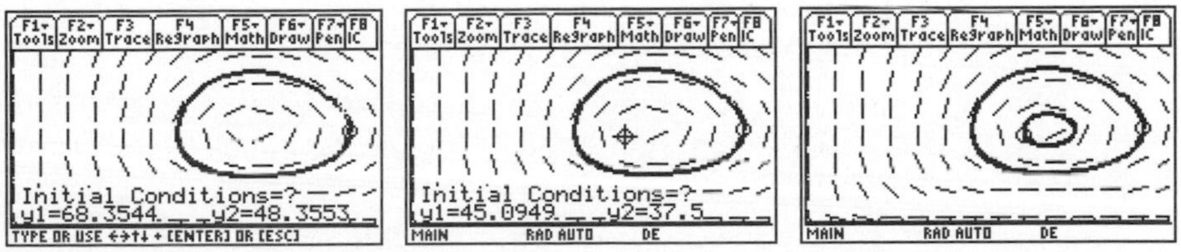

Figure 15-23: Graphing additional phase portraits.

Dealing with ODEs of large order

The abilities of the calculator appear to limit you to solving only first or second order ODEs, to solving only linear systems of first order ODEs, and to graphing only first or second order ODEs. But a mathematical solution to these apparent limitations does exist — just convert the higher order ODEs to a system of first order ODEs! Here's how you do this.

A higher order ODE has the form $y^{(n)} = f(t, y, y', y'', \ldots, y^{(n-1)})$, where $y^{(k)}$ denotes the kth derivative of y with respect to t. This ODE is converted to a system of n first order ODEs by setting $y_1 = y$, $y_2 = y'$, $y_3 = y''$, . . . , $yn = y^{(n-1)}$. These settings result in $y'_1 = y' = y_2$, $y'_2 = y'' = y_3$, . . . , $y'_n = y^{(n)}$, where $y^{(n)} = f(t, y, y', y'', \ldots, y^{(n-1)}) = f(t, y_1, y_2, y_3, \ldots, y_n)$. This gives the system:

$$
\begin{aligned}
y'_1 &= y_2 \\
y'_2 &= y_3 \\
&\vdots \\
y'_{n-1} &= y_n \\
y'_n &= f(t, y_1, y_2, \ldots, y_n)
\end{aligned}
$$

As an example, the third order ODE $y''' + 3y'' + 2y' - 5y = \sin(2t)$ can be written as $y''' = f(t, y, y', y'') = 5y - 2y' - 3y'' + \sin(2t)$. Substituting $y_1 = y$, $y_2 = y'$, and $y_3 = y''$ results in the system:

$$
\begin{aligned}
y'_1 &= y_2 \\
y'_2 &= y_3 \\
&\vdots \\
y'_3 &= 5y_1 - 2y_2 - 3y_3 + \sin(2t)
\end{aligned}
$$

This method can also be used to transform a system of ODEs of order greater than one to a system of first order ODEs by converting each equation in the system to a system of first order ODEs. This comes in quite handy if you want to use the calculator to solve for or graph the solutions to the original system of higher order ODEs.

As an example, the system

$$
\begin{aligned}
x'' &= -3x + y \\
y'' &= 2x - 2y + 40\sin 3t
\end{aligned}
$$

of second order ODEs can be transformed to a system of first order ODEs by setting $x_1 = x$ and $x_2 = x'$ to transform the first equation and setting $y_1 = y$ and $y_2 = y'$ to transform the second equation. This results in the equivalent system

$$
\begin{aligned}
x'_1 &= x'_2 \\
x'_2 &= -3x'_1 + y_1 \\
y'_1 &= y_2 \\
y'_2 &= 2x_1 - 2y_1 + 40\sin 3t
\end{aligned}
$$

of first order differential equations where x_1 and y_1 are the solutions to the original system of ODEs.

Part VI
Dealing with Matrices

The 5th Wave By Rich Tennant

In this part . . .

In this part, I show you how to use matrices in arithmetic expressions and how to find the inverse, transpose, and determinant of a matrix. I also show you how to use matrices to solve a system of linear equations.

Chapter 16

Creating and Editing Matrices

In This Chapter

- Defining a matrix
- Editing and resizing a matrix
- Displaying the contents of a matrix
- Augmenting two matrices
- Making a copy of a matrix
- Deleting a matrix from the calculator's memory

A matrix is a rectangular array of elements arranged in rows and columns. The dimensions, $r \times c$, of a matrix are defined by the number of rows and columns in the matrix. The calculator allows you to define as many matrices as the memory of the calculator can accommodate, and that's most likely more than you'll ever need. Even better than that, each matrix can be huge — dimensions up to 999×99 — and can contain just about any type of element you desire, such as a real or complex number, an expression like cos(*x*), or a name like Phido.

Defining a Matrix

You can define a matrix in two places: on the Home screen or in the Data/Matrix editor. The Data/Matrix editor is by far the best place to define a matrix, especially if you plan to edit matrix elements at a later time or if you're defining several large matrices. On the other hand, if you're defining only one or two rather small matrices (such as a 2×2 matrix), it's faster to define them on the Home screen. This section tells you how to define a matrix on the Home screen and how to define matrices in the Data/Matrix editor.

Defining a matrix on the Home screen

When you define matrices on the Home screen, you must enclose them in square brackets, as illustrated on the command line in Figure 16-1. The elements in the rows of the matrix are separated by commas, and the rows are delineated by semicolons. You enter the square brackets by pressing [2nd][,] and [2nd][÷], and you enter the semicolon by pressing [2nd][9].

If you want to reference your matrix at a later time, store it in a variable, as in Figure 16-1. This is quite convenient if, for example, you want to add or multiply two matrices. (I explain storing and recalling variables in Chapter 2, and in Chapter 17 you find out about adding and multiplying matrices.) For example, to enter the first matrix in Figure 16-1 and store it in variable *a*, you use these keystrokes: [2nd][,][1][,][2][2nd][9][3][,][4][2nd][÷][STO▸][ALPHA][=][ENTER].

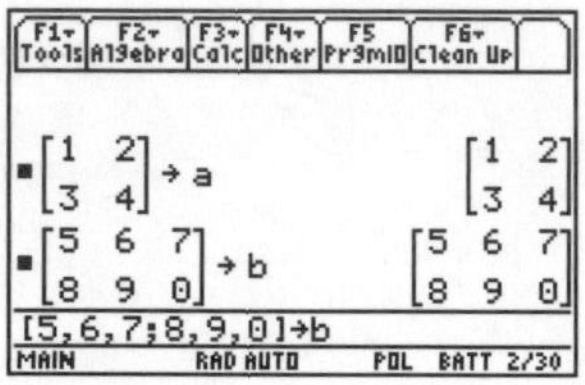

Figure 16-1: Defining matrices on the Home screen.

Using the Data/Matrix editor to define and edit matrices

The Data/Matrix editor is a handy place to define and edit matrices, especially if you have several large matrices to define. And if you're prone to keystroke errors, you'll appreciate how convenient it is to edit matrices in the Data/Matrix editor.

Entering a matrix in the Data/Matrix editor

To define one or more matrices in the Data/Matrix editor, follow these steps:

1. **Press [APPS].**
2. **On the TI-89 Titanium, use the Arrow keys to highlight Data/Matri and then press [ENTER]. On the TI-89, press [6].**

   On the TI-89 Titanium, you see the first picture in Figure 16-2. On the TI-89, you see a screen displaying the same information but in a different location.

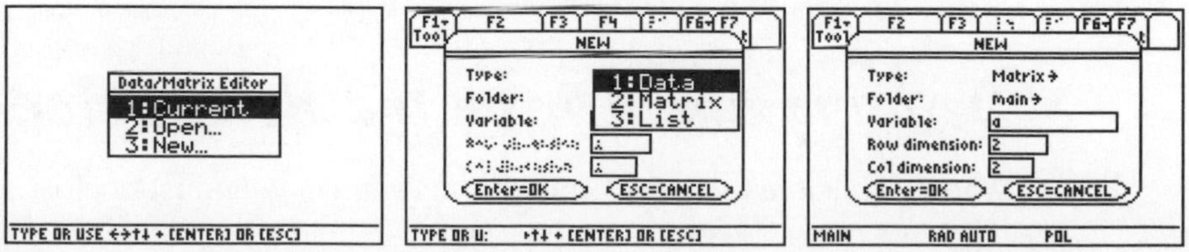

Figure 16-2: Setting up the Data/Matrix editor.

3. **Press [3] to enter a new matrix and then press ⊙ to display the types of data you can enter, as illustrated in the second picture in Figure 16-2.**

4. **Press [2] to tell the calculator that you want to enter a matrix.**

5. **Press ⊝ and select the folder in which the matrix is to be stored.**

   If you haven't defined a special folder for your matrix, accept the default (main) folder and go to the next step. If you have defined a special folder for your matrix, press ⊙ and then press the number key corresponding to that folder.

6. **Press ⊝ and key in a name for your matrix.**

   The name of your matrix can consist of no more than eight characters, and the first character must be a letter. The calculator is already in Alpha mode and is expecting the first entry you key in to be a letter. So, for example, if you want to name your matrix "a," press [=] because the letter *a* is above this key. (I explain Alpha mode and entering text in Chapter 1.)

   Although giving your matrix a multicharacter name like "amatrix" is far more descriptive than giving it a one-letter name such as "a," when it comes time to delete the matrix from the memory of the calculator, it's much easier to delete matrices that have one-letter names than it is to delete those that have multi-character names. So if you don't need your matrices to remain in the memory of the calculator for a long period of time, give them one-letter names. I let you know how to delete matrices from the memory of the calculator later in this chapter.

7. **If the last character in the name of your matrix is a letter, press [ALPHA] to take the calculator out of Alpha mode so that you can enter numbers in the following steps.**

8. **Press ⊝ and key in the number of rows in the matrix, and then press ⊝ and key in the number of columns in the matrix.**

   You see a screen similar to the one in the third picture in Figure 16-2 (shown earlier).

   If you try to key in a number but get a letter, press [CLEAR][ALPHA] and then enter the number.

9. **Press ⊝[ENTER] to define your matrix.**

   If you get an error message after pressing [ENTER], this means that the calculator already contains a matrix that has the same name you entered in Step 6. To rectify this situation, press [ENTER]⊝⊝, key in a new name for your matrix, and then press [ENTER] twice.

   You see a screen similar to the first picture in Figure 16-3, in which the calculator has filled in the elements of your matrix with zeros and has highlighted the first-row, first-column element.

TIP

You can adjust the width of the columns in your matrix to display more than the default three columns or to accommodate large matrix entries. The column width tells you how many characters can be displayed in each cell of the matrix. So, for example, if the entries in your matrix range from –10 to 10, you need a column width of 3 to accommodate the negative sign and the two digits in the number 10. The smallest allowable width is 3, and the largest is 12. The default width is 6. To adjust the column width in your matrix, press [F1][↑][ENTER][→] to display the allowable column widths. Then move down (by using the Arrow keys) to the desired width and press [ENTER] two times.

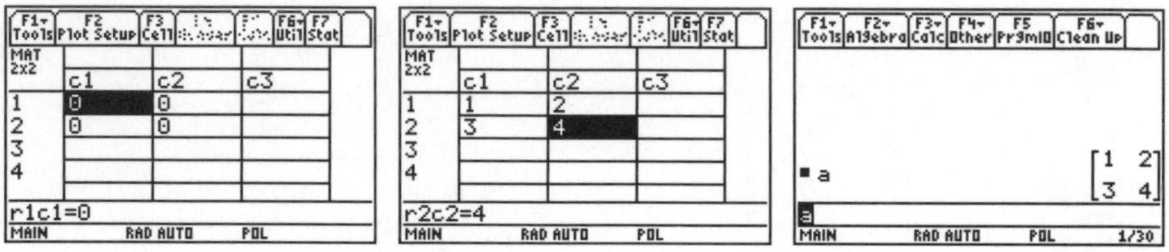

Figure 16-3: Defining a matrix in the Data/Matrix editor and viewing it on the Home screen.

10. **Key in the value of the first-row, first-column element, and then press [ENTER] to store that value in the matrix.**

    As you key in your value, the cursor moves to the command line and overwrites the zero entry with the value you key in.

11. **Key in the values of the next element in the matrix and press [ENTER] to store that value in the matrix.**

    Continue to do this until all elements in the matrix are defined to your specifications.

    After you press [ENTER] to store one element of the matrix in the calculator's memory, the cursor highlights the next element in the matrix that needs to be defined. Key in the value of this element and press [ENTER]. When there are no more elements to be defined, the last-row, last-column element that you entered remains highlighted, as in the second picture in Figure 16-3.

    After defining the elements in your matrix, the matrix is automatically stored in the memory of the calculator under the name you declared in Step 6. So, for example, you can view the matrix on the Home screen, as in the third picture in Figure 16-3. To do this, press [HOME][ALPHA], key in the name of the matrix, and then press [ENTER]. Press [2nd][APPS] to return to the Data/Matrix editor.

12. **To define another matrix, press [F1][3] and follow Steps 3 through 11.**
13. **When you finish defining matrices, press [2nd][ESC] to exit (quit) the Data/Matrix editor.**

TIP

You don't necessarily have to press [2nd][ESC] to exit (quit) the Data/Matrix editor. Instead, you can press [2nd][APPS] to return to the application you were using prior to the Data/Matrix editor, or you can enter a new application without exiting the Data/Matrix editor. Then, when you want to return to the Data/Matrix editor, press [2nd][APPS]. For example, if you want to go to the Home screen after using the Data/Matrix editor, press [HOME]. Then if you want to return to the Data/Matrix editor, press [2nd][APPS].

Recalling a matrix defined in the Data/Matrix editor

You usually want to recall a matrix so you can edit its elements, resize it, or add (or delete) rows or columns. I let you know how to complete these tasks after recalling a matrix later in this chapter. To recall a matrix not currently displayed in the Data/Matrix editor, follow these steps:

1. **If you're in the Data/Matrix editor but the matrix you want isn't currently displayed, press [F1][1].**

   If you aren't already in the Data/Matrix editor, follow Steps 1 and 2 in the preceding section and then press [2] to recall the matrix.

   You see a screen similar to the first picture in Figure 16-4.

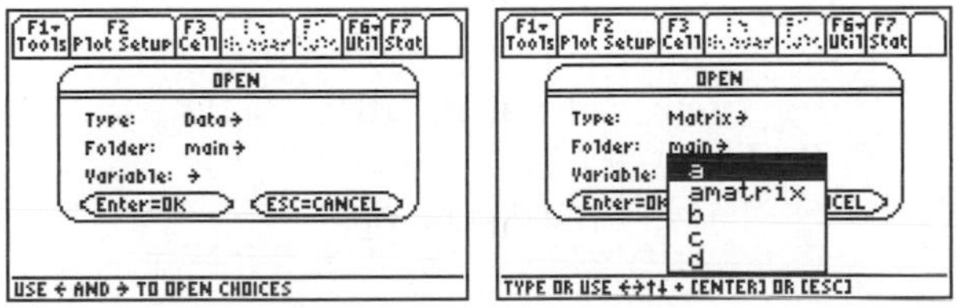

Figure 16-4: Recalling a matrix to the Data/Matrix editor.

2. **Press (▶)[2] to tell the calculator that you want to open an existing matrix.**
3. **Press (▼) to advance to the Folder option.**
4. **If the matrix isn't stored in the main folder, press (▶) and then press the number of the folder in which the matrix is stored.**
5. **Press (▼)(▶) to display the matrices stored in that folder, as illustrated in the second picture in Figure 16-4.**
6. **If necessary, repeatedly press (▼) to highlight the matrix you want to edit.**
7. **Press [ENTER] to select the highlighted matrix and press [ENTER] again to display the matrix.**

Editing a matrix in the Data/Matrix editor

When a matrix is displayed in the Data/Matrix editor, you can edit or redefine the elements in the matrix. (I tell you how to recall a matrix in the preceding section.)

To redefine a matrix element, use the Arrow keys to highlight the element and key in the new value. As you key in the new value, the cursor moves to the command line and replaces the existing value with the value you key in. When you're finished, press [ENTER] to save that value in the matrix.

To edit a matrix element, use the Arrow keys to highlight the element and press [ENTER] to place the value of that element on the command line. Then press ◀ or ▶ to place the cursor at the beginning or end of the value, whichever is more convenient for your editing needs. Finally, edit the value and then press [ENTER] to save the edited value in the matrix. For example, to change a matrix element from 12345 to 12305, press ▶ to place the cursor at the end of 12345 after the 5. Then press ◀ to place the cursor after the 4. Finally, press [←][0] to change the 4 to a zero and then press [ENTER].

Adding and deleting rows and columns

You can add or delete rows or columns in a matrix that you previously created in the Data/Matrix editor. To do so, that matrix must currently be displayed in the Data/Matrix editor.

To add more rows or columns (or both) after the last row or last column of the matrix, use the Arrow keys to place the cursor in the row or column you want to add, as illustrated for the 2×2 matrix in the first picture in Figure 16-5, in which the cursor is placed in the third row, third column. Then key in a value for the new matrix element you're creating. The calculator resizes the matrix (as illustrated under the Toolbar in the second picture in Figure 16-5, where the size of this matrix changed to 3×3) and fills in the remaining elements in the new rows and columns with zeros.

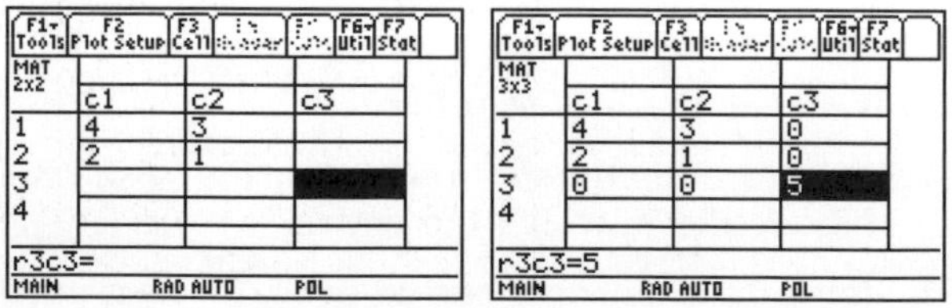

Figure 16-5: Adding additional rows and columns to a matrix.

To insert a row in a matrix, place the cursor in an existing row, as illustrated in the first picture in Figure 16-6, and press [2nd][F1][1][2]. The new row is inserted in the row of the cursor location, and the other rows are moved down, as illustrated in the second picture in Figure 16-6. To insert a column, press [2nd][F1][1][3]. The new column is inserted in the column where the cursor is located, and the other columns are moved to the right.

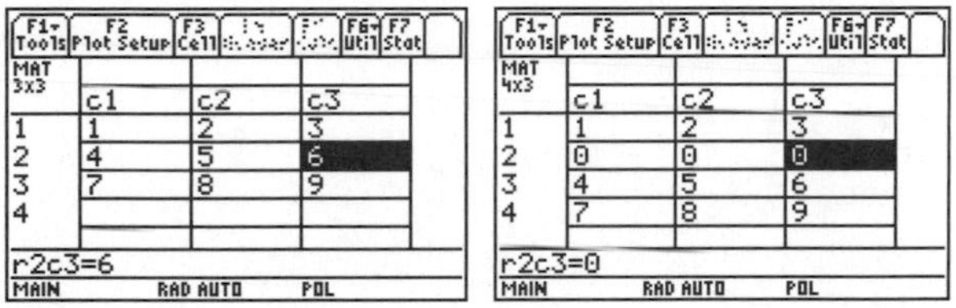

Figure 16-6: Inserting a row in a matrix.

To delete a row or column of a matrix, place the cursor in the row or column and press [2nd][F1][2][2] to delete the row containing the cursor or press [2nd][F1][2][3] to delete the column.

Augmenting Two Matrices

Augmenting two matrices allows you to append one matrix to another matrix. Both matrices must be previously defined and have the same number of rows. To augment two matrices, follow these steps:

1. **If necessary, press [HOME] to access the Home screen.**
2. **Press [2nd][5][4][7] to select the augment command from the MATH Matrix menu.**
3. **Enter the name of the first matrix and then press [,].**

   The first matrix is the matrix that appears on the left in the augmented matrix. This is illustrated in the third picture in Figure 16-7.
4. **Enter the name of the second matrix and then press [)].**
5. **Store the augmented matrix under a specified name.**

   To do so, press [STO▸], enter the name of the matrix in which you plan to store the augmented matrix, and then press [ENTER].

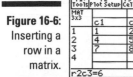

   If the name you give the augmented matrix is the same name as an existing matrix or other stored variable, that matrix or variable will be erased and replaced with the augmented matrix. If you don't know what variable names the calculator is currently using, press [2nd][−] and use the ⊝ key to view the names in current use. When you finish viewing variable names, press [2nd][ESC].

Figure 16-7: Augmenting two matrices.

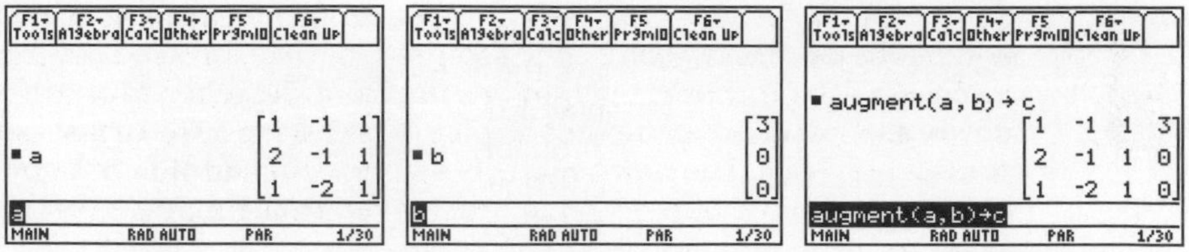

Matrix a — Matrix b — Augmented matrix

Copying One Matrix to Another

There are several situations in which you might want to copy the contents of one matrix to another. One of these is when you want to define a new matrix by adding rows or columns to an existing matrix and still keep the existing matrix in memory. To do so, copy the existing matrix to a new matrix and then redefine the new matrix to have the number of rows and columns you desire.

You can make a copy of a matrix in two places:

- **On the Home screen:**
  1. Key in the name of the matrix you want to copy.
  2. Press [STO▸].
  3. Key in a name for the copy of the matrix.
  4. Press [ENTER] to save the copy under the new name.
- **In the Data/Matrix editor:**
  1. Recall the matrix you want to copy.

     For more on how to do this, see the section, “Recalling a matrix defined in the Data/Matrix editor,” earlier in this chapter.
  2. Press [F1][2].
  3. Select a folder for the copy of the matrix.
  4. Press [▼].
  5. Key in a name for the copy (the calculator is already in Alpha mode).
  6. Press [ENTER] twice.

WARNING! Independent of where you made a copy of a matrix (on the Home screen or in the Data/Matrix editor), if the name you give to the copy of the matrix is the same name as an existing matrix or other stored variable, that matrix or variable will be erased and replaced with the matrix you’re copying. If you aren’t sure what variable names are currently being used by the calculator, press [2nd][−] and use the [▼] key to view the names in current use. When you’re finished viewing variable names, press [2nd][ESC].

After using [F1][2] (Save Copy As) in the Data/Matrix editor to make a copy of a matrix, the calculator displays the original matrix, *not the copy*. This can be confusing because the calculator doesn't tell you the name of the displayed matrix and because you're probably used to having a word processor display the copy of a document after using the same command (Save Copy As). So if you want to make changes to the copy of your matrix, you have to first recall (open) it in the Data/Matrix editor.

Deleting a Matrix from Memory

If you used one-letter names for your matrices, you can quickly delete them from the memory of the calculator by pressing [HOME] if you aren't already on the Home screen and then pressing [2nd][F1][1][ENTER]. The disadvantage of this method is that it deletes *all* variables that have one-letter names whether they be matrices or stored constants.

To delete a single matrix, whether it has a one- or multi-character name, press [2nd][–] to display the VAR-LINK screen. Then repeatedly press ↓ to highlight the name of the matrix you want to delete. (Matrices have the extension MAT displayed to the right of their names.) Finally, press [F1][1][ENTER] to delete that matrix from the memory of the calculator.

To delete several matrices on the VAR-LINK screen, use the ↓ key to highlight one matrix and press [F4] to place a check next to it. Continue this process until you have checked all the matrices you want to delete. Then press [F1][1][ENTER] to delete the checked variables.

Chapter 17

Using Matrices

In This Chapter

- Using matrices in arithmetic expressions
- Finding a scalar multiple of a matrix
- Negating a matrix
- Using the identity matrix in an arithmetic expression
- Using matrices to solve a system of equations
- Converting a matrix to reduced row-echelon form

Do you know how to use the calculator to perform arithmetic operations with matrices? And do you know how to use matrices to solve a system of equations? If not, you're in luck because this chapter tells you how.

Matrix Arithmetic

When evaluating arithmetic expressions that involve matrices, you usually want to perform the following basic operations:

- Scalar multiplication
- Negation (additive inverse)
- Addition
- Subtraction
- Multiplication
- Inversion (multiplicative inverse)
- Raising a matrix to an integral power
- Finding the transpose of a matrix
- Using the identity matrix in an arithmetic expression

Here's how you enter these matrix operations in an arithmetic expression:

1. **Define the matrices on the Home screen or in the Data/Matrix editor.**

   You find out how to do this in Chapter 16.

2. **Press [HOME] to access the Home screen.**

   If you want to clear the Home screen, press [F1][8]. To clear the command line, press [CLEAR].

   You perform all matrix operations on the Home screen.

3. **Enter the operations you want to perform and press [ENTER] when you're finished.**

   As with algebraic expressions, the Home screen is where you evaluate arithmetic expressions that involve matrices. To enter the name of a matrix into an expression, press [ALPHA] and then press the key corresponding to the name you gave the matrix. (For a matrix that has a multi-character name, press [2nd][ALPHA] to enter more than one letter, and then press [ALPHA] to take the calculator out of Alpha mode.) Here's how you enter the various operations into the arithmetic expression:

   - **Entering the scalar multiple of a matrix:** To enter the scalar multiple of a matrix in an arithmetic expression, enter the value of the scalar and then enter the name of the matrix, as shown in the first picture in Figure 17-1.
   - **Negating a matrix:** To negate a matrix, press [(-)] and then enter the name of the matrix, as shown in the second picture in Figure 17-1.

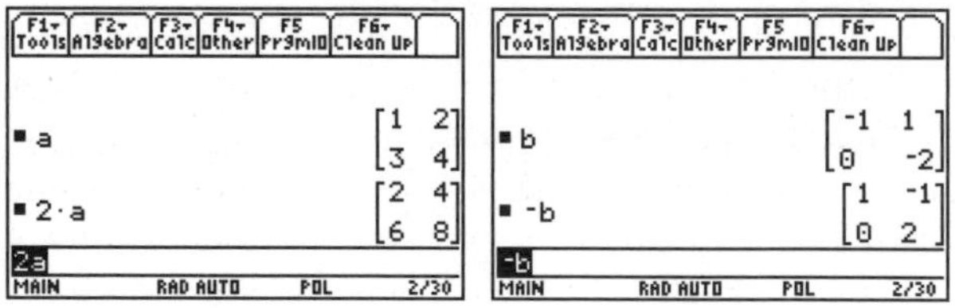

Scalar multiple Negation

Figure 17-1: The scalar multiple and the negation of a matrix.

   - **Entering the identity matrix:** You don't have to define an identity matrix on the Home screen or in the Data/Matrix editor in order to use it in an algebraic expression. To enter an identity matrix in an expression, press [2nd][5][4][6] to select the **identity** command from the MATH Matrix menu. Then enter the size of the identity matrix and press [)] to close the parentheses. For example, enter **2** for the 2×2 identity matrix, as in the first picture in Figure 17-2.

- **Adding or subtracting matrices:** When adding or subtracting matrices, the matrices must have the same dimensions. If they don't, you get an error message.

  Entering the addition and subtraction of matrices is straightforward: Just combine the matrices by pressing [+] or [−], as appropriate. The second picture in Figure 17-2 illustrates this process.

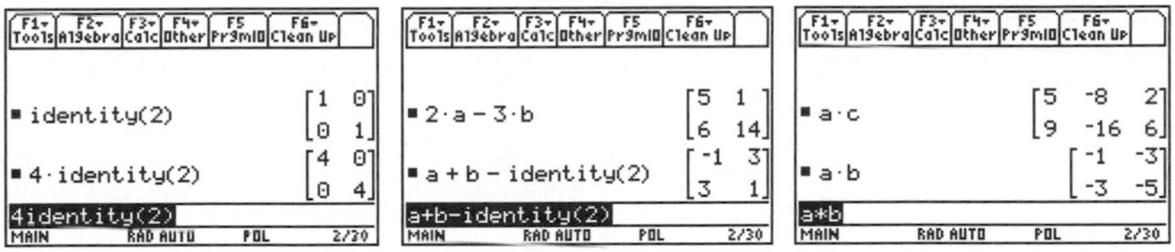

Identity matrix | Addition and subtraction | Multiplication

Figure 17-2: The identity matrix and addition, subtraction, and multiplication of matrices.

- **Multiplying two matrices:** When finding the product $A*B$ of two matrices, the number of columns in the first matrix, A, must equal the number or rows in the second matrix, B. If this condition isn't satisfied, you get an error message.

  The multiplication of matrices is simple: Just indicate the product by pressing [×], as in the third picture in Figure 17-2.

  WARNING! Do not use juxtaposition (*AB*) to denote the product of two matrices *A* and *B*. If you do, you will most likely get an error message because the calculator interprets *AB* as being a single entity with a two-letter name.

- **Finding the inverse of a matrix:** When finding the inverse of a matrix, the matrix must be *square* (number of rows = number of columns) and *nonsingular* (nonzero determinant). If it's not, you get an error message. (I explain evaluating the determinant of a matrix in the next section.)

  You enter the inverse of a matrix by entering the name of the matrix and then pressing [^][(-)][1], as in the first picture in Figure 17-3.

- **Raising a matrix to an integral power:** When finding the power of a matrix, the matrix must be square. If it isn't, you get an error message.

  To enter a power of a matrix, just enter the name of the matrix, press [^], and enter the power, as in the second picture in Figure 17-3.

- **Transposing a matrix:** To transpose a matrix in an arithmetic expression, enter the name of the matrix and then press [2nd][5][4][1] to select the **Transpose (T)** command from the MATH Matrix menu, as illustrated in the third picture in Figure 17-3.

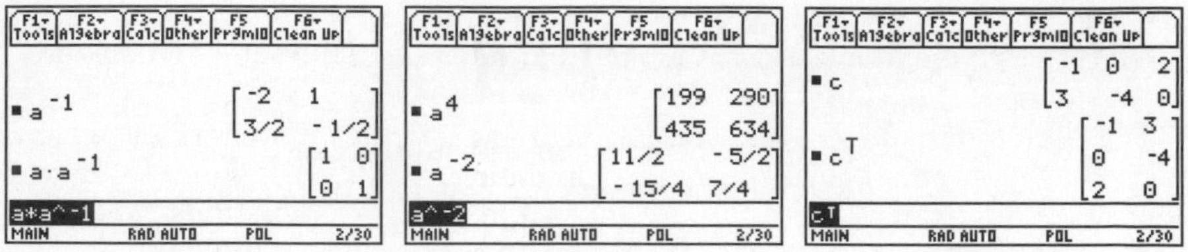

Figure 17-3: The inverse, power, and transpose of a matrix.

Finding Determinants and Eigenvalues

When finding the determinant, eigenvalues, or eigenvectors of a matrix, the matrix must be square (the number of rows = the number of columns). If it isn't, you get an error message.

To find the determinant, eigenvalues, or eigenvectors of a square matrix, follow these steps:

1. **If necessary, press [HOME] to access the Home screen.**
2. **Press [2nd][5][4] to display the MATH Matrix menu.**
3. **Press [2] to find the determinant of the matrix, press [9] to find the eigenvalues, or press [ALPHA][=] to find the eigenvectors.**

   TIP: Instead of pressing the number or letter of the desired option, you can repeatedly press ⊙ to highlight the option and then press [ENTER].
4. **Enter the name of the matrix and then press [)].**

   To enter a letter in the name of the matrix, press [ALPHA] and then press the key corresponding to the appropriate letter.
5. **Press [ENTER] to evaluate the option you selected in Step 3.**

   This procedure is illustrated in Figure 17-4. In the first picture, the determinant is displayed. In the second picture in this figure the eigenvalues (–1 and –2) are displayed in a list and the corresponding eigenvectors are displayed in columns. For example, in this picture, [1, 0] is an eigenvector corresponding to the eigenvalue –1.

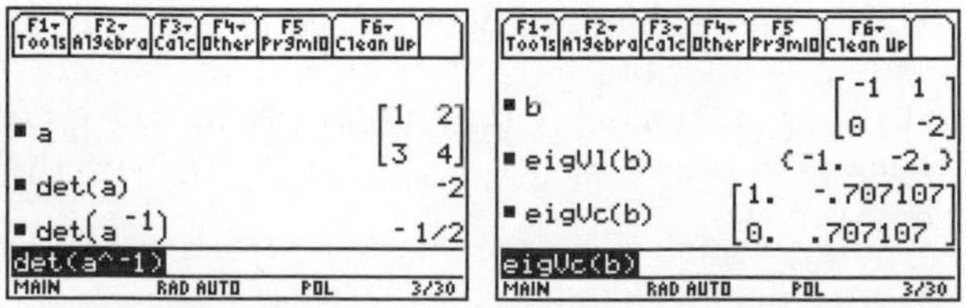

Figure 17-4: The determinant, eigenvalues, and eigenvectors of a matrix.

Solving Systems of Equations

In the olden days (15 to 20 years ago) when computers and calculators were very expensive and limited in their capabilities, if you wanted to find the inverse of a matrix or solve a large system of equations, you had to perform row operations on a matrix (such as adding a multiple of one row to another row) to put the matrix in what is called *reduced row-echelon form* (rref). From the rref form of the matrix, you could then read the inverse of the matrix or the solutions to the system of equations (provided these entities existed).

With a TI-89 graphing calculator, there are, for the most part, easier ways of finding the inverse of a matrix (as I explain earlier in this chapter) or of solving a system of equations (as I explain in Chapter 3). But when confronted with the task of solving a system of three or more equations, it's easier to use matrices than it is to use the method explained in Chapter 3. This section tells you how to do this.

$$a_{11}x + a_{12}y + a_{13}z = b_1$$
$$a_{21}x + a_{22}y + a_{23}z = b_2$$
$$a_{31}x + a_{32}y + a_{33}z = b_3$$

Three matrices are associated with a system of linear equations: the coefficient matrix, the solution matrix, and the augmented matrix. For example, *A*, *B*, and *C*, are (respectively) the coefficient matrix, solution matrix, and augmented matrix for the system of equations just given.

$$A = \begin{bmatrix} a_{11} & a_{12} & a_{13} \\ a_{21} & a_{22} & a_{23} \\ a_{31} & a_{32} & a_{33} \end{bmatrix} \quad B = \begin{bmatrix} b_1 \\ b_2 \\ b_3 \end{bmatrix} \quad C = \begin{bmatrix} a_{11} & a_{12} & a_{11} & b_1 \\ a_{21} & a_{22} & a_{23} & b_2 \\ a_{31} & a_{32} & a_{33} & b_3 \end{bmatrix}$$

Systems of linear equations can be solved by first putting the augmented matrix for the system in rref form. The mathematical definition of rref isn't important here. It is simply an equivalent form of the original system of equations, which, when converted back to a system of equations, gives you the solutions (if any) to the original system of equations.

For example, when the reduced row-echelon matrix in the first picture in Figure 17-5 is converted to a system of equations, it gives the solutions $x = -3$, $y = 3$, and $z = 9$. The matrix in the second picture in Figure 17-5 converts to the system $x - z = 0$ and $y - z = -2$. This arrangement indicates that the system has an infinite number of solutions — namely, all solutions in which $x = z$ and $y = z - 2$, where z is any real number. The third picture in Figure 17-5 illustrates a system that has no solution — the last line of the matrix says that $0 = 1$, which is clearly impossible!

Figure 17-5: Finding the solutions given by the reduced row-echelon matrix.

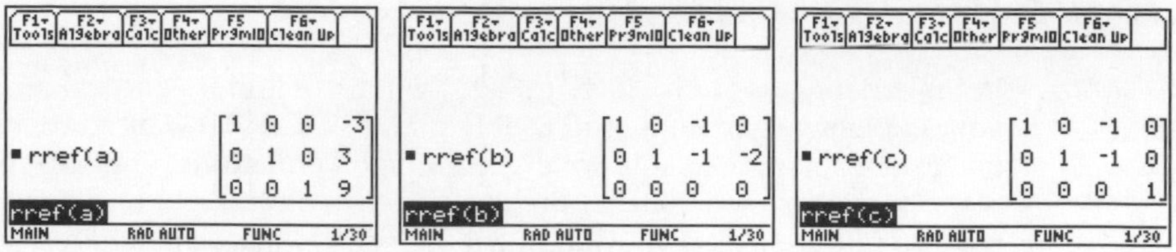

Unique solution | Infinite solutions | No solution

To solve a system of equations, follow these steps:

1. **Define the augmented matrix in the Data/Matrix editor.**

   The augmented matrix for the system of equations is explained at the beginning of this section. Chapter 16 explains how to define a matrix in the Data/Matrix editor.

   TIP: You can define the coefficient and solution matrices for the system of equations and then augment these matrices to form the augmented matrix. (For more about augmenting matrices, see Chapter 16.)

2. **Press [HOME] to access the Home screen.**
3. **Press [2nd][5][4][4] to select the rref command from the MATH Matrix menu.**
4. **Enter the name of the augmented matrix and then press [)].**

   To enter a letter in the name of the matrix, press [ALPHA] and then press the key corresponding to the appropriate letter.

5. **Press [ENTER] to put the augmented matrix in reduced row-echelon form.**
6. **To find the solutions (if any) to the original system of equations, convert the reduced row-echelon matrix to a system of equations.**

   The beginning of this section describes converting a reduced row-echelon matrix to a system of equations.

Although the calculator allows you to solve systems of equations without reducing them to reduced row-echelon form, if you have two or more systems of equations that have the same coefficient matrix, you can solve them simultaneously by finding the rref form of the matrix formed by augmenting the coefficient matrix and the solution matrices. Here's an example:

$$\begin{array}{ll} x-y+z=3 & x-y+z=2 \\ 2x-y+z=0 & 2x-y+z=3 \\ x-2y+z=0 & x-y+z=0 \end{array}$$

The two systems of equations just given have the same coefficient matrix A but different solution matrices B and C.

$$A = \begin{bmatrix} 1 & -1 & 1 \\ 2 & -1 & 1 \\ 1 & -2 & 1 \end{bmatrix} \quad B = \begin{bmatrix} 3 \\ 0 \\ 0 \end{bmatrix} \quad C = \begin{bmatrix} 2 \\ 3 \\ 0 \end{bmatrix}$$

The matrix formed by augmenting A, B, and C appears in the first picture in Figure 17-6. The reduced row-echelon form of this matrix appears in the second picture in the figure. The fourth column of the matrix in the second picture tells us that the solution to the first system of equations is $x = -3$, $y = 3$, and $z = 9$; the fifth column shows that the solution to the second system is $x = 1$, $y = 2$, and $z = 3$.

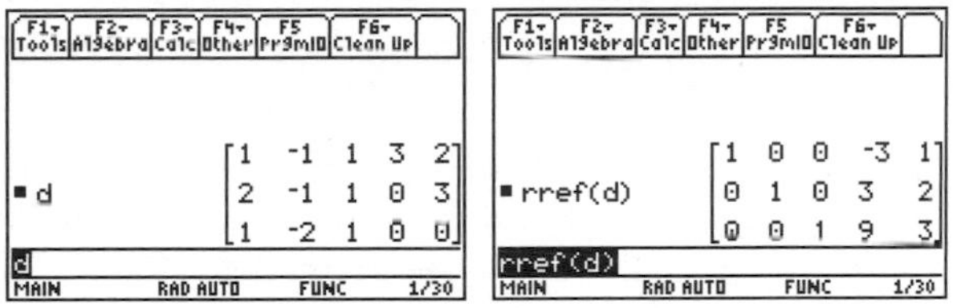

Figure 17-6: Simultaneously solving two systems of equations.

Part VII
Dealing with Probability and Statistics

"Okay – let's play the statistical probabilities of this situation. There are 4 of us and 1 of him. Phillip will probably start screaming, Nora will probably faint, you'll probably yell at me for leaving the truck open, and there's a good probability I'll run like a weenie if he comes toward us."

In this part . . .

This part gives you a look at calculating permutations and combinations, as well as generating random numbers. I also show you how to graph and analyze one- and two-variable statistical data sets. And if you want to do regression modeling (curve-fitting) — hey, who doesn't? — I show you how to do that, too.

Chapter 18

Probability

In This Chapter

- Evaluating permutations and combinations
- Generating random numbers

Do you need to calculate the number of ways you can arrange six people at a table or the number of ways you can select four people from a group of six people? Or do you just need an unbiased way of selecting people at random? If so, this is the chapter for you.

Permutations and Combinations

A *permutation,* denoted by **nPr**, answers the question "From a set of *n* different items, how many ways can you select *and* order (arrange) *r* of these items?" A *combination,* denoted by **nCr**, answers the question "From a set of *n* different items, how many ways can you select (independent or order) *r* of these items?" To evaluate a permutation or combination, follow these steps:

1. **If necessary, press [HOME] to go to the Home screen.**
2. **Press [2nd][5][7] to access the MATH Probability menu.**
3. **Press [2] to evaluate a permutation or press [3] to evaluate a combination.**
4. **Enter the total number (*n*) of items in the set.**

   Use the number keys to enter a positive integer or press [ALPHA] and enter a letter, as illustrated in Figure 18-1.
5. **Press [,] and enter the number (*r*) of items to be selected from the set.**

   As with *n*, *r* can be a positive number or a letter.
6. **Press [)][ENTER] to display the result (as shown in Figure 18-1).**

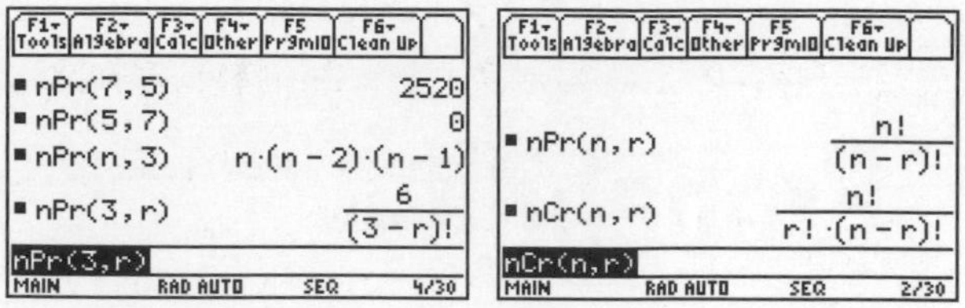

Figure 18-1: Evaluating permutations and combinations.

Generating Random Numbers

When generating random numbers, you usually want to generate numbers that are integers contained in a specified range, or decimal numbers that are strictly between 0 and 1.

Generating random integers

To generate random integers that fall between the integers 1 and *n* or between –*n* and –1, follow these steps:

1. **If necessary, press [HOME] to go to the Home screen.**
2. **Press [2nd][5][7][4] to select rand from the MATH Probability menu.**
3. **Enter a value for *n* and then press [)].**

   To generate random integers between 1 and *n*, use the number keys to enter a positive integer for *n*. To generate integers between –*n* and –1, enter a negative integer.

4. **Press [)][ENTER] to generate the first random integer, and to generate more random integers, repeatedly press [ENTER].**

   This is illustrated in the two pictures in Figure 18-2.

Pressing any key except [ENTER] stops the calculator from generating random integers.

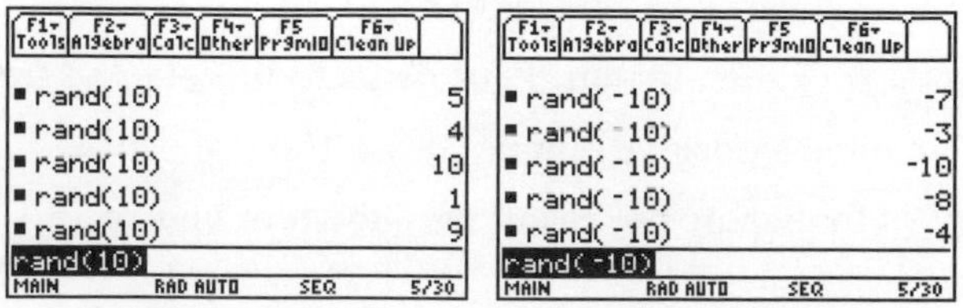

Positive integers Negative integers

Figure 18-2: Generating random integers.

Seeding the random number generator

If you've never used your calculator to generate random numbers, it spews forth the same random numbers as a new calculator or a calculator that has been reset to the factory defaults. This defeats the purpose of generating random numbers. To prevent this from happening, press [2nd][5][7][6] to select **RandSeed** from the MATH Probability menu, enter any number, and press [ENTER]. The number you enter is then used by the numerical routine in the calculator to generate random numbers. This process is called seeding the random number generator.

However, occasionally you might want to generate the same random numbers as another calculator or which your calculator previously generated. Replicating an experiment is an example of such an occasion. Setting **RandSeed** to the same number each time you run the experiment means that your calculator will generate the same random numbers.

Generating random decimals

To generate random decimal numbers that are strictly between 0 and 1, press [2nd][5][7][4] to select the **rand** command from the MATH Probability menu. Then press [)][ENTER] to generate the first random decimal and repeatedly press [ENTER] to generate more random decimals. Figure 18-3 illustrates this process.

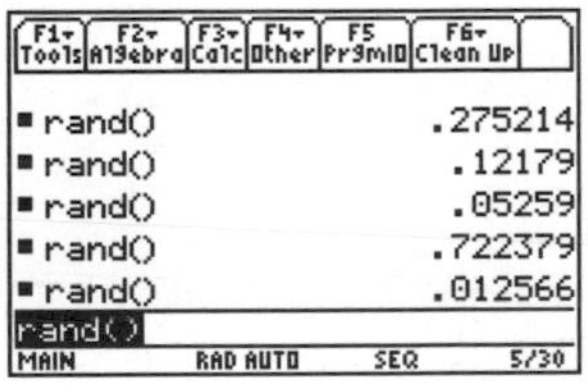

Figure 18-3: Generating random decimals.

Chapter 19

Dealing with Statistical Data

In This Chapter

- Entering data into the calculator
- Editing and sorting data
- Saving and recalling data
- Deleting data sets from memory

The calculator has many features that provide information about the data that you've entered. The calculator can graph data as a scatter plot, histogram, or box plot. It can calculate the median and quartiles. It can even find a regression model (curve fitting) for your data. It can do this and much, much more. This chapter tells you how to enter your data into the calculator; Chapter 20 shows you how to use the calculator to analyze that data.

Entering Data

What you use to enter statistical data into the calculator is the Data/Matrix editor — a relatively large spreadsheet that can accommodate up to 99 columns (*data lists*). And each data list can handle a maximum of 999 entries. Pictures of the Data/Matrix editor appear in Figure 19-1.

To use the Data/Matrix editor to enter and save your data, follow these steps:

1. **Press [APPS].**
2. **On the TI-89 Titanium, use the Arrow keys to highlight Data/Matri and then press [ENTER]. On the TI-89, press [6].**

   On the TI-89 Titanium, you see the first picture in Figure 19-1; on the TI-89, you see a screen displaying the same information but in a different location.

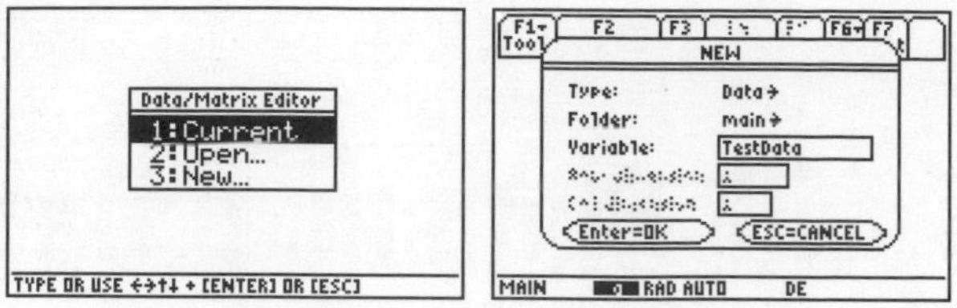

Figure 19-1: Setting up the Data/Matrix editor.

3. **Press [3] to enter a new data set. Then press ▽ to select the default Data type and select the folder in which to save your data.**

   If you haven't defined a special folder for your data set, accept the default (Main) folder, as illustrated in the second picture in Figure 19-1, and go to the next step. If you have defined a special folder for your data, press ▷ and then press the number key that corresponds to that folder.

4. **Press ▽ and key in a name for your data.**

   The name of your data set can consist of no more than eight characters, and the first character must be a letter. The calculator is already in Alpha Lock mode and is expecting your entries to be letters. I explain Alpha mode and entering text in Chapter 1.

   If you have a non-Titanium TI-89 calculator and you get symbols instead of letters when you try to name your data set, you need to upgrade your operating system. Chapter 21 tells you how to do this, but you don't need to do it right now. To continue entering your data, press [CLEAR] to erase any symbols you entered, and then press [2nd][ALPHA] to put your calculator in Alpha Lock mode.

5. **Press [ENTER][ENTER] to enter the Data/Matrix editor.**

   If you get an error message after pressing [ENTER], this means that the calculator already contains a data set with the same name you entered in Step 4. To rectify this situation, press [ENTER]▽▽, key in a new name for your data set, and then press [ENTER] twice.

   You see a screen similar to the first picture in Figure 19-2, in which the calculator has provided a spreadsheet and has highlighted the first-row, first-column element.

   You can adjust the width of the columns to display more than the default three columns or to accommodate large entries. The column width tells you how many characters can be displayed in each cell of the spreadsheet. So, for example, if the entries in your data range from –10 to 10, you need a column width of 3 to accommodate the negative sign and the two digits in the number –10. The smallest allowable width is 3, and the largest is 12; the default width is 6. To adjust the column width, press [F1]△[ENTER]▷ to display the allowable column widths. Then arrow to the desired width and press [ENTER] twice.

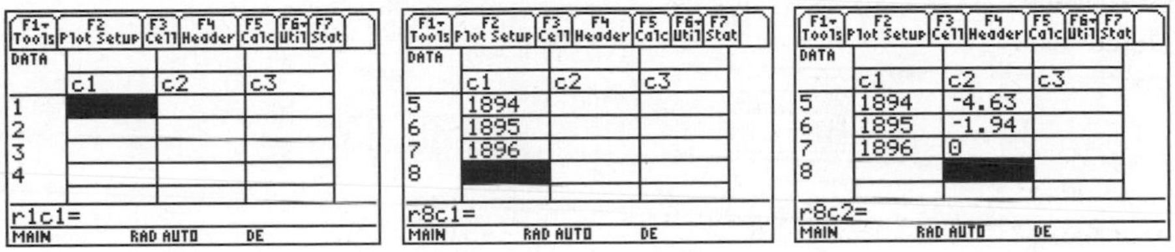

Figure 19-2: Entering data in the Data/Matrix editor.

6. **Key in the value of the first entry in the first column and press ENTER to store that value in the calculator. Key in the values of the next entry in the first column and press ENTER to store that value in the calculator. Continue to do this until all entries in the first column are defined to your specifications, as illustrated in the second picture in Figure 19-2.**

   As you key in your value, the cursor moves to the command line. After you press ENTER, the cursor moves to the next cell in the column.

   TIP: To quickly enter data that follows a sequential pattern (such as 1, 2, 3, . . . , 100 or 1, 4, 9, 25, . . . , 81) or data that is derived from a formula (such as 2∗c1 or ln(c2)), see the next section, "Using formulas to enter data."

7. **Press ◆▲▶ to move to the first cell in the next column and key in the entries in the next column. Continue to do this until you've defined all columns to your specifications, as illustrated in the third picture in Figure 19-2.**

As you defined each entry in your data set, it was automatically stored in the memory of the calculator under the name you declared in Step 4. So you don't need to do anything else to ensure that your data has been saved. To exit (quit) the Data/Matrix editor, press 2nd ESC or press 2nd APPS to return to the application you were using prior to using the Data/Matrix editor.

Using formulas to enter data

If you had to enter data for each year in the 20th century, would you key in all 100 years of the century? Of course not — you'd use a formula to enter the data for you. To use a formula to define your data, follow these steps:

1. **After setting up the Data/Matrix editor, as I explain in the first five steps in the preceding section, use the ◀▶▲▼ keys to highlight the column heading of the column in which your data is to appear, as illustrated in the first picture in Figure 19-3, where c1 is highlighted.**

2. **Enter the formula for that column and press ENTER.**

   As you enter your formula, the cursor moves to the command line. After you press ENTER, the data determined by your formula is entered in the column, and the first entry in the column is highlighted.

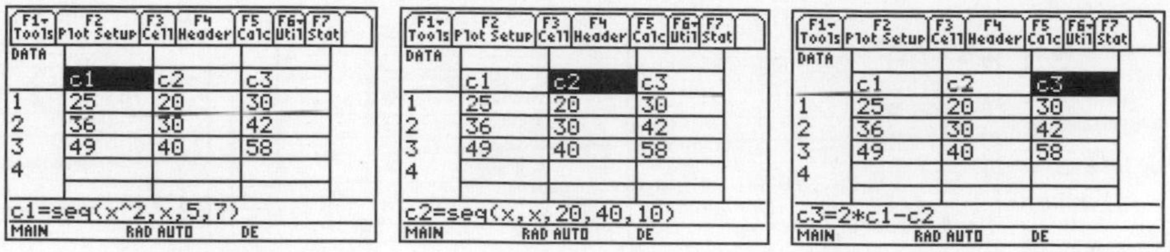

Figure 19-3: Examples of using formulas to enter data.

To enter a formula defining a sequence, as illustrated on the command lines in the first two pictures in Figure 19-3, press [2nd][5][3][ENTER] to select the **seq** command from the Math List menu and enter the formula as a function of one variable. Then press [,] and enter the variable you used to define the formula. Then press [,], enter the first value of the variable in the formula, press [,], and enter the last value of the variable. If the variable is incremented by a number other than 1, press [,] and enter the value of the increment, as illustrated on the command line in the second picture in Figure 19-3, in which the variable *x* is incremented by 10. Finally, press [)] to close the parentheses and then press [ENTER] to enter the data determined by this formula.

You can also enter a formula by referencing data housed in other columns, as illustrated on the command line in the third picture in Figure 19-3. For example, if you want the entries in column c2 to be twice those in column c1, define c2 as 2∗c1. (To enter c1, press [ALPHA][)][1].) If your formula references more than one column, as illustrated on the command line in the third picture in Figure 19-3, those columns must contain the same number of entries. If they don't, you get an error message.

Creating column titles

Imagine your dismay when you return to the data set you stored in the calculator last week and realize that you haven't the slightest idea what the data in the columns represent. You can avoid this problem easily by giving your columns a title.

To create a column title, use the ◀▶▲▼ keys to place the cursor in the blank cell above the column heading, as illustrated in the first picture in Figure 19-4. Then press [2nd][ALPHA] to put the calculator in Alpha mode and key in your title, as illustrated in the second picture in this figure. Finally, press [ENTER] to insert the title above the column heading, as illustrated in the third picture in this figure, and then press [ALPHA] to take the calculator out of Alpha mode.

TIP

Column titles can consist of as many characters as you want. The calculator might not be able to fit the whole title in the title area, but if you place the cursor on the title, the full title appears on the command line.

Figure 19-4: Creating column titles.

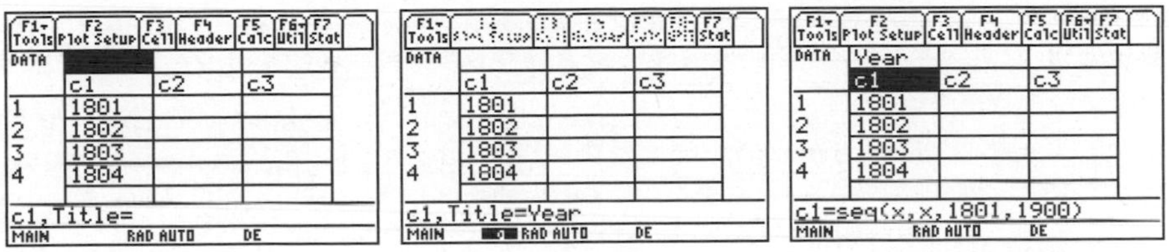

Highlight title area | Enter title | Press [ENTER]

Recalling, Editing, and Sorting Data

Everyone makes mistakes now and then, so knowing how to recall and edit data is a good idea. And if you don't like sorting data by hand, you might as well find out how to get the calculator to do it for you.

Recalling data

To recall a data set when you aren't currently in the Data/Matrix editor, follow these steps:

1. **Press [APPS]. On the TI-89 Titanium, use the Arrow keys to highlight Data/Matri and press [ENTER]; on the TI-89, press [6].**

   You're confronted with three options, as illustrated in the first picture in Figure 19-5 for the TI-89 Titanium. On the TI-89, you see a screen displaying the same information but in a different location.

Figure 19-5: Recalling a data set.

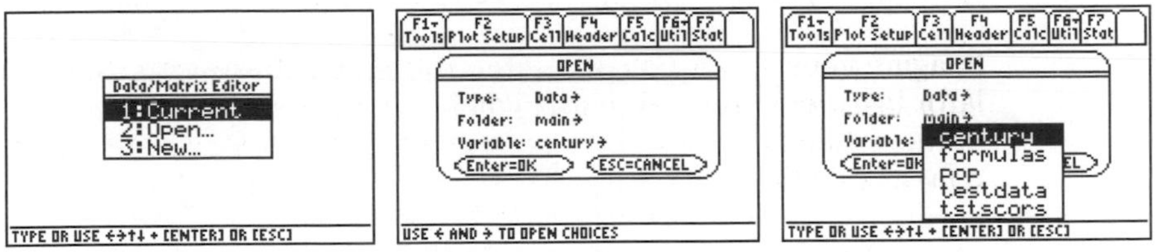

2. **If the data set you're recalling is the last data set you entered in the Data/Matrix editor, press [1] and skip the remaining steps; otherwise, press [2].**

   The OPEN menu appears, as shown in the second picture in Figure 19-5.

3. **Press ⊙. If you saved your data set in the Main folder, go to the next step; otherwise, press ⊙ to display the folders housed in the memory of the calculator, and then press the number key for the folder in which your data set is stored.**

4. **Press ⊝⊙ to display the list of data sets housed in the folder you selected in Step 3 (see the third picture in Figure 19-5).**

   You most likely don't have the same data sets stored in your calculator as I have in mine, so the third picture in Figure 19-5 doesn't look anything like what you see on your calculator's screen. But it does give you an idea of what to expect after completing this step.

5. **Repeatedly press ⊝ to highlight the name of your data set, and then press [ENTER][ENTER] to recall the data set.**

To recall a data set when you're currently in the Data/Matrix editor, press [F1][1] to select the **Open** command from the Tools menu. Then follow Steps 3 through 5 in this section.

Editing data

The calculator, with one exception, allows you to perform the following editing procedures:

- Changing the value of an entry
- Erasing the contents of a column
- Inserting a cell, row, or column
- Deleting a cell, row, or column

The exception is: You cannot mess with a column that is defined by a formula. (I explain using formulas to define a column earlier in this chapter.) But there is an easy solution to this exception — just erase the formula that defines the column. To do this, place the cursor on the column heading (c*n* where *n* is the number of the column), press [ENTER], and then press [CLEAR][ENTER]. The entries in the column that were defined by the formula remain but are no longer governed by the formula. So you can now edit them as you please.

Here's how you edit entries that are not (or are no longer) defined by a formula:

- **Change the value of an entry:** To do this, place the cursor on the entry and press [ENTER] to place the entry on the command line. Then enter a new value or press ⊙ and edit the original entry. When you're finished, press [ENTER] to save the new value in the calculator. (See Chapter 1 for details on editing expressions.)
- **Erase the contents of a column:** To do this, place the cursor anywhere in the column and then press [2nd][F1][5] to execute the **Clear Column** command in the Util menu.
- **Insert a cell:** To do this, place the cursor in the cell that will appear *below* the cell you want to insert and press [2nd][F1][1][1] to execute the

Insert cell command in the Util menu. Then enter the value you want to place in this cell and press [ENTER].

- **Insert a row or column:** To do this, place the cursor in the row or column that will appear *after* the row or column you want to insert and press [2nd][F1][1][2] to insert a row or press [2nd][F1][1][3] to insert a column. If you insert a row, the calculator places "undef" in the entries of the row. Simply edit these entries, as I explain earlier in this section.
- **Deleting a cell, row, or column:** To do this, place the cursor in the cell, row, or column, and then press [2nd][F1][2][1] to delete the cell, [2nd][F1][2][2] to delete the row, or [2nd][F1][2][3] to delete the column.

Sorting data

There are two ways in which you can sort data:

- Sort a single column so that numbers are in numeric order and characters are in alphabetical order.
- Sort all columns in the data set by using one column as the "key" column.

One exception exists: You can't sort data that is defined by a formula. (I explain using formulas to define a column earlier in this chapter.) But an easy solution to this exception does exist — just erase the formula that defines the column. To do this, place the cursor on the column heading (c*n* where *n* is the number of the column), press [ENTER], and then press [CLEAR][ENTER]. The entries in the column that were defined by the formula remain but are no longer governed by the formula. So you can now sort the data in the column.

Sorting a column

To sort a single column, place the cursor anywhere in the column, as illustrated in the first picture in Figure 19-6, and then press [2nd][F1][3] to execute the **Sort Column** command in the Util menu. As illustrated in the second picture in Figure 19-6, the **Sort Column** command sorts numbers in numeric order and characters in alphabetical order, with the numbers appearing first.

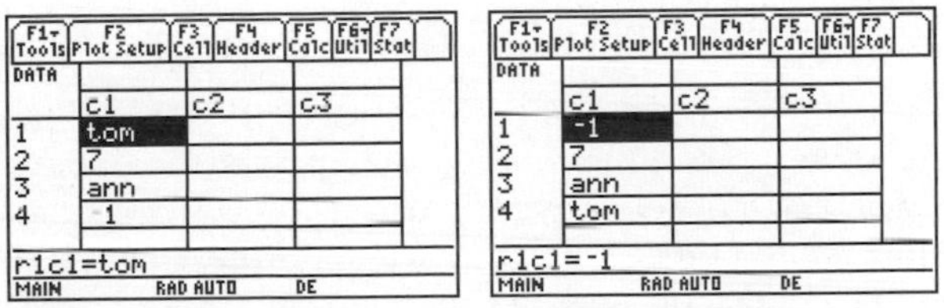

Unsorted Sorted

Figure 19-6: Sorting a single column in a data set.

Sorting all columns using one column as the "key"

Suppose that your data set is a grade book in which the first column contains students' first names, the second contains their last names, and the remaining columns contain their test scores. You want to sort the data based on the students' last names. That is, you want to sort all columns by using the second column as the "key" column. This method of sorting sorts the "key" column and maintains the integrity of the row structure — that is, after sorting, the row containing a student's last name in the second column still has his first name in the first column and his test scores in the remaining columns.

To sort all columns of your data set by using one column as the "key" column, first place the cursor anywhere in the "key" column, as illustrated in the first picture in Figure 19-7. Then press [2nd][F1][4] to execute the **Sort Col, adjust all** command in the Util menu, as illustrated in the second picture in this figure.

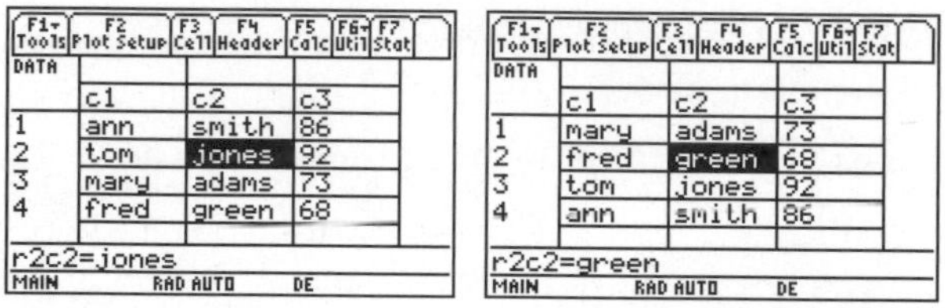

Unsorted   Sorted

Figure 19-7: Sorting the whole data set based on one "key" column.

Saving and Deleting Data Sets

Saving your data set is no big deal because the calculator automatically saves it as you construct it. But after a while, those data sets build up in the memory of the calculator. When it comes time to weed out some of those data sets housed in memory, you can do it in two ways: You can transfer a data set to your PC so you can recall it at a later time, or you can banish it to cyberspace by deleting it from your calculator.

Chapter 21 tells you how to transfer a data set to your PC and how to send it back to your calculator at a later date. Chapter 22 tells you how to transfer a data set from one calculator to another calculator.

To delete data sets from your calculator's memory, press [2nd][−] to display the VAR-LINK screen. Then repeatedly press ⊙ to highlight the name of the data set you want to delete. (Data sets have the extension DATA displayed to the right of their names.) Then press [F4] to place a checkmark next to it. Continue this process until you have checked all the data sets you want to delete. Then press [F1][1][ENTER] to delete the checked data sets.

Chapter 20

Analyzing Statistical Data

In This Chapter

- Plotting statistical data
- Creating histograms and box plots to describe one-variable data
- Creating scatter and line plots to describe two-variable data
- Tracing statistical data plots
- Finding the mean, median, standard deviation, and other neat stuff
- Finding a regression model for your data (curve fitting)

In descriptive statistical analysis, you usually want to plot your data and find the mean, median, standard deviation, and so on. You might also want to find a regression model for your data (a process also called *curve fitting*). This chapter tells you how to get the calculator to do these things for you.

Plotting One-Variable Data

The most common plots used to graph one-variable data are histograms and box plots. In a *histogram,* the data is grouped into classes of equal size; a bar in the histogram represents one class. The height of the bar represents the quantity of data contained in that class, as in the first picture in Figure 20-1.

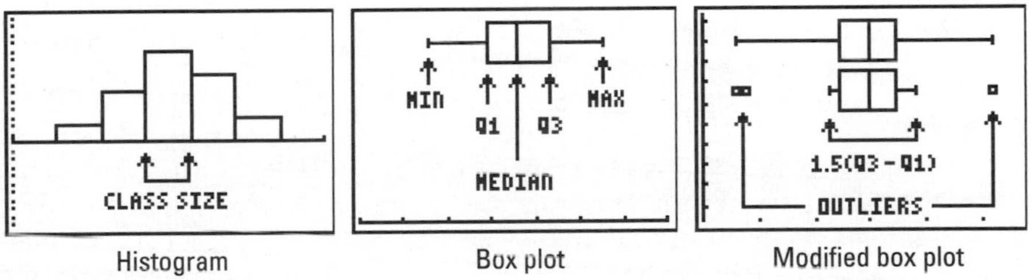

Figure 20-1: One-variable statistical plots.

A *box plot* (as in the second picture in Figure 20-1) consists of a box-with-whiskers. The box represents the data that exists between the first and third quartiles. The box is divided into two parts, with the division line defined by the median of the data. The whiskers represent the locations of the minimum and maximum data points.

The third picture in Figure 20-1 illustrates both a standard box plot and a modified box plot of the same data. In a modified box plot, the whiskers represent data in the range defined by 1.5(q3 – q1), and the outliers are plotted as points beyond the whiskers.

TIP

If your data has *outliers* (data values that are much larger or smaller than the other data values), consider constructing a modified box plot instead of a box plot.

To construct a histogram, box plot, or modified box plot, follow these steps:

1. **If necessary, press [MODE][▶][1][ENTER] to put the calculator in Function Graph mode.**
2. **Press [♦][F1] to enter the Y= editor, and then press [F5][1] to turn off all functions and Stat Plots.**
3. **If you haven't already done so, store your data in the calculator. If your data is already stored in the calculator, recall it in the Data/Matrix editor.**

   Chapter 19 tells you how to store data in the calculator and how to recall it in the Data/Matrix editor.

4. **Press [F2] to display the Plot Setup screen, use the [▼] key to highlight the Plot number you want to define, and then press [F1] to define the plot.**

   Your screen looks similar to the first picture in Figure 20-2 but without the drop-down menu displayed. At the top of the screen is the folder and data set you want to plot and the plot number in which this data is to be plotted.

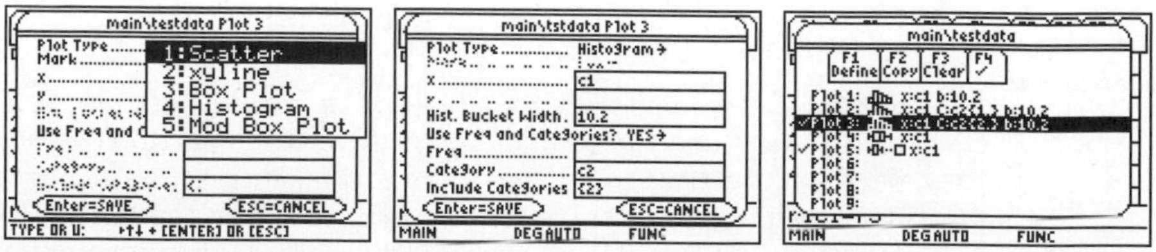

Figure 20-2: Setting up one-variable statistical plots.

5. **Press [▶] to display the Plot Type drop-down window, as illustrated in the first picture in Figure 20-2, and then press [3], [4], or [5] to tell the calculator you want to construct a box plot, histogram, or modified box plot, respectively.**

6. **If you're constructing a box plot or histogram, skip this step and go to Step 7. If you're constructing a modified box plot, press ⊝⊙ and then key in the number of the character you want to have the calculator use to plot the outliers.**

   Here's what your choices look like: Box (□), Cross (×), Plus (+), Square (■), and Dot (•). My recommendation is to go with Box. The dot is barely visible on the screen.

7. **Press ⊝ and enter the column heading (c*n*) that contains your data.**

   For example, if your data is in the first column, enter c1. The calculator is already in Alpha mode, so, for example, you enter c1 by pressing [)] to enter the letter c, then pressing [ALPHA] to take the calculator out of Alpha mode, and finally pressing [1].

8. **If you're graphing a box plot or a modified box plot, skip this step and go to Step 9. If you're graphing a histogram, press ⊝ and enter a value for the bucket width.**

   The *bucket width* is the actual width of each bar in the graph of the histogram. That is,

   $$\text{Hist. Bucket Width} = \frac{\textbf{xmax} - \textbf{xmin}}{\text{Number of bars}}$$

   where **xmax** and **xmin** are respectively the largest and smallest values of *x* appearing on the *x*-axis in the graph of the histogram.

   If you set the bucket width equal to the *class size* (the difference between consecutive lower class limits), some of your data points might occur on the boundary between two bars. As an example, if your data consists of test scores ranging from 50 to 100, and you group them into the typical data classes (F: 50 – 59, D: 60 – 69, C: 70 – 79, B: 80 – 89, A: 90 – 100), then when the bucket width equals the class size (which is 10), a test score of 60 is on the boundary between the first and second bars in the histogram.

   The only potential problem created by having a data point on the boundary between two bars is that when you trace the histogram to determine how many data points are in each bar, you must remember that a data point on the boundary between two bars is counted in the bar on the right. (I give you the lowdown on tracing a histogram later in this chapter, in the section "Tracing Statistical Data Plots.")

   For the histograms in Figure 20-3, which plot test scores ranging from 50 to 100, I avoided the problem of data points occurring on the boundary between two bars by setting **xmin** to 49.5 and setting **xmax** to 100.5. Because the histogram consisted of 5 bars, I set the bucket width to (100.5 – 49.5)/5 = 10.2, as illustrated in the second picture in Figure 20-2. (**xmin** and **xmax** are Window settings, which you set in Step 12.)

9. **Press ⊝⊙. Do you want to use frequencies or categories? If NO, then press [1]; if YES, press [2].**

The *frequency* of a data point is the number of times it occurs. This can also be viewed as a weighted value indicating that data vary in their degrees of importance. If the frequency of each data point is 1, you don't have to use frequencies because the calculator uses the default frequency value of 1 when you don't tell it to do otherwise.

Categories allow you to group your data into smaller subgroups, which you can then plot. As an example, the first picture in Figure 20-3 shows a histogram of test scores for all students, and the next two pictures in this figure show the histograms for the two subgroups consisting of girls and boys.

To create categories, assign a positive integer to each category and store these values in a column in the Data/Matrix editor. For example, to create the histograms in Figure 20-3 I stored the test scores in column **c1**, and in column **c2** I placed the number 1 next to the test score of a female and the number 2 next to the score of a male student.

You can have as many categories as you want, and you can have the calculator plot more than one category at a time. For example, if you have test scores for two classes of students, you could assign the following categories: 1 (first class, female), 2 (first class, male), 3 (second class, female), and 4 (second class, male). Then to plot the test scores for just the first class, you plot categories 1 and 2; to plot the test scores for the male students in both classes, plot categories 2 and 4.

TIP

I put the letters and words in the pictures in Figure 20-3 by using the **Text Pen** tool in the Graph window. If you'd like to do the same, the sidebar in this chapter tells you how.

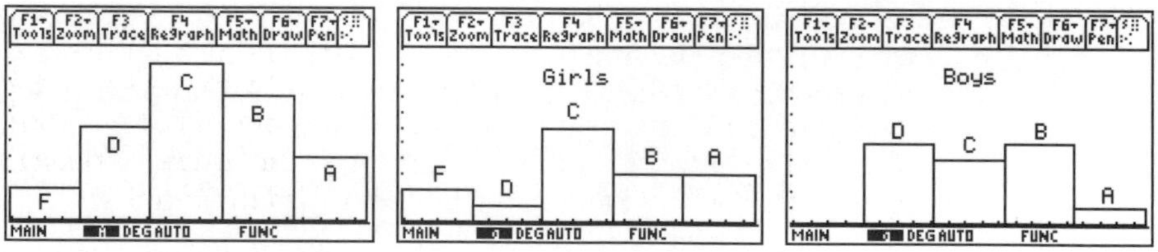

Figure 20-3: An example of plotting categories.

10. **If you answered NO in the preceding step, skip this step and go to the next. If you answered YES, press ⊙ and then do the following:**

    a. **If you aren't using frequencies, skip this step and go to Step b. If you are using frequencies, enter the column heading (c*n*) that contains the frequencies.**

    b. **If you aren't using categories, go to Step 11. If you are using categories, press ⊙ and enter the column heading (c*n*) containing the categories.**

c. **Press ▼►◄ to place the cursor inside the brackets in the Include Categories field and then enter the numbers of the categories you want to plot.**

If you're plotting more than one category, the numbers of the categories must be separated by commas. For example, to plot categories 1, 2, and 3, you place 1,2,3 inside the brackets.

11. **Press [ENTER] twice.**

You see a screen similar to the third picture in Figure 20-2, which shows a summary of the plots stored in the calculator.

12. **Put a checkmark next to the plots that you want to graph and leave the others unchecked.**

The calculator graphs only the plots that have checkmarks next to them. For example, in the third picture in Figure 20-2, plots 3 and 5 are checked, so they'll be graphed; the others won't be graphed. To change the checked status of a plot, use the ▲▼ keys to highlight the plot and then press [F4] to change the checked status of the plot.

13. **Press [◆][F2] to enter the Window editor and then press [F2][9] to have ZoomData graph the data for you.**

ZoomData doesn't adjust the settings for the *y*-axis, so you might get a graph as unpleasant-looking as the one in the first picture in Figure 20-4. In addition, **ZoomData** might not give you the appropriate number of bars in a histogram. You rectify these problems in the following step.

14. **If necessary, press [◆][F2] to reenter the Window editor and then adjust the settings in this editor to get a better view of your graph, as illustrated in the second picture in Figure 20-4.**

If you're graphing only box plots and modified box plots, you need to adjust only the settings for **ymin**, **ymax**, and **yscl**. If you're graphing a histogram and didn't get the expected number of bars in your graph, you need to also adjust the settings for **xmin**, **xmax**, and **xscl** so they agree with the way you set the bucket width in Step 8.

If in Step 8 you set the bucket width equal to the class size, set **xmin** equal to the lower limit of the first class, **xmax** equal to the upper limit of the last class, and **xscl** equal to the class size. If you used the formula in Step 8 to determine the bucket width, set **xmin** and **xmax** to the values you used in this formula and set **xscl** equal to either the class size or the bucket width.

TIP

If you're planning to trace your Stat Plot, decrease the setting for **ymin** so there's room at the bottom of the screen for the calculator to display its findings as it traces the graph. The basic rule is that the calculator uses the bottom one-fourth of the graphing area when tracing your Stat Plot, as illustrated in the third picture in Figure 20-4. If you don't leave room at the bottom of the screen, the calculator displays its findings on top of your stat plot — and you might find this very annoying.

TECHNICAL STUFF

Writing on a graph

Would you like to enhance your graph with words describing what's being graphed? If so, here's how you do it. While the graph is displayed on the screen, press [2nd][F2][7] to activate the **Text Pen** tool. Then move the cursor to the location on the screen where the upper right of the letter or symbol is to appear and press the key that corresponds to the letter or symbol. (Don't forget to press [ALPHA] to insert a letter.) The **Text Pen** tool remains active until you press [ESC], so while it's active, you can move the cursor to another location to write elsewhere on the graph.

If you make a mistake while using this tool or if you don't like the location of your letter or symbol, you can use the space key to erase it. To do this, move the cursor to the upper right of the letter or symbol and press [ALPHA][(-)] (or just [(-)] if you're in Alpha Lock mode) to overwrite the letter or symbol with a blank space.

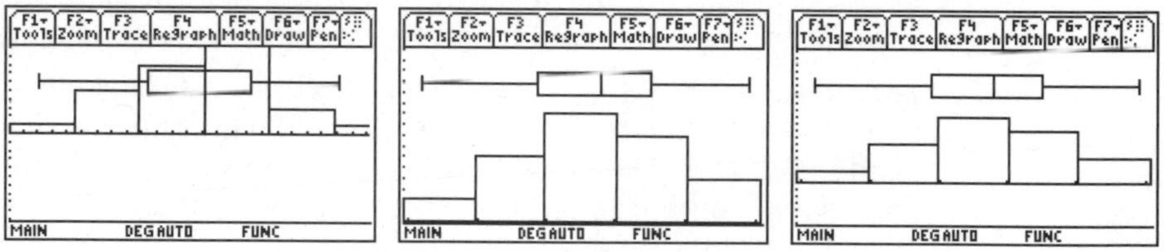

Figure 20-4: Adjusting the viewing window for stat plots.

Plotting Two-Variable Data

The most common plots used to graph two-variable data sets are the scatter plot and the *xy*-line plot. A *scatter plot* plots the points (*x*, *y*) where *x* is a value from one data list and *y* is the corresponding value from the other data list. The *xy-line plot* is simply a scatter plot with consecutive points joined by a straight line.

To construct a scatter plot or an *xy*-line plot, follow these steps:

1. **Follow Steps 1 through 4 in the preceding section with the following difference:**

   If you're graphing an *xy*-line plot, in Step 3, sort your data by using the data for the *x*-coordinates of the data points as the "key" column. If you don't, your *xy*-line plot might look like the one in the third picture in Figure 20-5. (I explain sorting data by using a "key" column in Chapter 19.)

2. **Press ⊙ to display the Plot Type drop-down window, as illustrated in the first picture in Figure 20-2, and then press [1] or [2] to tell the calculator you want to construct a scatter plot or an *xy*-line plot.**

Figure 20-5: Two-variable statistical data plots.

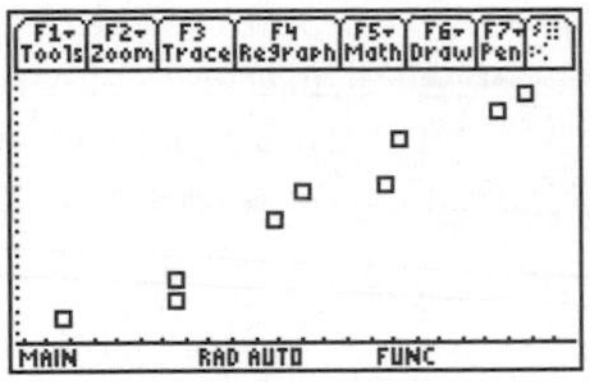

Scatter plot

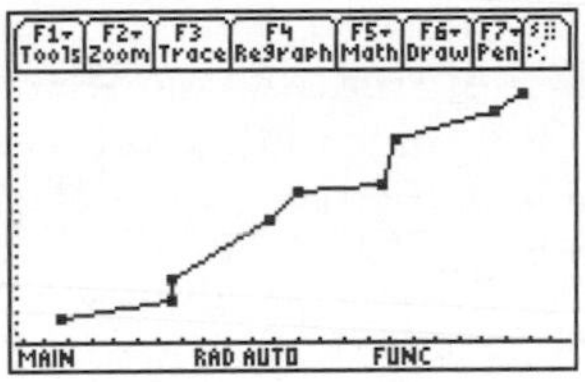

xy-line, sorted

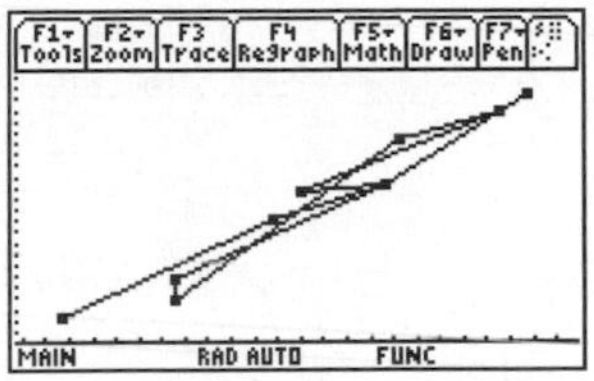

xy-line, unsorted

3. **Press ▼▶ and then key in the number of the character you want the calculator to use to plot the data points.**

   Here's what your choices look like: Box (□), Cross (×), Plus (+), Square (■), and Dot (•). I recommend using Box or Square. Don't use Dot unless you enjoy squinting.

4. **Press ▼ and enter the column heading (c*n*) that contains the data for the *x*-coordinates of the data points. Then press ▼ and enter the column heading (c*n*) that contains the data for the *y*-coordinates of the data points.**

   For example, if your data is in the first and second columns, enter c1, press ▼, and enter c2. The calculator is already in Alpha mode, so, for example, you enter c1 by pressing [)] to enter the letter *c*, then press [ALPHA] to take the calculator out of Alpha mode, and finally press [1]. After pressing ▼, press [ALPHA] to put the calculator back in Alpha mode, press [)] to enter the letter c, then press [ALPHA] to take the calculator out of Alpha mode, and finally press [2].

5. **Follow Steps 9 through 14 in the preceding section.**

Tracing Statistical Data Plots

To trace a statistical data plot, press [F3] while the plot is displayed on the screen. In the upper-right corner, you see the Stat Plot number (P1, P2, and so on). If you have more than one Stat Plot on the screen, repeatedly press ▲ until the plot you want to trace appears in the upper-right corner.

Use the ◀▶ keys to trace the plot. What you see depends on the type of plot:

- **Tracing a histogram:** As you trace a histogram, the cursor moves from the top center of one bar to the top center of the next bar. At the bottom of the screen, you see the values of **min**, **max**, and **n**. This tells you that there are *n* data points *x* such that min $\le x <$ max. This is illustrated in the first picture in Figure 20-6.

- **Tracing a box plot:** As you trace a box plot from left to right, the values that appear at the bottom of the screen are **minX** (the minimum data value), **q1** (the value of the first quartile), **Med** (the value of the median), **q3** (the value of the third quartile), and **maxX** (the maximum data value).
- **Tracing a modified box plot:** As you trace a modified box plot from left to right, the values that appear at the bottom of the screen are **minX** (the minimum data value) and then the other outliers, if any, to the left of the interval defined by 1.5(q3 – q1). The next value you see at the bottom of the screen is the value of the left bound of the interval defined by 1.5(q3 – q1). Then, as with a box plot, you see the values of the first quartile, the median, and the third quartile. After that, you see the value of the right bound of the interval defined by 1.5(q3 – q1), the outliers to the right of this, if any, and finally **maxX** (the maximum data value). This is illustrated in the second picture in Figure 20-6.
- **Tracing a scatter plot or an *xy*-line plot:** As you trace a scatter plot or an *xy*-line plot, the coordinates of the cursor location appear at the bottom of the screen, as illustrated in the third picture in Figure 20-6.

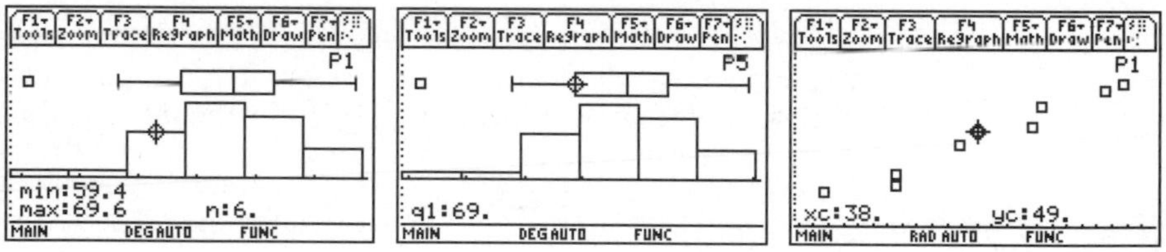

Histogram | Modified box plot | Scatter plot

Figure 20-6: Tracing statistical data plots.

Analyzing Your Data

The calculator can perform one- and two-variable statistical data analysis. For one-variable data analysis, the statistical data variable is denoted by *x*. For two-variable data analysis, the data variable for the first data list is denoted by *x*, and the data variable for the second data list is denoted by *y*. Table 20-1 lists the variables calculated by using one-variable data analysis (**One-Var**), as well as those calculated by using two-variable analysis (**Two-Var**).

Table 20-1 One- and Two-Variable Data Analysis

One-Var	*Two-Var*	*Meaning*
$\bar{x}$	$\bar{x}$, $\bar{y}$	Mean of data values
Σx	Σx, Σy	Sum of data values
Σx^2	Σx^2, Σy^2	Sum of squares of data values

One-Var	*Two-Var*	*Meaning*
Sx	Sx, Sy	Sample standard deviation
σx	σx, σy	Population standard deviation
nStat	nStat	Total number of data points
minX	minX, minY	Minimum data value
maxX	maxX, maxY	Maximum data value
q1		First quartile
medStat		Median
q3		Third quartile
	Σxy	Sum of x∗y

Analyzing one- and two-variable data

To analyze one- or two-variable data, follow these steps:

1. **If you haven't already done so, store your data in the calculator. If your data is already stored in the calculator, recall it in the Data/Matrix editor.**

   Chapter 19 tells you how to store data in the calculator and how to recall it in the Data/Matrix editor.

2. **Press [F5] to display the Calculate screen and then press ⊙ to display the Calculation Type drop-down window, as illustrated in the first picture in Figure 20-7.**

   At the top of the screen, you see the folder and name of the data set you want to analyze.

3. **If necessary, repeatedly press ⊙ to highlight the appropriate item and then press [ENTER] to select that item.**

   To analyze one-variable data, highlight item 1; to analyze two-variable data, highlight item 2.

4. **Press ⊙ and enter the column heading (c*n*) that contains the data for the *x*-values of your data. If you're analyzing one-variable data, go to the next step. If you're analyzing two-variable data, press ⊙ and then enter the column heading (c*n*) that contains the data for the *y*-value of your data.**

   For example, if you're analyzing two-variable data and your data are in the first and second columns of the Data/Matrix editor, enter **c1**, press

↓, and enter **c2.** The calculator is already in Alpha mode, so you enter c1, for example, by pressing [)] to enter the letter c, then press [ALPHA] to take the calculator out of Alpha mode, and finally press [1]. After pressing ↓, press [ALPHA] to put the calculator back in Alpha mode, press [)] to enter the letter c, and then press [2].

5. **Press ↓→. Do you want to use frequencies or categories? If NO, then press [1]; if YES, then press [2].**

   For a detailed explanation of frequencies and categories, see Step 9 in the first section, "Plotting One-Variable Data," in this chapter.

6. **If you answered NO in the preceding step, skip this step and go Step 7. If you answered YES, press ↓ and then do the following:**

   a. **If you aren't using frequencies, skip ahead to Step b. If you are using frequencies, enter the column heading (c*n*) that contains the frequencies.**

   b. **If you aren't using categories, go to Step 7. If you are using categories, press ↓ and enter the column heading (c*n*) that contains the categories.**

   c. **Press ↓→← to place the cursor inside the brackets in the Include Categories field and then enter the numbers of the categories you want to use.**

      If you're analyzing more than one category, the numbers of the categories must be separated by commas. For example, to analyze categories 1, 2, and 3, you place 1,2,3 inside the brackets.

   d. **Press [ENTER] to save the last entry you made.**

7. **Press [ENTER] to display the analysis of your data, as illustrated in the second picture in Figure 20-7. When you're finished viewing this display, press [ENTER] to return to the Data/Matrix editor.**

   After you press [ENTER] the first time, you see a screen similar to the second picture in Figure 20-7, which — with the exception of the population standard deviations (σx and σy) — shows the values of statistical variables listed in Table 20-1. Not all the statistical variables fit on this screen, as indicated by the down arrow in the last entry in the second picture in Figure 20-7. To see the statistical variables not displayed, repeatedly press ↓ until they appear.

   The next section tells you how to find the values of the population standard deviations. It also tells you how to use all statistical variables listed in Table 20-1 in a mathematical expression.

TIP

While the calculator is in the Data/Matrix editor, you can press [2nd][F2] to redisplay the results of the last one- or two-variable statistical analysis displayed by the calculator. If the calculator no longer has the results of that analysis stored in its memory, you get an error message when you press [2nd][F2].

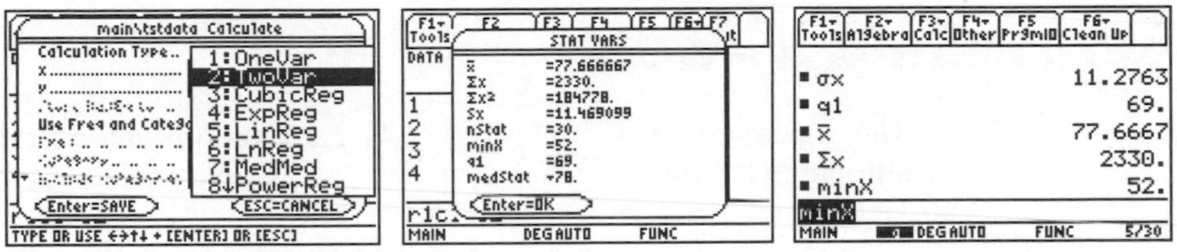

Figure 20-7: Analyzing one-variable data.

Using statistical variables in calculations

Perhaps you need to use the value of a statistical variable in a mathematical expression (see the third picture in Figure 20-7), or maybe you want to display the values of the population standard deviations on the Home screen. If so, you must be able to enter the symbols σ, x̄, ȳ, and Σ into the calculator, and you need to know how to enter lower- and uppercase letters. No sweat! Chapter 1 tells you how to use the [ALPHA] and [↑] keys to enter variables such as q1 and minX. The following tells you how to enter the symbols σ, x̄, ȳ, and Σ:

- **To enter σ:** Press [2nd][+][ENTER] to display the Greek CHAR menu, repeatedly press ▽ to highlight item G, and then press [ENTER].

  To, for example, display the value of σx on the Home screen, as illustrated in the first entry in the third picture in Figure 20-7, enter σ and then press [X][ENTER].

  The shortcut keystrokes for entering σ are [◆][(][↑][3]. Although these keystrokes give you no clue as to why you're pressing them, they're well worth remembering if you frequently find yourself in need of knowing the values of the population standard deviations σx and σy, which (for reasons that baffle me) don't appear on the Calculate screen, as illustrated in the second picture in Figure 20-7.

- **To enter Σ:** On the Home screen, press [F3][4] to enter Σ(and then press [←] to erase the left parenthesis.

  For example, to display the value of Σx on the Home screen, as illustrated in the fourth entry in the third picture in Figure 20-7, press [F3][4][←][X][ENTER].

  To enter Σ when you aren't on the Home screen, you can use the shortcut keystrokes [◆][(][ALPHA][3].

- **To enter x̄ and ȳ:** Press [2nd][+][2] to enter the Math CHAR menu, repeatedly press ▽ to highlight item A for x̄ or item B for ȳ, and then press [ENTER].

  As an example, the keystrokes I used to display the value of x̄ in the third entry in the third picture in Figure 20-7 are [2nd][+][2][ALPHA][=][ENTER].

Regression Models

Regression modeling is the process of finding a function that approximates the relationship between the two variables in a two-variable data set. (An example appears in Figure 20-8, in which a straight line approximates the relationship between the two variables.) Table 20-2 shows the types of regression models the calculator can compute.

Table 20-2 Types of Regression Models

TI-Command	*Model Type*	*Equation*
MedMed	Median-median	$y = ax + b$
LinReg	Linear	$y = ax + b$
QuadReg	Quadratic	$y = ax^2 + bx + c$
CubicReg	Cubic	$y = ax^3 + bx^2 + cx + d$
QuartReg	Quartic	$y = ax^4 + bx^3 + cx^2 + dx + e$
LnReg	Logarithmic	$y = a + b*\ln(x)$
ExpReg	Exponential	$y = a*b^x$
PowrReg	Power	$y = a*x^b$
Logistic	Logistic	$y = c/(1 + a*e^{-bx})$
SinReg	Sinusoidal	$y = a*\sin(bx + c) + d$

To compute a regression model for your two-variable data, follow these steps:

1. **Follow Steps 1 through 4 in the section titled "Analyzing one- and two-variable data," earlier in this chapter. In Step 3 select the appropriate regression model.**

   The first picture of Figure 20-8 illustrates that the linear regression model was selected in Step 3.

2. **Press ⊙⊙, use the ⊙ key to highlight a function in which to store the regression equation, and then press [ENTER] to select that function.**

   If the function you select is already defined in the Y= editor, that definition is replaced with the regression equation.

3. **If you aren't using frequencies or categories, press [ENTER] to display the regression formula, as illustrated in the second picture in Figure 20-8. If you are using frequencies or categories, follow Steps 5 through 7 in the section titled "Analyzing one- and two-variable data," earlier in this chapter, to display the regression formulas.**

   For a detailed explanation of frequencies and categories, see Step 9 in the first section ("Plotting One-Variable Data") in this chapter.

   As illustrated in the second picture in Figure 20-8, the calculator displays the equation for the model you selected and gives the values of the coefficients in this equation. It also gives the values of the correlation coefficient (**Corr**) and the coefficient of determination (**R^2**).

4. **Graph your data and regression model, as illustrated in the third picture in Figure 20-8.**

   To do this, follow the directions given in the earlier section, "Plotting Two-Variable Data." If the regression model does not display with your data, press [◆][F1] to enter the Y= editor, use the ⊖ key to highlight the function you selected in Step 2, and press [F4] to place a checkmark next to it. Then press [◆][F3] to graph the data and regression model.

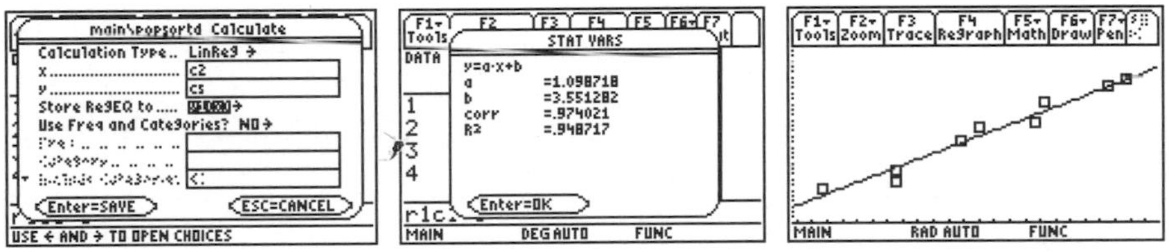

Figure 20-8: Calculating and graphing a regression model.

Part VIII

Communicating with PCs and Other Calculators

The 5th Wave By Rich Tennant

"WHAT EXACTLY ARE WE SAYING HERE?"

In this part . . .

This part gets you ready to transfer files between your calculator and a PC, or between your calculator and another calculator. I also tell you how to download and install the free TI Connect software you can use to (among other things) transfer files to and from your PC.

Chapter 21

Communicating with a PC Using TI Connect™

In This Chapter

- Downloading the TI Connect software
- Installing and running the TI Connect software
- Connecting your calculator to your computer
- Transferring files between your calculator and your computer
- Upgrading the calculator's operating system

You need two things to enable your calculator to communicate with your computer: TI Connect (software) and a TI-Graph Link cable. TI Connect is free; the TI-Graph Link cable isn't. If the cable didn't come bundled with your calculator, you can purchase one at the Texas Instruments online store at `www.education.ti.com`.

Downloading TI Connect

The TI Connect software is on the TI Resource CD that most likely came with your calculator. However, the version on this CD might not be current. The following steps tell you how to download the current version of TI Connect from the Texas Instruments Web site, as it existed at the time this book was published:

1. **Go to the Texas Instruments Web site (`www.education.ti.com`).**
2. **In the Know What You're Looking For? drop-down list, select TI Connect.**
3. **Click Downloads in the column on the left side of the screen.**
4. **Click either the Download Latest TI Connect for Windows or the Download Latest TI Connect for Mac link.**

   If you haven't already logged into the TI Web site, the Login page appears. (If you've logged in, go to Step 6.)

5. **If you're already a member, enter your e-mail address and password and then click the Login button. If you're not a member, sign up — it's free.**

   To sign up as a member, click New User Registration and follow the directions. TI might ask you a lot of questions, but registering gives you access to a plethora of free stuff — such as TI Connect and the calculator programs I tell you about in Chapter 23.

6. **Click the appropriate language.**

7. **Follow the directions given during the downloading process.**

   Make a note of the directory in which you save the downloaded TI Connect file on your PC so you'll know where to find it when you want to install it on your PC, as I explain in the next section.

Installing and Running TI Connect

After you've downloaded TI Connect, you install it by double-clicking the downloaded TI Connect file you saved on your computer. Then follow the directions given by the installation program you just launched.

When you start the TI Connect program, you see the many subprograms it contains. To see what these subprograms are used for, click the Help button in the lower-right corner of the screen. In this chapter, I explain how to use TI Device Explorer to transfer files between your calculator and your PC.

TIP: Each of the subprograms housed in TI Connect has excellent Help menus that tell you exactly how to use the program.

Connecting Calculator and Computer

To connect your calculator to your computer, you use the TI-Graph Link cable that came with your calculator. If you don't have a TI-Graph Link cable, you can purchase one at the Texas Instruments online store at `www.education.ti.com`.

There are four types of Link cables. Three of them work with all TI-89 calculators: the gray serial cable, the black serial cable, and the silver USB cable. The fourth type, the black USB-to-USB cable, works only with the TI-89 Titanium and came bundled with the calculator. (It's the USB-to-USB cable that has differently sized ends.) Because the ends of all four of these cables have different shapes and sizes, you can easily figure out how to connect your calculator to your computer. The small end fits in a slot located at the bottom of the TI-89 or in a slot at the top of a TI-89 Titanium. The other end plugs into one of your computer's serial or USB ports.

Transferring Files

After you've connected the calculator to your computer, the TI Device Explorer program housed in TI Connect can transfer files between the two devices. This allows you to archive calculator files on your computer.

To transfer files between your calculator and PC, start the TI Connect software and click the TI Device Explorer program. A directory appears, listing the folders housed in your calculator. Expanding these folders works the same on your calculator as on your computer. From this directory, you can do the following:

- **To copy or move files from your calculator to your PC,** highlight the files you want to transfer, click File, and then choose either Copy to PC or Move to PC. When the Choose Folder window appears, select the location to which your files will be transferred and click Select.
- **To copy files to the calculator from a PC running Windows,** you don't need to be in the TI Device Explorer program. Instead, here's what you do:
    1. Open Windows Explorer.
    2. Highlight the files you want to copy.
    3. Right-click the highlighted files and then click Send To TI Device.
    4. When asked whether you want the files sent to RAM or Archive, select RAM if you plan to edit the file; otherwise, select Archive, where it cannot be edited or inadvertently deleted.

TIP

The Help menu in TI Device Explorer is packed with useful information. In it, you can find directions for editing and deleting calculator files and directions for backing up all the files on your calculator.

Upgrading the OS

Texas Instruments periodically upgrades the operating systems of the TI-89 family of calculators. To upgrade the operating system, start the TI Connect software and click the TI Device Explorer program. Click the Help menu at the top of the screen, and then click TI Device Explorer Help. In the column on the left, double-click Updating TI Software, and then click Updating Operating Systems with TI OS Downloader. After that, just follow the on-screen directions.

Chapter 22

Communicating between Calculators

In This Chapter

- Linking calculators to transfer files
- Determining what files you can transfer
- Selecting files to transfer
- Transferring files between calculators

You can transfer data lists, programs, matrices, and other such files from one calculator to another if you link the calculators with the unit-to-unit Calculator Link cable that came bundled with your calculator. This chapter describes how to make such transfers.

Linking Calculators

Linked calculators can share information — a real bonus if you want to transfer data from one calculator to another. You can link calculators by using the unit-to-unit Calculator Link cable that came bundled with the calculator. If you're no longer in possession of the cable, you can purchase one at the Texas Instruments online store at `www.education.ti.com`.

Two types of unit-to-unit Calculator Link cables exist: One has two cylindrically shaped plugs on each end (I/O-to-I/O), and the other has two small rectangular ends of the same size (USB-to-USB). The one that comes bundled with the TI-89 Titanium is the USB-to-USB cable. At the time this book was written, the USB-to-USB cable can link only two TI-89 Titanium calculators. The other cable (I/O-to-I/O) can be used to link any two calculators of the following type: TI-89, TI-89 Titanium, TI-92, TI-92 Plus, and Voyage 200.

To link two TI-89 Titanium calculators, plug the USB-to-USB cable into the ports on the top right of each calculator. To link the other types of calculators, plug each end of the I/O-to-I/O cable into the only hole in which they fit — this hole is located either at the top or bottom of the calculator.

If you get an error message when transferring files from one calculator to another using the I/O-to-I/O cable, the most likely cause is that the unit-to-unit cable isn't fully inserted into the port of one calculator.

Transferring Files

You can transfer files between your TI-89 and another TI-89, TI-89 Titanium, TI-92, TI-92 Plus, or Voyage 200, provided you have the appropriate unit-to-unit Calculator Link cable (as I explain in the preceding section). To transfer files from the sending calculator to the receiving calculator, follow these steps:

1. **Press [2nd][–] on the sending calculator to access the VAR-LINK menu.**

   The VAR-LINK menu appears in the first picture in Figure 22-1.

2. **Use the ⊙ and ⊙ keys to highlight the file you want to send and press [F4] to place a checkmark to the left of that file.**

   Repeat this procedure for each file you want to send.

   Only those files that have been checked will be sent to another calculator. The second picture in Figure 22-1 shows three selected files that are waiting to be sent to another calculator.

   To place a checkmark to the left of all files in the calculator, press [F5][1]. To check all files in a single folder, as in the third picture in Figure 22-1, use the ⊙ and ⊙ keys to highlight the name of the folder and press [F4].

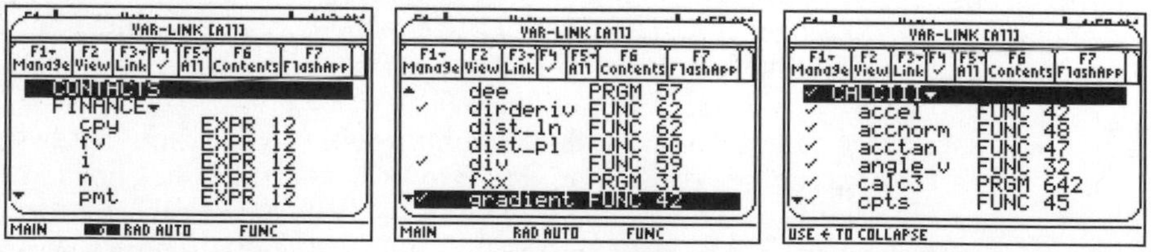

VAR-LINK menu — Checked files — Checked folder

Figure 22-1: Selecting files for transmission between calculators.

3. **On the receiving calculator, press [2nd][−][F3][2].**

   At the bottom of the screen you see the message VAR-LINK: WAITING TO RECEIVE.

   Always put the receiving calculator in Receiving mode before you transfer files from the sending calculator.

4. **On the sending calculator, press [F3][1] to send the files to the receiving calculator.**

   As files are transferred, on the receiving calculator you might see a screen similar to the first picture in Figure 22-2 indicating that the receiving calculator already has a file with the same name as the one being transferred. If you want to overwrite this file with the one being transferred, press [ENTER]. If not, press ▶ and then press the number of the desired option in the Duplicate File menu, as illustrated in the second picture in Figure 22-2. Then press [ENTER] to execute your selection.

   The third option (NO) in the Duplicate File menu affords you the opportunity to rename the file that is being transmitted, as illustrated in the third picture in Figure 22-2. To do this, press ▼ and key in a new name. Because the calculator is already in Alpha mode, you don't need to press [ALPHA] before entering each letter — just press the key that has the appropriate letter above it. After entering the new filename, press [ENTER] twice; once to confirm the new filename, and again to transfer the newly renamed file. If there is already a file in the calculator with the same name as your newly named file, you get the Duplicate File menu again. In this case, just give the file a different name.

5. **When you're finished transferring files, press [ESC] to exit the VAR-LINK menu.**

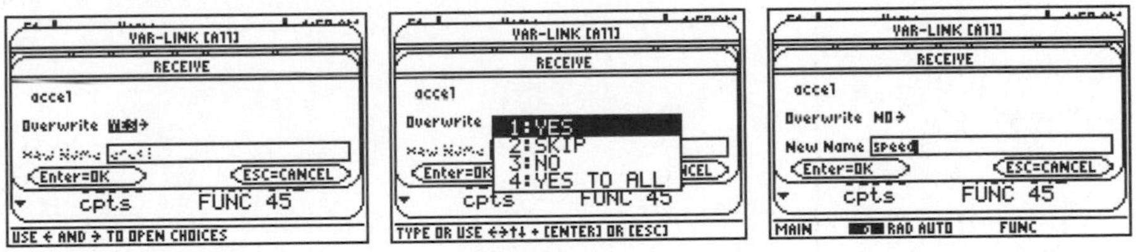

Figure 22-2: Dealing with duplicate file names.

Duplicate File menu — Menu options — Renaming a file

If you want to terminate the transfer of files while it is in progress, press [ON] on either calculator. Then press [ESC] when you're confronted with the Link transmission error message. If you put one calculator in Receiving mode and then decide not to transfer any files to that calculator, press [ON] to take it out of Receiving mode.

Transferring Files to Several Calculators

After transferring files between two calculators, you can then use the sending calculator to transfer the same files to a third calculator without having to reselect the files. To do this, simply link the two calculators, press [2nd][-][F3][2] on the third calculator, and then press [F3][1] on the sending calculator. And if you first select the newly transmitted files on the third calculator (as I describe in Step 2 in the preceding section), you can then use that calculator to send files to yet another calculator.

Part IX
The Part of Tens

"You can sure do a lot with a TI-89, but I never thought dressing one up in G.I. Joe clothes and calling it your little desk commander would be one of them."

In this part . . .

This part wraps up some handy items in packages of (approximately) ten. I tell you how to download and install application programs that enhance the capabilities of your calculator, and briefly describe ten of my favorite applications. I also list the most common errors that crop up while using the calculator, and explain the most common error messages the calculator might give you.

Chapter 23

Ten Great Applications

In This Chapter

- Reviewing ten applications for the calculator
- Finding and downloading applications

The Texas Instruments Web site has about two dozen applications that you can download and install on your calculator. Most of these application programs are free, and those that aren't free are very inexpensive.

Texas Instruments already might have installed some of these applications on your calculator — even some that aren't available on its Web site for free. To see what applications are already on your calculator, press [APPS].

In the following section, I briefly describe ten great applications you can download from the Texas Instruments Web site. At the end of this chapter, I tell you how to find, download, and install those programs.

Ten Great Applications

Here are ten great applications you can download from the Texas Instruments Web site (`www.education.ti.com`):

- **Cabri Jr.:** This is an interactive, dynamic geometry program that allows you to export geometric figures between the calculator and the Cabri Geometry II Plus software for Windows.
- **Calculus Tools:** This is a must-have application for calculus students — and teachers, too. Not only does it contain operations not found on the calculator (like **grad**, **div**, and **curl**), but it also allows you to interactively investigate calculus applications (such as Riemann sums and Newton's method). This application can even do the Ratio test.
- **CellSheet:** This application turns your calculator into a spreadsheet. If you also download the free TI CellSheet Converter software, you can transfer your spreadsheet files between your calculator and Microsoft Excel or Appleworks.

- **The Geometer's Sketchpad:** This application is the calculator version of the well-known PC software. It doesn't have all the bells and whistles that the PC version has, but it will do the geometry you need to do.
- **Notefolio:** Download this application if you want to turn your calculator into a word processor. And while you're at it, download the Notefolio Creator software so you can transfer files between your calculator and Microsoft Word. This application is most useful on the TI-92 and Voyage 200 because these calculators have a keyboard. If you have a TI-89 or fat fingers, I suggest purchasing a TI Keyboard so you can key in words the way you do on a computer.
- **Organizer:** This application is a personal organizer that you can use to schedule events, create to-do lists, and save phone numbers and e-mail addresses.
- **Statistics with List Editor:** The stat options built into the TI-89 are wimpy compared to what you get with the Statistics with List Editor application. For a college-level stat course, this is a must-have application.
- **StudyCards:** This application creates electronic flash cards. Be sure also to download the free TI StudyCard Creator software that allows you to create the flash cards on your PC.
- **Symbolic Math Guide:** This application helps students with various topics (including algebra, pre-calculus, and calculus) by having them interactively do various problems step by step.
- **TI-Reader:** Want to read *The Scarlet Letter* on your calculator? Or would you like your calculator to function as a dictionary? If so, download this application. And while you're about it, download the TI-Reader Converter — it gets your PC to convert ebooks into something your calculator can display. And after you've done that, Google *"free ebooks"* so you can download something to read.

Downloading an Application

The following steps tell you how to download application programs from the Texas Instruments Web site as of the time this book was published. To download and install applications, follow these steps:

1. **Go to the Texas Instruments Web site at `www.education.ti.com`.**
2. **In the Know What You're Looking For? drop-down list, click Apps and OS Versions.**
3. **Click the link that matches the type of calculator you have.**

4. **Click the application you want to download.**
5. **Click Download Instructions in the left column of the screen and read the instructions on how to download your application.**

   The download instructions are the same for each application, so you need read the instructions only once. After reading them, close the window containing these instructions.
6. **Click Download under the picture of the calculator screen and follow the directions you're given.**

   As you follow these directions, you are asked to accept the License Agreement and log in. If you aren't a member of the site, sign up — it's free.

   Make a note of the directory in which you save the downloaded application on your PC so you'll know where to find it when you want to install it on your calculator, as I explain in the next section.

Installing an Application

To install applications on your calculator, you need the TI Connect software and a TI-Graph Link cable. See Chapter 21 for information on downloading and installing the software and connecting your calculator to your PC by using the TI-Graph Link cable. You can also find directions for copying the application file to your calculator in Chapter 21.

Chapter 24

Ten Common Errors and Messages

In This Chapter

- Common errors made by first-time users
- Common errors made by experienced users
- Common error messages that leave you baffled

Even the best calculating machine is only as good as its input. Everyone makes mistakes, and when you make a mistake on the calculator, you usually get an error message. This chapter tells you how to avoid eight of the most common errors made on the calculator as well as how to interpret ten of the most common and baffling error messages.

Eight Great Ways to Mess Up

Here's a list of eight common errors made when using the calculator, as well as explanations of what happens when you make such errors:

- **Using [−] instead of [(-)] to indicate that a number is negative:** If you press [−] instead of [(-)] at the beginning of an entry, the calculator assumes that you want to subtract what comes after the minus sign from the previous answer. If you use [−] instead of [(-)] in the interior of an expression to denote a negative number, the calculator responds with the "Syntax" error message.
- **Not properly indicating the order of operations:** When evaluating expressions, the order of operations is crucial. To the calculator, for example, -3^2 equals –9. If you were expecting the answer to be 9, I suggest consulting Chapter 2 for a refresher on the order of operations. According to this order of operations, the calculator, where appropriate, first simplifies what's in parentheses, then squares the result, and then performs the operation of negation. So to the calculator, $-3^2 = -9$, whereas $(-3)^2 = 9$.

Also, when graphing rational functions, users who are new to the calculator often make the basic mistake of omitting the parentheses that must be used to set the numerator apart from the denominator.

- **Improperly using implied multiplication:** Implied multiplication is when you omit the times sign in an expression such a 2sin(π). There are times when this is okay with the calculator, such as with 2sin(π), and there are times when it isn't, such as with *x*sin(π). The reason is because the calculator allows names of objects to consist of more than one letter. So to the calculator, *xsin* is a four-letter name. To be on the safe side, use the times sign instead of implied multiplication.
- **Not clearing the contents of one-letter variables before starting a new problem:** If, for example, you previously stored the number 4 in the variable *a* and then at a later date try to solve the equation $x^2 - 4x + a = 0$, the calculator gives the unexpected answer of 2 if you don't first clear the value 4 from the one-letter variable *a*. Because we quickly forget what we've stored in the one-letter variables, be sure to always press [2nd][F1][1] (which clears *a* through *z*) before starting a new problem on the Home screen.
- **Not properly entering the arguments of a function:** If the arguments are improperly entered, you get an error message. As an example, if you enter ∫(2*x*) to integrate 2*x*, you get the "Too few arguments" error message because this function requires two arguments: the function being integrated and the independent variable. The Cheat Sheet attached at the front of this book shows the proper form for entering the arguments of some of the most common functions that require more than one argument.
- **Entering an angle in degrees when the calculator is in Radian mode:** Actually, you *can* do so legitimately, but you have to let the calculator know that you're overriding the Angle mode by pressing [2nd][|] to place a degree symbol after your entry.
- **Improperly setting the Window:** If you get the "Window variables domain" error message when graphing functions, this is most likely caused by setting **xmin** ≥ **xmax** or by setting **ymin** ≥ **ymax** in the Window editor. Chapter 5 has the details on setting the Window editor.
- **Improperly setting the Complex Format in the Mode menu:** The Complex Format item in the Mode menu, accessed by pressing [MODE]⊙⊙⊙⊙⊙⊙, gives you three choices: Real, Rectangular, and Polar. If you don't want numbers displayed in polar form, set the Complex Format to Rectangular — even if you deal exclusively with real numbers. Reason: The rectangular format displays real numbers in the normal fashion, and it also displays complex numbers in the form $a + bi$.

Ten Baffling Error Messages

Most error messages you get make perfect sense, such as those that say "Argument must be a decimal number" or "No solution found." In a perfect world, all error messages would make sense. But in the world of the small TI-89 screen, some error messages are abbreviated to the point where they might leave you baffled. Or you might understand the message but be clueless about how to solve the problem. Here are ten baffling messages you might encounter:

- **Bound:** You get this message when you incorrectly set the Lower Bound to be larger than the Upper Bound when using a function in the Graph Math menu. You can rectify the problem by starting over and properly setting these bounds.
- **Break:** You get this message when you press [ON] to stop a calculation or to stop graphing. All this message really tells you is that the calculator has successfully stopped doing whatever it was doing.
- **Data type:** This type of error occurs if, for example, you enter a negative number when the calculator requires a positive number.
- **false:** This isn't actually an error message, but it is baffling. Basically, it says that the calculator can't do what you asked it to do. For example, you get this result when you use the **solve** command to solve an equation that contains complex numbers or to solve an equation with complex solutions. To solve such equations, use the **cSolve** command.
- **Link transmission:** This message pops up when you're attempting to transfer data between two calculators. You get this message for one of two reasons: The cable linking the calculators isn't firmly set, or you had one calculator send the data *before* you instructed the other calculator to receive the data. To rectify this problem, check the cable connecting the calculators and then resend the data, making sure to set the receiving calculator to **Receive** *before* you set the sending calculator to **Send**.
- **Missing (:** This message usually means exactly what it says — you haven't properly closed all parentheses. But you also get this message when you try to store an expression in a function name used by the calculator. For example, you get this message when you try to store a number in the variable *max*. When this happens, just use a different name for your variable.
- **Non-real result:** Although this message makes perfect sense, it might leave you baffled about fixing the problem. This message usually occurs when the Complex Format in the Mode menu is set to Real. If you set it to Rectangular — by pressing [MODE]⊙⊙⊙⊙⊙⊙ — and reevaluate your expression, you should get the desired results.

- **Singular matrix:** You get this message when your expression explicitly or implicitly refers to the inverse of a singular matrix — that is, a matrix whose determinate is zero. Because the inverse of such a matrix doesn't exist, whatever you asked the calculator to do before getting this message is mathematically impossible.
- **Syntax:** This is the catch-all of all catch-all error messages. When evaluating expressions, this message usually means that you didn't properly enter the expression — maybe you did something haphazard like pressing [−] instead of [(-)] to indicate that a quantity is negative. When storing a number in a variable name, this message usually means that the name you chose is reserved by the calculator. For example, storing a number in *disp* gives you this message. In this case, just use a different name.
- **Window variables domain:** This error message tells you that the current Window settings are preventing the calculator from doing what you're asking it to do. If you get this message when you're trying to graph something, the problem is usually caused by setting **xmin** to be greater than **xmax** or setting **ymin** greater than **ymax**.

  If you get this message when using a command in the Graph Math menu, the problem is caused by entering an argument that isn't contained within the Window settings. For example, using the **value** command to evaluate the graphed function at $x = 11$ when **xmin** = –10 and **xmax** = 10 in the Window editor gives you this error message because 11 is not between –10 and 10.

Appendix

Creating Custom Menus

In This Chapter

- Creating and using a Custom menu
- Installing and uninstalling a Custom menu
- Editing a Custom menu
- Creating functions to put in a Custom menu

Are you sick of all those keystrokes needed to find often-used commands lurking deep in the submenu structure of the MATH menu? Have you often felt that your life would be easier if TI had a single command for the keystrokes that you use to complete the same task over and over again? Or are you just tired of having to remember where things are located on the calculator? If so, a Custom menu is the solution to your problems.

As the name indicates, a Custom menu is a menu that replaces the Toolbar menu on the Home screen with the commands of your choice. It can even house the programs you write to save keystrokes when performing an often-repeated task. I create a Custom menu for each class I teach so that I don't have to keep reminding students where to find things on the calculator. Figure A-1 shows one of my Custom menu creations.

The calculator has a Default Custom menu, as illustrated in the second picture in Figure A-1. Press [2nd][HOME] to check it out and then press [2nd][HOME] again to return to the Toolbar menu, as illustrated in the first picture in Figure A-1. If you'd like to create your own Custom menu, be sure to check out this appendix.

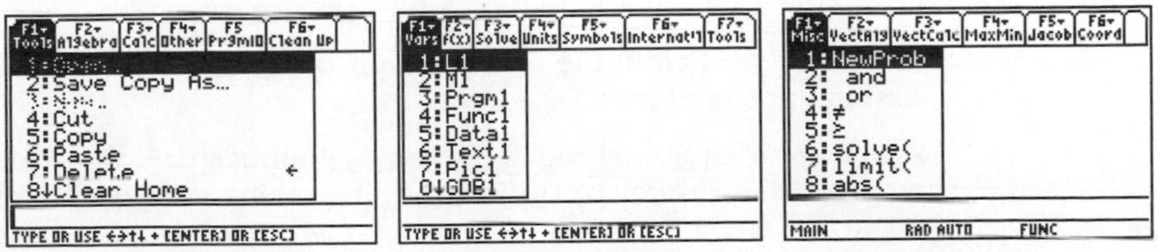

Toolbar menu — Default Custom menu — User Custom menu

Figure A-1: Toolbar and Custom menus.

Creating and Using a Custom Menu

As Figure A-2 shows, a Custom menu is a program that has a specified format consisting of blocks of code. The first block gives the name of the menu and tells the calculator that it is a program for a Custom menu. The interior blocks tell the calculator the names to give the Function keys (F1, F2, and so on) and which commands (items) to place in the submenus accessed by these keys. The last block (EndCustm and EndPrgm) tells the calculator that you're finished constructing the Custom menu.

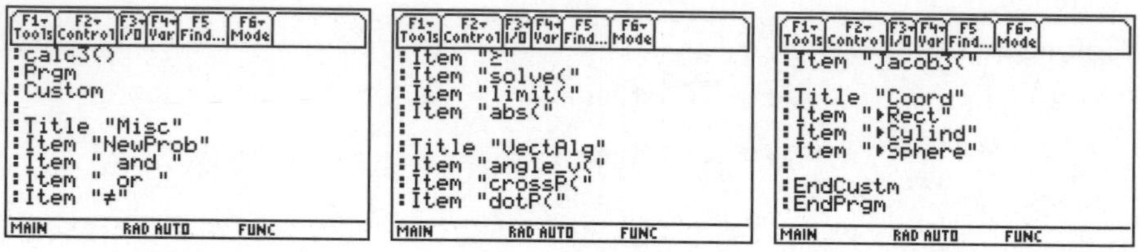

First screen — Second screen — Last screen

Figure A-2: Screen shots from a user-created Custom menu.

Creating a Custom menu isn't as difficult as it looks, but it does take time to key in all those words on a TI-89. A timesaving alternative is to create the menu on a PC and then transfer it to your calculator. The sidebar in this chapter tells you how to download and use the (free) software needed to create the menu on your PC, and Chapter 21 tells you how to transfer that menu from your PC to your calculator. Of course, you need a cable that links your calculator to your PC. Such a cable most likely came with your calculator. If not, you can purchase one at the Texas Instruments online store at `http://education.ti.com/us/product/main.html`.

Creating a Custom menu on the calculator

To create a Custom menu on your calculator, follow these steps:

1. **Press APPS and select the Program editor. Then press 3 to tell that calculator that you're creating a new program.**
2. **Press ⊙⊙ and enter the name of your program, as illustrated in the first picture in Figure A-3.**

   A calculator equipped with a current operating system is in Alpha Lock mode waiting for you to enter letters. The name you give your Custom menu can consist of no more than eight letters or numbers, the first of which must be a letter. You find out about entering letters, numbers, and symbols in Chapter 1.

TIP

If you get symbols when you try to enter letters, you need to upgrade your operating system. Chapter 21 tells you how to do this, but you don't need to do it right now. Simply press [2nd][ALPHA] to put the calculator in Alpha Lock mode and upgrade the operating system when convenient.

3. **Press [ENTER] twice.**

   You see a screen similar to the second picture in Figure A-3.

4. **Press [↓][↓][F2][7] to tell the calculator you're creating a Custom menu and then press [ENTER] to leave a space between lines, as shown in the third picture in Figure A-3.**

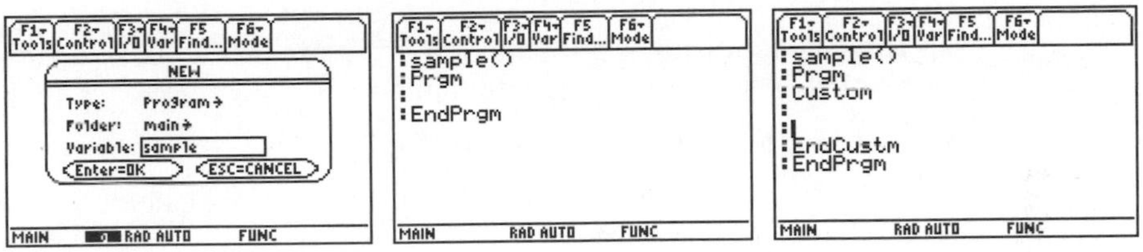

Figure A-3: Creating the first block of a Custom menu.

5. **While the calculator is in Alpha Lock, enter the word *title,* press [(-)] to enter a space, press [2nd][1] to start a quote, enter the name of the submenu that will be accessed when you press [F1], press [2nd][1] to end the quote, and then press [ENTER] (see the first picture in Figure A-4).**

6. **With Alpha Lock engaged, enter the word *item,* press [(-)] to enter a space, and then press [2nd][1] to start a quote, as illustrated in the second picture in Figure A-4).**

TIP

   If you so choose, you can select the words *title* and *item* from the CATALOG. Chapter 1 tells you how to do this.

7. **Enter the command for this item, press [2nd][1] to end the quote, and then press [ENTER], as illustrated in the third picture in Figure A-4.**

   To enter a command from a calculator menu, simply select the command from the appropriate calculator menu and then press [ENTER]. For example, the command **abs(** in the third picture in Figure A-4 was placed there by pressing [2nd][5][1][2].

   To enter a command that is a function or program of your creation, simply key in its name. I explain creating functions and programs later in this chapter.

8. **Repeat Step 7 until you've defined all items that appear in the submenu.**

   An example of this procedure appears in the first two pictures of Figure A-2, in which the submenu titled Misc has eight items.

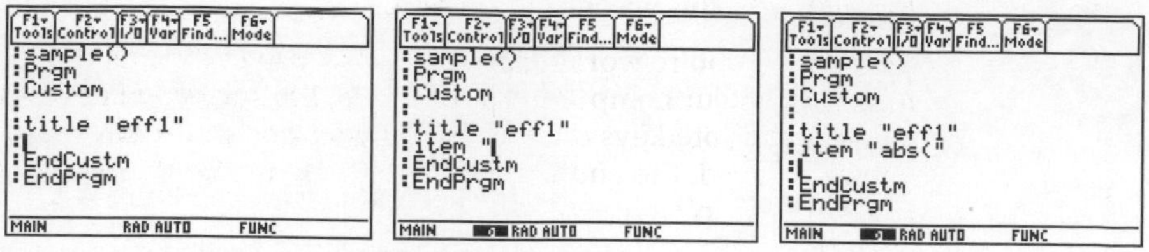

Figure A-4: Defining the submenus (Function keys) in a Custom menu.

9. **Repeat Steps 5 through 8 to define the other submenus that will be accessed by the other Function keys.**

   The calculator has eight Function keys, so you can have a maximum of eight submenus in your Custom menu.

10. **Press [2nd][ESC] to exit the Program editor.**

    Your Custom menu is automatically saved in the calculator under the name you gave it in Step 2.

Creating a Custom menu on a PC

To create a Custom menu on a PC and then transfer it to your calculator, follow these steps:

1. **Start the appropriate TI-Graph Link software.**

   The TI-89, TI-92 Plus, and Voyage 200 are on speaking terms, so the "appropriate" TI-Graph Link software can be for either the TI-89 or TI-92 Plus. The sidebar in this chapter tells you how to download this free software from Texas Instrument's Web site.

2. **In the far-right panel, click in the Name text box and enter the name of your Custom menu.**

   The name can consist of no more than eight characters, the first of which must be a letter.

3. **Click in the blank space between Prgm and EndPrgm.**
4. **Click the drop-down box in the left panel and select Control.**
5. **In the left column of this panel, click Special, and then in the right column, double-click Custom . . . EndCustm to place this command in your program.**

   With the exception of the first line, your program should look like the third picture in Figure A-3.

6. **Enter the submenus for your Custom menu by following Steps 5 through 9 in the preceding section.**

   Remember, you're working on a PC — so you can do all those keystrokes directly on your computer. For example, I'm sure you know where the space and quote keys are located on your computer keyboard. When you're finished, the end of your custom menu should look similar to the third picture in Figure A-4.

7. **To save your Custom menu, choose File⇨Save.**

8. **In the File Name text box, change UNTITLED to the name of your Custom menu.**

9. **In the box to the right of the Save As panel, select the directory on your PC where you want to save the Custom menu and then click OK.**

   Sorry folks, but you're dealing with an ancient piece of software that recognizes only eight-character names for the directories. TI has updated the TI-Graph Link software so that it works with current TI calculators, but it hasn't done the same for the PC. Why? I don't know. Perhaps the TI creators think their calculators are so good that no one would ever need to program them!

10. **Transfer your Custom menu to your calculator.**

    See Chapter 21 for details.

Installing and using a Custom menu

After storing your Custom menu in the calculator, you have to install it before you can use it. To do this, follow these steps:

1. **If you aren't already on the Home screen, press [HOME].**

2. **Press [2nd][-] to display the VAR-LINK screen and then repeatedly press ⊝ to highlight your Custom menu, as illustrated in the first picture in Figure A-5.**

3. **Press [ENTER] to place the Custom menu on the command line of the Home screen and then press [)] to close the parentheses.**

   TECHNICAL STUFF

   If you're wondering why the calculator inserts the parentheses, it's because your Custom menu is a calculator program, and programs expect you to input information in the parentheses the same way you, for example, place information in the parentheses when using the calculator to evaluate an integral or to factor a polynomial.

4. **Press [ENTER] to install your Custom menu, as illustrated in the second picture in Figure A-5.**

Figure A-5: Installing and using a Custom menu.

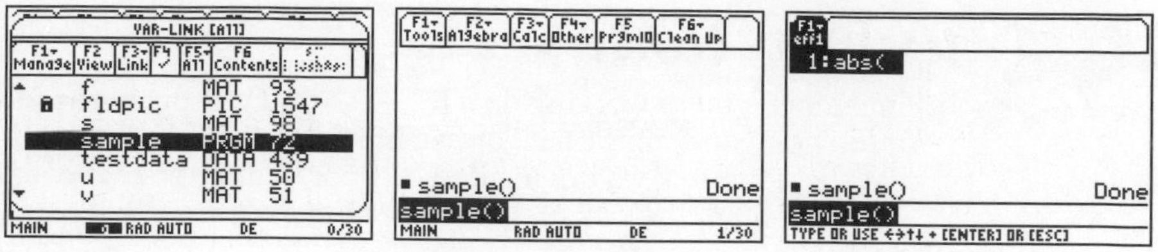

To display your Custom menu, press [2nd][HOME]. The third picture in Figure A-5 displays the Custom menu created in the third picture in Figure A-4. You select items from your Custom menu the same way you select items from any menu, as I explain in Chapter 1. To return to the Toolbar menu, shown in the second picture in Figure A-5, press [2nd][HOME] again. This key combination toggles between the Toolbar menu and your Custom menu.

To uninstall your Custom menu, first display the Toolbar menu on the Home screen. Then press [2nd][F1][3][ENTER] to execute the **Restore custom default** command in the Clean Up menu. Finally, press [2nd][HOME] to display the Toolbar menu. The key combination of [2nd][HOME] on the TI-89 now toggles between the Toolbar menu and the Default Custom menu, as illustrated in the first two pictures in Figure A-1, shown earlier. To use your Custom menu again, you have to reinstall it.

Editing a Custom menu

To edit your Custom menu, you must first display it on either a PC or the calculator. Then you can edit it the same way you created it. Chapter 1 tells you how to edit on a calculator, and editing on a PC is . . . well, I'm sure you know how to do that.

- **To display and edit your Custom menu on a calculator,** press [APPS] and select the Program Editor. Then press [2] to open an existing program and press ⊙⊙⊙ to display the programs housed in your calculator's memory. Repeatedly press ⊙ to highlight your Custom menu and then press [ENTER] twice to display it. As you edit your Custom menu, the calculator automatically saves your changes to the menu.
- **To display and edit your Custom menu on a PC,** start the TI-Graph Link software, choose File⇨Open, locate your Custom menu on the PC, and then click OK. After editing your Custom menu, transfer it to your calculator.

If you edit a Custom menu that is already installed on your calculator, you don't have to reinstall it — the changes made to the menu should automatically appear. If they don't, toggle between the Toolbar menu and your Custom menu to make them appear.

Creating Functions

Are you fed up with pressing many keys to perform the same calculator task over and over again? Or are you disappointed that the calculator doesn't have a command that executes a mathematical formula you frequently use? If so, you can create your own function or program that houses multiple keystrokes in a single command — and then you can place that command in a Custom menu so you have easy access to it. (You find out how to create that Custom menu earlier in this chapter.) This section tells you how to create a keystroke-saving function. To create a more elaborate program, consult the manual that came with your calculator.

Are you wondering what the difference is between a function and a program? Well, a *program* has multiple steps, as illustrated in the example in the first picture in Figure A-6. A *function* does only one thing — it takes the information you give it and spits out an answer, as illustrated in the second picture in Figure A-6. Most of the menu items housed in the calculator are functions, and none are programs. For example, the **solve** command is a function — you give it an equation and a variable, and it spits out the solution for the variable.

TIP

If you're adept at creating programs, you won't have any trouble taking the directions for creating a function and adapting them to creating a program. The programming commands used by the calculator are housed in the Control and I/O menus, which appear in the Toolbar, as illustrated in the third picture in Figure A-6.

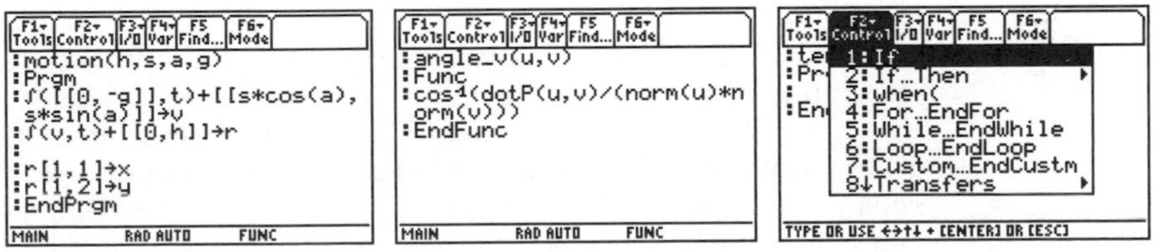

Program | Function | Control menu

Figure A-6: The difference between a program and a function.

Creating a function on the calculator

To create a function on the calculator, follow these steps:

1. **Press [APPS] and select the Program editor. Then press [3] to tell the calculator that you're creating a new program or function.**
2. **Press ⓡ[2] to tell the calculator you're creating a function.**

3. **Press ⊙⊙ and enter the name of your function, as illustrated in the first picture in Figure A-7.**

   A calculator endowed with a current operating system is in Alpha Lock mode waiting for you to enter letters. The name you give your Custom menu can consist of no more than eight letters or numbers, the first of which must be a letter. I explain entering letters, numbers, and symbols in Chapter 1.

   TIP: If you get symbols when you try to enter letters, you need to upgrade your operating system. Chapter 21 tells you how to do this, but you don't need to do it right now. Simply press [2nd][ALPHA] to put the calculator in Alpha Lock mode and upgrade the operating system when convenient.

4. **Press [ENTER] twice.**

   You see a screen similar to the second picture in Figure A-7 with the cursor placed between the parentheses after the name of the function.

5. **Enter the variables used in the formula.**

   The variables you enter are dummy variables — this means that the calculator doesn't care what you call them. For example, if one of the variables you enter is *a* and the calculator has a number stored in *a*, the calculator ignores this and treats *a* as a variable (instead of equating *a* to the value stored in it).

   If your formula has more than one variable, these variables must be separated by commas, as shown on the first line of the third picture in Figure A-7.

6. **Press ⊙⊙ and enter the formula for your function, as illustrated in the third picture in Figure A-7.**

   As an example, the third picture in Figure A-7 shows the formula for a function designed to find the *n*th column of matrix *a*; the second picture in Figure A-6 is that of a function which computes the angle between two vectors, **u** and **v**. The commands in these formulas were entered by using the menus on the calculator ([2nd][5][4][1] for the transpose ^{T} of a matrix) or the keypad ([◆][Z]for $\cos^{-1}$ on the TI-89).

7. **Press [2nd][ESC] to exit the Program editor.**

   Your function is automatically saved in the calculator under the name you gave it in Step 3.

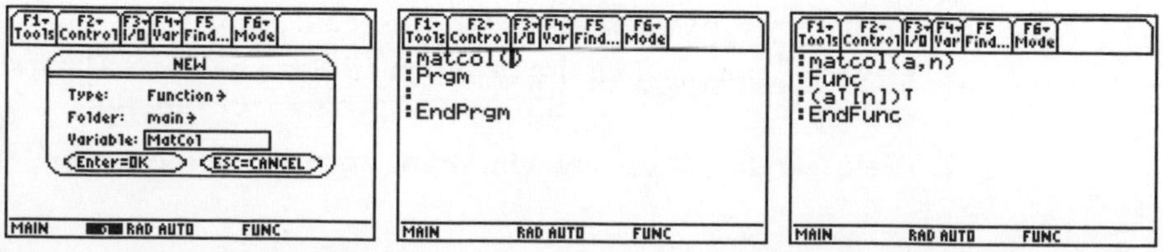

Figure A-7: Creating a function to find the column of a matrix.

Creating a function on a PC

To create a function on your PC and then transfer it to your calculator, follow these steps:

1. **Start the appropriate TI-Graph Link software.**

   The TI-89, TI-92 Plus, and Voyage 200 are on speaking terms, so the "appropriate" TI-Graph Link software can be for either the TI-89 or TI-92 Plus. The sidebar in this chapter tells you how to download this free software from the Texas Instruments Web site.

2. **In the far-right panel, click in the Name text box and enter the name of your function.**

   The name can consist of no more than eight characters, the first of which must be a letter.

3. **Place the cursor between the parentheses on the first line in the right screen.**

4. **Follow Steps 5 and 6 in the preceding section.**

   Ignore the fact that the ancient TI-Graph Link software doesn't recognize the difference between a function and a program — to any regeneration of calculator or software, a function is a special form of a program.

   Keyboard strokes, such as those producing $\cos^{-1}$, can always be found on the left screen by selecting Catalog from the drop-down box and then clicking the appropriate letter.

   TIP: Did the left screen of the TI-Graph Link software mysteriously disappear? No problem, just click **Window** and then click **Function List** to make it reappear.

5. **To save your function, choose File⇨Save.**

6. **In the File Name text box, change UNTITLED to the name of your Custom menu.**

7. **In the box on the right side of the Save As panel, select the directory on your PC where you want to save the Custom menu and then click OK.**

   Unfortunately, you're dealing with an ancient piece of software that recognizes only eight-character names for the directories. TI has updated the TI-Graph Link software so that it works with current TI calculators, but it hasn't done the same for the PC.

8. **Transfer your Custom menu to your calculator.**

   You find out how to do this in Chapter 21.

Downloading and using TI-Graph Link

TI-Graph Link is a computer software program that allows you to create calculator programs on your PC and then transfer them to your calculator. To download this free software, do the following from your PC while it is connected to the Internet:

1. **Go to `http://education.ti.com`.**
2. **In the Search text box in the upper-right corner, type TI Graph Link and then click the orange Search button.**
3. **Click the link for TI-Graph Link.**
4. **Click your flavor of PC: Windows, Macintosh, or MS-DOS.**
5. **Click your flavor of calculator: TI-89 or TI-92 Plus. (Voyage 200 users, click TI-92 Plus.)**
6. **Log in.**

   If you haven't registered, do it now — it's free and well worth it.
7. **Click your language of choice. (English is the third selection.)**
8. **Click the software you selected.**

   I know this is redundant, but do it anyway.
9. **Follow the given directions. (Run installs the software on your PC, and Save saves the software so you can manually install it later.)**

After installing TI-Graph Link, you have to restart your PC before you can use it. When you have TI-Graph Link up and running, it works very much like the calculator. On the left side of the screen, you see the double-column Function List menu and a drop-down box with Catalog as the default entry. Click a letter in the left column, and all the Catalog items beginning with that letter appear in the right column. Click the down arrow on the drop-down box to see the other menus that appear on the calculator. When you click one of these menus, its contents appear in the right column.

On the right side of the screen, you see the Program editor all set up and ready for you to write your program. This side of the screen works just as you would expect it to work, but with one exception — to place a function in the right column of the Function List in the Program editor, position the cursor in the proper place in the Program editor and then double-click the function in the Function List. After you've written your program, you can transfer it to your calculator.

Index

• Numbers & Symbols •

• A •

• D •

• F •

• G •

• H •

• I •

• J •

• K •

• L •

• M •

• N •

• O •

• Q •

• R •

• T •

• U •

• V •

• W •

• X •

• Y •

• Z •

BUSINESS, CAREERS & PERSONAL FINANCE

0-7645-5307-0

0-7645-5331-3 *†

Also available:

- Accounting For Dummies † 0-7645-5314-3
- Business Plans Kit For Dummies † 0-7645-5365-8
- Cover Letters For Dummies 0-7645-5224-4
- Frugal Living For Dummies 0-7645-5403-4
- Leadership For Dummies 0-7645-5176-0
- Managing For Dummies 0-7645-1771-6
- Marketing For Dummies 0-7645-5600-2
- Personal Finance For Dummies * 0-7645-2590-5
- Project Management For Dummies 0-7645-5283-X
- Resumes For Dummies † 0-7645-5471-9
- Selling For Dummies 0-7645-5363-1
- Small Business Kit For Dummies *† 0-7645-5093-4

HOME & BUSINESS COMPUTER BASICS

0-7645-4074-2

0-7645-3758-X

Also available:

- ACT! 6 For Dummies 0-7645-2645-6
- iLife '04 All-in-One Desk Reference For Dummies 0-7645-7347-0
- iPAQ For Dummies 0-7645-6769-1
- Mac OS X Panther Timesaving Techniques For Dummies 0-7645-5812-9
- Macs For Dummies 0-7645-5656-8
- Microsoft Money 2004 For Dummies 0-7645-4195-1
- Office 2003 All-in-One Desk Reference For Dummies 0-7645-3883-7
- Outlook 2003 For Dummies 0-7645-3759-8
- PCs For Dummies 0-7645-4074-2
- TiVo For Dummies 0-7645-6923-6
- Upgrading and Fixing PCs For Dummies 0-7645-1665-5
- Windows XP Timesaving Techniques For Dummies 0-7645-3748-2

FOOD, HOME, GARDEN, HOBBIES, MUSIC & PETS

0-7645-5295-3

0-7645-5232-5

Also available:

- Bass Guitar For Dummies 0-7645-2487-9
- Diabetes Cookbook For Dummies 0-7645-5230-9
- Gardening For Dummies * 0-7645-5130-2
- Guitar For Dummies 0-7645-5106-X
- Holiday Decorating For Dummies 0-7645-2570-0
- Home Improvement All-in-One For Dummies 0-7645-5680-0
- Knitting For Dummies 0-7645-5395-X
- Piano For Dummies 0-7645-5105-1
- Puppies For Dummies 0-7645-5255-4
- Scrapbooking For Dummies 0-7645-7208-3
- Senior Dogs For Dummies 0-7645-5818-8
- Singing For Dummies 0-7645-2475-5
- 30-Minute Meals For Dummies 0-7645-2589-1

INTERNET & DIGITAL MEDIA

0-7645-1664-7

0-7645-6924-4

Also available:

- 2005 Online Shopping Directory For Dummies 0-7645-7495-7
- CD & DVD Recording For Dummies 0-7645-5956-7
- eBay For Dummies 0-7645-5654-1
- Fighting Spam For Dummies 0-7645-5965-6
- Genealogy Online For Dummies 0-7645-5964-8
- Google For Dummies 0-7645-4420-9
- Home Recording For Musicians For Dummies 0-7645-1634-5
- The Internet For Dummies 0-7645-4173-0
- iPod & iTunes For Dummies 0-7645-7772-7
- Preventing Identity Theft For Dummies 0-7645-7336-5
- Pro Tools All-in-One Desk Reference For Dummies 0-7645-5714-9
- Roxio Easy Media Creator For Dummies 0-7645-7131-1

*** Separate Canadian edition also available**

† Separate U.K. edition also available

Available wherever books are sold. For more information or to order direct: U.S. customers visit www.dummies.com or call 1-877-762-2974. U.K. customers visit www.wileyeurope.com or call 0800 243407. Canadian customers visit www.wiley.ca or call 1-800-567-4797.

WILEY

SPORTS, FITNESS, PARENTING, RELIGION & SPIRITUALITY

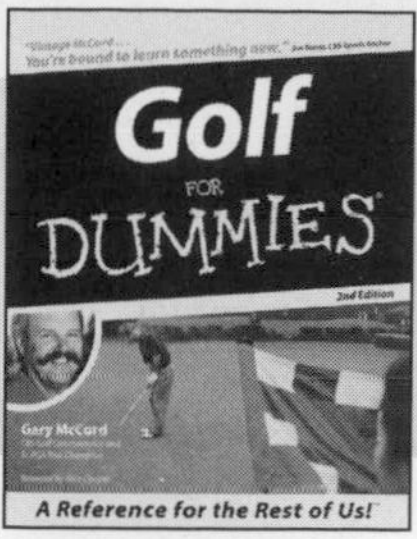

0-7645-5146-9

0-7645-5418-2

Also available:

- Adoption For Dummies 0-7645-5488-3
- Basketball For Dummies 0-7645-5248-1
- The Bible For Dummies 0-7645-5296-1
- Buddhism For Dummies 0-7645-5359-3
- Catholicism For Dummies 0-7645-5391-7
- Hockey For Dummies 0-7645-5228-7
- Judaism For Dummies 0-7645-5299-6
- Martial Arts For Dummies 0-7645-5358-5
- Pilates For Dummies 0-7645-5397-6
- Religion For Dummies 0-7645-5264-3
- Teaching Kids to Read For Dummies 0-7645-4043-2
- Weight Training For Dummies 0-7645-5168-X
- Yoga For Dummies 0-7645-5117-5

TRAVEL

0-7645-5438-7

0-7645-5453-0

Also available:

- Alaska For Dummies 0-7645-1761-9
- Arizona For Dummies 0-7645-6938-4
- Cancún and the Yucatán For Dummies 0-7645-2437-2
- Cruise Vacations For Dummies 0-7645-6941-4
- Europe For Dummies 0-7645-5456-5
- Ireland For Dummies 0-7645-5455-7
- Las Vegas For Dummies 0-7645-5448-4
- London For Dummies 0-7645-4277-X
- New York City For Dummies 0-7645-6945-7
- Paris For Dummies 0-7645-5494-8
- RV Vacations For Dummies 0-7645-5443-3
- Walt Disney World & Orlando For Dummies 0-7645-6943-0

GRAPHICS, DESIGN & WEB DEVELOPMENT

0-7645-4345-8

0-7645-5589-8

Also available:

- Adobe Acrobat 6 PDF For Dummies 0-7645-3760-1
- Building a Web Site For Dummies 0-7645-7144-3
- Dreamweaver MX 2004 For Dummies 0-7645-4342-3
- FrontPage 2003 For Dummies 0-7645-3882-9
- HTML 4 For Dummies 0-7645-1995-6
- Illustrator CS For Dummies 0-7645-4084-X
- Macromedia Flash MX 2004 For Dummies 0-7645-4358-X
- Photoshop 7 All-in-One Desk Reference For Dummies 0-7645-1667-1
- Photoshop CS Timesaving Techniques For Dummies 0-7645-6782-9
- PHP 5 For Dummies 0-7645-4166-8
- PowerPoint 2003 For Dummies 0-7645-3908-6
- QuarkXPress 6 For Dummies 0-7645-2593-X

NETWORKING, SECURITY, PROGRAMMING & DATABASES

0-7645-6852-3

0-7645-5784-X

Also available:

- A+ Certification For Dummies 0-7645-4187-0
- Access 2003 All-in-One Desk Reference For Dummies 0-7645-3988-4
- Beginning Programming For Dummies 0-7645-4997-9
- C For Dummies 0-7645-7068-4
- Firewalls For Dummies 0-7645-4048-3
- Home Networking For Dummies 0-7645-42796
- Network Security For Dummies 0-7645-1679-5
- Networking For Dummies 0-7645-1677-9
- TCP/IP For Dummies 0-7645-1760-0
- VBA For Dummies 0-7645-3989-2
- Wireless All In-One Desk Reference For Dummies 0-7645-7496-5
- Wireless Home Networking For Dummies 0-7645-3910-8